REGENERATION OF VERTEBRATE SENSORY RECEPTOR CELLS

The Ciba Foundation is an international scientific and educational charity. It was established in 1947 by the Swiss chemical and pharmaceutical company of CIBA Limited—now CIBA-GEIGY Limited. The Foundation operates independently in London under English trust law.

The Ciba Foundation exists to promote international cooperation in biological, medical and chemical research. It organizes about eight international multidisciplinary symposia each year on topics that seem ready for discussion by a small group of research workers. The papers and discussions are published in the Ciba Foundation symposium series. The Foundation also holds many shorter meetings (not published), organized by the Foundation itself or by outside scientific organizations. The staff always welcome suggestions for future meetings.

The Foundation's house at 41 Portland Place, London W1N 4BN, provides facilities for meetings of all kinds. Its Media Resource Service supplies information to journalists on all scientific and technological topics. The library, open five days a week to any graduate in science or medicine, also provides information on scientific meetings throughout the world and answers general enquiries on biomedical and chemical subjects. Scientists from any part of the world may stay in the house during working visits to London.

Ciba Foundation Symposium 160

REGENERATION OF VERTEBRATE SENSORY RECEPTOR CELLS

A Wiley-Interscience Publication

1991

JOHN WILEY & SONS

Chichester · New York · Brisbane · Toronto · Singapore

Published in 1991 by John Wiley & Sons Ltd.
Baffins Lane, Chichester
West Sussex PO19 1UD, England

Other Wiley Editorial Offices

John Wiley & Sons, Inc., 605 Third Avenue,
New York, NY 10158-0012, USA

Jacaranda Wiley Ltd, G.P.O. Box 859, Brisbane,
Queensland 4001, Australia

John Wiley & Sons (Canada) Ltd, 5353 Dundas Road West,
Fourth Floor, Etobicoke, Ontario M9B 6H8, Canada

John Wiley & Sons (SEA) Pte Ltd, 37 Jalan Pemimpin 05-04,
Block B, Union Industrial Building, Singapore 2057

Suggested series entry for library catalogues:
Ciba Foundation Symposia

Ciba Foundation Symposium 160
ix + 341 pages, 71 figures, 11 tables

Library of Congress Cataloging-in-Publication Data

Regeneration of vertebrate sensory receptor cells.
p. cm.—(Ciba Foundation symposium: 160)
Editors, Gregory R. Bock (organizer) and Julie Whelan.
"A Wiley-Interscience Publication."
"Symposium on Regeneration of Vertebrate Sensory Receptor Cells, held at the Ciba Foundation, London, 4–6 December 1990"—Contents p.
Includes bibliographical references and indexes.
ISBN 0 471 92960 3
1. Sensory receptors—Regeneration—Congresses. 2. Cochlea—Regeneration—Congresses. 3. Retina—Regeneration—Congresses. 4. Vertebrates—Physiology—Congresses. I. Symposium on Regeneration of Vertebrate Sensory Receptor Cells (1990: Ciba Foundation) II. Series.
[DNLM: 1. Receptors, Sensory—physiology—congresses. 2. Regeneration—congresses. 3. Vertebrates—physiology—congresses. W3 C161F v. 160/WL 102.9 R333 1990]
596′.0182-dc20
DNLM/DLC
for Library of Congress 91-21619
CIP

British Library Cataloguing in Publication Data

A catalogue record for this book is available from the British Library

ISBN 0 471 92960 3

Phototypeset by Dobbie Typesetting Limited, Tavistock, Devon.
Printed and bound in Great Britain by Biddles Ltd., Guildford.

Contents

Symposium on Regeneration of Vertebrate Sensory Receptor Cells, held at the Ciba Foundation, London 4–6 December 1990

The topic of the symposium was proposed by Professor Edwin W. Rubel

Editors: Gregory R. Bock (Organizer) and Julie Whelan

Participants

G. D. Burd Department of Molecular & Cellular Biology, University of Arizona, Life Sciences South, Room 444, Tucson, AZ 85721, USA

A. L. Calof Department of Biology, University of Iowa, Iowa City, IA 52242, USA

J. T. Corwin Department of Otolaryngology, Head & Neck Surgery and Department of Neuroscience, University of Virginia Medical Center, Charlottesville, VA 22908, USA

R. M. Costanzo Department of Physiology, Medical College of Virginia, Virginia Commonwealth University, Richmond, VA 23298-0551, USA

D. A. Cotanche Department of Anatomy, Boston University School of Medicine, Boston, MA 02118, USA

A. I. Farbman Department of Neurobiology & Physiology, Section of Biological Sciences, Northwestern University, O. T. Hogan Hall, 2153 Sheridan Road, Evanston, IL 60208-3520, USA

R. D. Fernald Departments of Neuroscience and Psychology, Stanford University, CA 94305, USA

J. M. Jørgensen Department of Zoophysiology, Building 131, Institute of Zoology & Zoophysiology, Aarhus University, DK-8000 Aarhus, Denmark

J. Lewis ICRF Developmental Biology Unit, Department of Zoology, University of Oxford, South Parks Road, Oxford, OX1 3PS, UK

G. V. Lopashov Institute of Gene Biology, USSR Academy of Sciences, 3 Vavilov Street 34/5, Moscow 117334, USSR

F. L. Margolis Laboratory of Chemosensory Neurobiology, Roche Institute of Molecular Biology, 340 Kingsland Street, Nutley, NJ 07110-1199, USA

B. Oakley Department of Biology, Natural Science Building, University of Michigan, Ann Arbor, MI 48109-1048, USA

S. L. Palay *The Journal of Comparative Neurology*, 78 Temple Road, Concord, MA 01742, USA

C. S. Potten Department of Epithelial Biology, Paterson Institute of Cancer Research, Christie Hospital & Holt Radium Institute, Wilmslow Road, Manchester M20 9BX, UK

M. K. Powers Department of Psychology, Vanderbilt University, Nashville, TN 37240, USA

J. C. Presson Department of Zoology, University of Maryland, College Park, MD 20742, USA

R. Pujol INSERM Unit 254, Laboratory of Auditory Neurobiology, Hopital St Charles, F-34059 Montpellier cedex 1, France

P. A. Raymond Department of Anatomy & Cell Biology, University of Michigan, Ann Arbor, MI 48109-0616, USA

D. S. Reasner (*Ciba Foundation Bursar*) Worcester Foundation for Experimental Biology, 222 Maple Avenue, Shrewsbury, MA 01545, USA

T. A. Reh Department of Biostructure SM-20, University of Washington, Seattle, WA 98195, USA

E. W. Rubel (*Chairman*) Hearing Development Laboratories, Department of Otolaryngology, Head & Neck Surgery, RL-30, University of Washington, Seattle, WA 98195, USA

B. M. Ryals Department of Audiology & Speech Pathology, James Madison University, Education Building, Room 333, Harrisonberg, VA 22807, USA

R. H. Steinberg Beckman Vision Center, University of California, School of Medicine, San Francisco, CA 94143-0730, USA

J. Voyvodic Department of Biology, University College, Gower Street, London WC1E 6BT, UK

F. M. Watt Room 602, Imperial Cancer Research Fund, PO Box 123, 44 Lincoln's Inn Fields, London WC2A 3PX, UK

H. H. Zakon Department of Zoology, University of Texas, Patterson Laboratory, Austin, TX 78712-1064, USA

Introduction

E. W. Rubel

Hearing Development Laboratories, Department of Otolaryngology, Head and Neck Surgery, RL-30, University of Washington, Seattle, WA 98195, USA

This symposium originated from the notion that it would be fun to bring together a group of people who study the regeneration of different cell populations but who do not usually have the chance to discuss common themes and problems. Subsequently, Greg Bock visited Seattle, the idea was discussed, and a group of participants was evolved. After that, the Ciba Foundation took over. The result is the present symposium.

Sensory receptor cells are our link with the environment: they connect us to the environment and allow us to act upon changes in the environment as well as perceive the consequences of these actions. Sensory receptors vary markedly between species and between receptor cell populations. In mammals, particularly in humans, many receptor cell populations are permanent; they are not replaced when they are damaged. Although the condition is not life threatening like heart disease or cancer, it represents a major alteration in the quality of life that can be a permanent change for as long as a person is alive. These changes have an impact on our society in major ways. About one in every 700–800 babies is born with significant hearing loss that severely compromises the development of communication and the number of people with serious hearing loss increases to 50% by the age of 65. A rough calculation suggests that something like 600 million people in the world suffer significant impairment of their hearing. More than 100 million people around the world are thought to have significant taste and smell disorders. Similar numbers can be found with respect to permanent vision disorders. Most receptor cell populations—although this isn't necessarily true of the olfactory and taste cells—aren't replaced after injury resulting from disease or other kinds of damage. Many investigators are attempting to understand why sensory receptors are replaced in some other groups of vertebrates, and why they cannot be replaced similarly in humans. Part of the impetus behind this symposium was to see if we can learn from these species and apply this knowledge to the replacement of receptors in our own species.

The organization of the meeting was an attempt to bring together people who work on the regeneration of receptor cell populations of various kinds, to determine the state of knowledge in various systems, and what we can learn from each other. In addition, by bringing in cell biologists who work on the control of mitosis, on the differentiation of cells, and on the factors regulating these processes, we hoped to be able to discover new approaches and new experiments that will provide new directions for our research. For three days we shared this information among ourselves and in this volume we hope to communicate what we've learned with our colleagues at large. My hope as chairman was that we would have very free-wheeling discussion periods, in which people working on sensory systems and receptor systems other than the one being immediately discussed would not hesitate to participate. This end was achieved, and I think the discussions that follow each chapter are at least as valuable as the chapters themselves. They provide a framework of the relative state of understanding in each sensory system. What is more, I think they are fun to read.

What I think we achieved from this meeting, and what I hope the reader will achieve, is increased information about sensory systems other than the one each of us works on, as well as an understanding of the present state of knowledge on regeneration in other systems. Even more important, from the meeting and, I hope, from the volume, we shall be able to return to our laboratories and do different, better, experiments. The major purpose is to be able to take what we learn here, go back, and then modify our research programmes in order to achieve a better understanding of the ways of Nature.

The general architecture of sensory neuroepithelia

Sanford L. Palay

Department of Anatomy and Cellular Biology, Harvard Medical School, Boston, MA 02115, USA and The Journal of Comparative Neurology, 78 Temple Road, Concord, MA 01742, USA

Abstract. All neuroepithelia are sheets of cells lining an internal or external surface of the body and resting on a basement membrane. They consist of at least two kinds of cell, receptor cells and sustentacular (supporting) cells. Some contain undifferentiated precursor cells and senescent or degenerating cells. The potential for plasticity and regeneration in different sensory neuroepithelia varies widely according to their origins and structure in any individual animal and according to the species in which they occur. Four sensory neuroepithelia are described as examples of the range of construction, complexity, and life history.

1991 Regeneration of vertebrate sensory receptor cells. Wiley, Chichester (Ciba Foundation Symposium 160) p 3–24

Although neuroepithelia differ in their embryological histories, in their specific structure, and in their capacity for regeneration, they display certain common architectural features. First of all, they are epithelial; that is, they are sheets of more or less similar cells resting upon a basal lamina and mesodermal connective tissue and fronting upon an external or internal open space. Second, they consist of at least two types of cell, specific sensorireceptor cells and supporting cells. Many of them contain a third cell type, precursors that give rise to the replacements of the other two cells. Third, the free or apical surfaces of the receptor cells, and in some cases also of the supporting cells, are decorated with elaborate motile or non-motile appendages—cilia or microvilli. Often these appendages protrude into or are attached to a dense extracellular secretory material, which may have an important role in the function of the ensemble as a sensory receptor organ. Fourth, the sensory elements are connected to the central nervous system by way of specific nerves, which either originate from the sensorireceptor cells themselves or synapse with them. In this way the neuroepithelia differ from a variety of other peripheral sense organs in which the sensory element is itself a terminal of a nerve fibre originating from a dorsal root ganglion cell. These general features accommodate a wide variety of different receptor organs responsive to photic, mechanical, kinetic, electrical, or chemical stimuli.

In order to cover this enormous variety of structure and function, I would have to go into much more detail than the space allotted to this communication allows. Therefore I have chosen to describe only four kinds of neuroepithelia, which illustrate both their common features and a range of problems involved in understanding their architecture. In addition I shall include some remarks about their development.

Taste buds

I begin with the taste bud because it appears to be a simple cluster of cells (Fig. 1). Actually its structure has been the centre of considerable controversy, and it exemplifies many of the problems associated with understanding the organization and life histories of neuroepithelia. A taste bud consists of a rounded cluster of epithelial cells mostly oriented in a vertical direction with their bases lying on the basement membrane and their apices extending microvilli into a narrow channel on the free surface, the taste pore. The cells are of at least two types, gustatory receptor cells and sustentacular (supporting) cells, but the categories of cells vary in different species and in different kinds of taste buds on different parts of the tongue (or different parts of the body in those animals that have taste buds on their exposed surfaces). The species variations are at least partly responsible for the vigorous contention that has arisen over the question of cell types in the gustatory organs during the past century and is still going on. Not only are there different numbers of cell types, but also the cell types are differently described in different species of animals and kinds of taste organs.

In taste buds on the rabbit's tongue, four cell types have been described (Murray 1986, Kinnamon 1987, Royer & Kinnamon 1991). Type I cells are fusiform cells extending from the basal lamina to the taste pore. The apical cytoplasm of these cells contains characteristic large dense granules and extends into fine short microvilli. Type I cells enclose their neighbouring cells and intragemmal nerve fibres in extended lamellar processes similar to those of Schwann cells around unmyelinated nerve fibres. Type II cells are also fusiform but lack the lamellar processes and the dense vesicles of the Type I cells. In addition, they are fairly large plump cells with vacuolated cytoplasm. Type III cells are generally thin, dark, and fusiform. They contain characteristic dense core vesicles, 80–140 nm in diameter, which crowd the basal and perinuclear cytoplasm, but decrease in concentration in the apical cytoplasm, where they are replaced by clear vesicles. The tips of Type III cells are crowned by elongated extensions that surpass the microvilli of the other cell types. Type IV cells lie at the base of the taste bud and do not possess apical processes. They are usually dark, triangular cells. In addition to these four cell types, other cells may be distinguished that have dark cytoplasm or that contain dense lamellar inclusions.

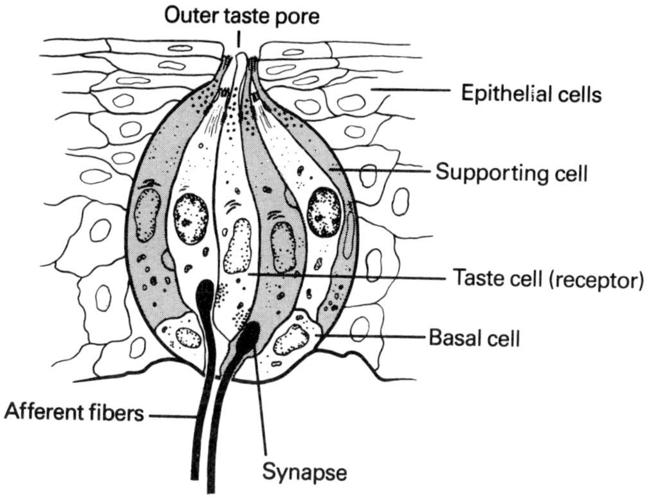

FIG. 1. Diagram of a taste bud. (From Kandel & Schwartz 1985, with permission.)

The appearance of these cells suggests that they are undergoing degeneration and that they may be end stages in the life cycle of Type III cells.

Which of these cell types should be regarded as sustentacular cells and which as receptor cells? In the rabbit, only Type III cells are in synaptic contact with nerve terminals. The presynaptic zones are usually round patches on the basal or perinuclear regions of the cell and are marked by prominent dense fibrillar material and dense projections attached to the cytoplasmic surface of the cell. Small clear vesicles aggregate in small or large clumps in this presynaptic cytoplasm. The tips of the afferent nerve fibres contain vesicles of various sizes, but no obviously specialized modification of the postsynaptic membrane. In the taste buds of mice, in contrast, all types of cells—dark, intermediate and light—engage in synapses with nerve fibres (Kinnamon et al 1985, Royer & Kinnamon 1988). Thus, if the presence of innervation is a minimal criterion for identification of the sensorireceptor cell in taste buds (Royer & Kinnamon 1991), different cell types can be sensorireceptor cells in different species, and indeed in different taste buds in any individual animal.

This inherent ambiguity has an important bearing on our understanding of the development of the cells in the taste bud. It has long been known that the cellular components of taste buds continually undergo renewal. The life cycle of any one cell occupies about fourteen to twenty-one days. Occasional mitotic figures seen among the basal cells suggests that these Type IV cells are stem cells that give rise to the other cell types. But there is considerable debate about whether the different types represent stages in a continuous cell line in which Type III is the final mature stage, or whether each cell type represents the product of a different cell line. Experiments in which taste buds are allowed to degenerate and regenerate after transection of the glossopharyngeal nerve have produced conflicting results. It is entirely possible that there are species differences in the generation of cells in taste buds corresponding to the differences in the distribution of innervation within the taste bud.

Both the development and persistence of taste buds require the presence of gustatory nerve fibres. When these specific nerve fibres are interrupted the taste buds atrophy and when the nerve fibres regenerate the taste buds reappear. As the gustatory cells are continually undergoing desquamation and replacement, the terminals of the nerve fibres must continually form fresh synaptic connections on the new cells. The discovery of how this is accomplished should be a fascinating study in neuroectodermal plasticity.

The olfactory epithelium

Taste buds are not the only neuroepithelial structures that undergo continuous renewal. The olfactory epithelium sustains a similar turnover of its component cells. Like other neuroepithelia, the olfactory epithelium is stratified and consists of three cell types—sustentacular cells, receptor cells, and basal cells (Fig. 2). The receptor cells, however, are unique in being true nerve cells, with dendritic processes extending to the surface and axonal processes extending to the brain, where they end in complicated synaptic formations known as the olfactory glomeruli. These are the only nerve cells in vertebrates that reside on the surface of the body, exposed to the outside world. They are also the only nerve cells in mammals that continue to be generated throughout the life of the animal (Graziadei & Monti-Graziadei 1979, Costanzo & Graziadei 1987). The receptor cells have a life cycle of 4–8 weeks. When degenerating cells die, their axons also disintegrate and are replaced by the growing axons of new receptor cells in the epithelium. These axons grow along the surviving axons in the olfactory nerves and must establish new connections in appropriate glomeruli in the olfactory bulb. The mechanisms that regulate and guide this natural and normal regenerative process might provide insights for ameliorating and enhancing plastic repair under pathological conditions in the central nervous system.

The olfactory epithelium originates from two embryological sources, the superficial ectoderm and the olfactory placode, a broad thickening in the ectoderm that forms very early in embryonic life almost simultaneously with the neural plate (Costanzo & Graziadei 1987). As development of the head end of the embryo proceeds, the cells of the placode and the superficial ectoderm intermingle, the latter providing the precursors of the sustentacular cells and the placode cells giving rise to the neuronal, sensory receptor cells of the olfactory epithelium. Undifferentiated cells, presumably from both cell lines, persist in the depths of the epithelium as the basal cells, which give rise to replacements of the other two, differentiated cells throughout the life of the organism.

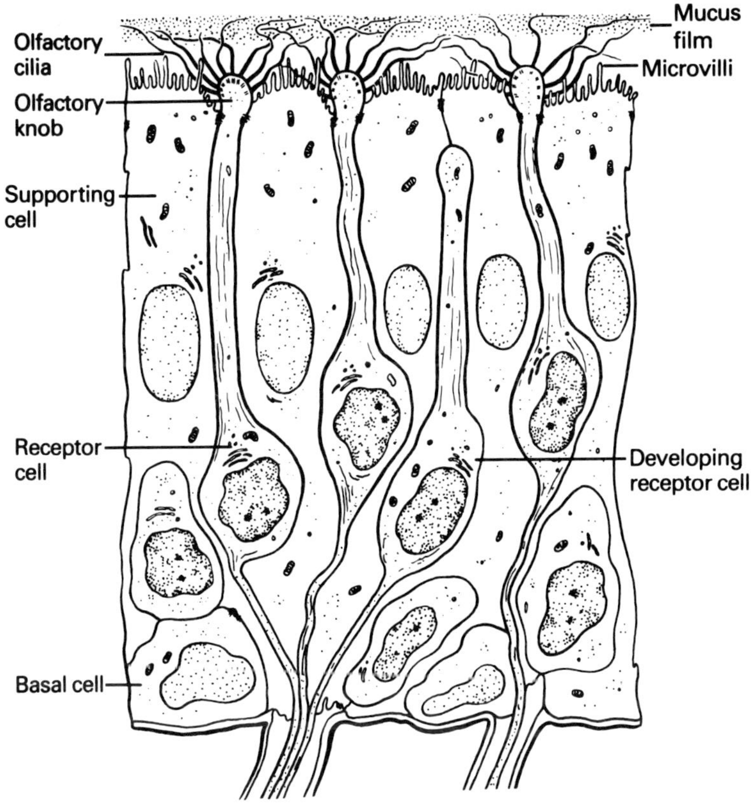

FIG. 2. Diagram of the structure of the olfactory epithelium. The palisade of nuclei in the middle third of the epithelium belong to the sustentacular (supporting) cells. Receptor cells are shown in two stages of maturation, and their axons leaving the epithelium are ensheathed in Schwann cells, whose nuclei are not shown. (From Kandel & Schwartz 1985, with permission.)

In the adult olfactory epithelium the sustentacular cells are arrayed in a sheet continuous with the ordinary pseudo-stratified columnar epithelium of the upper nasal passages (Morrison & Costanzo 1990). The ordinary columnar cells are ciliated, but the sustentacular cells of the olfactory sensory epithelium have only microvilli extending from their apices into the mucus overlying the free surface (Fig. 2). The nuclei of the sustentacular cells lie in the middle and upper thirds of the epithelium. The apical cytoplasm displays a thick terminal web and prominent desmosomes and tight junctions that bind neighbouring cells together. Axial intermediate filaments are prominent in the cytoplasm beneath the nuclei. The sustentacular cells, like epithelial cells elsewhere, have a brief lifespan, but their turnover rate is slower than that of the receptor cells in the same olfactory epithelium (Costanzo & Graziadei 1987).

The sensorireceptor cells, as mentioned, are true nerve cells. Their light, vesicular nuclei lie in the middle and lower thirds of the epithelium, below those of the sustentacular cells. The two cell types form a columnar unit with sustentacular cells in the centre and receptor cells arranged in a circle around them (Graziadei & Monti-Graziadei 1979, Costanzo & Graziadei 1987). Basal cells cluster at the bottoms of the columns. The receptor cell body is globular and contains a meagre complement of the usual cytoplasmic organelles, including some neurofilaments. The apical process, or dendrite, is a slender extension bound with junctional complexes to the sustentacular cells at the level of the terminal webs of the latter. Beyond this level, the dendrite protrudes as a rounded expansion, the so-called olfactory vesicle or dendritic knob, from which extend numerous cilia. In frogs (Reese 1965), each cell bears six to eight cilia, which start out as conventional cilia with basal bodies and the usual 9+2 array of axial microtubules. But, a short distance from their origin, the cilia suddenly narrow into slender tubes containing eleven single, parallel microtubules. The cilia are non-motile and splay into the mucous layer from the top of the olfactory vesicle like limp flower petals. In the human olfactory epithelium (Morrison & Costanzo 1990), each olfactory receptor cell can bear ten to thirty cilia, which taper gradually from their origins to their tips without alteration in their internal microtubular structure. The basal process of the cell is an unmyelinated axon, which extends through the basal lamina of the epithelium to be bundled together with its congeners into the filaments of the olfactory nerve. These axons are ensheathed by Schwann cells and terminate in the glomeruli of the olfactory bulb, where they synapse upon mitral cells.

In addition to receptor cells and sustentacular cells, a microvillar cell has recently been identified in the human olfactory epithelium (Moran et al 1982, Morrison & Costanzo 1990). This cell resembles the receptor cell, but displays only a dense brush of microvilli on its apical surface and no dendritic knob or ciliary crown. It is thought to have some receptor function.

Basal cells are pyramidal cells wedged among the basal processes of the sustentacular cells and the exiting axons of the receptor cells. They lack extensions that reach the free surface of the epithelium. Because mitoses are sometimes found among them, they have long been considered as stem cells for replacing the other cells in the olfactory epithelium. Studies with tritiated thymidine have confirmed this opinion (Hinds et al 1984). Stages in the life history of the receptor cells can be recognized in the epithelium, extending from the precursor basal cell through a cell resembling immature neurons in the central nervous system to the mature elongated receptor with its crown of peculiar cilia (Fig. 3; Graziadei & Monti-Graziadei 1978). Labelling with tritiated thymidine also shows that the distribution of dividing cells throughout the epithelium is uneven, with patches of mitotically active regions containing many immature receptors and large unlabelled areas containing mostly mature receptors (Graziadei & Monti-Graziadei 1979). The labelled regions are considered to be

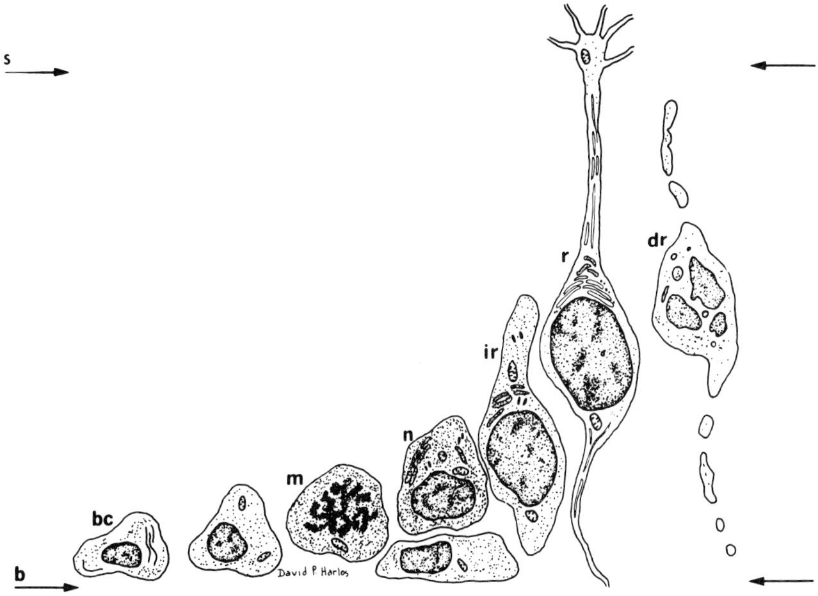

FIG. 3. Developmental stages in the life history of olfactory receptor cells. Arrows indicate the locations of the free surface (s) of the epithelium and the basal lamina (b) on which it rests. bc, basal cell; m, basal cell in mitosis; n, neuroblast; ir, immature receptor; r, mature receptor with dendritic knob and cilia; dr, degenerating receptor. (From Costanzo & Graziadei 1987, with permission.)

actively neurogenic whereas the others are thought to be quiescent. In addition, the numbers of basal cells vary according to their location in the olfactory epithelium. They are much more numerous in the active, neurogenic zones than in the quiescent ones.

The organ of Corti

As an example of a neuroepithelium concerned with the transduction of mechanical stimuli, I have chosen the mammalian organ of Corti, the peripheral organ of hearing. The organ of Corti is a specialized part of the cochlear duct, which courses through the spiral osseous cochlear canal. A thin fibrous sheet, the basilar membrane, stretches across the lumen of this osseous canal, dividing it along its entire length into two parallel tubular chambers. Another sheet, Reissner's membrane, further subdivides the upper chamber. Thus the lumen of the osseous canal is divided into three parallel spiral compartments, each lined by squamous or specialized epithelium and filled with fluid. The upper compartment is known as the scala vestibuli, the lower one as the scala tympani, and the one in between, bounded by Reissner's membrane and the basilar membrane, is the scala media or the cochlear duct. The scala vestibuli and the scala tympani communicate with each other at the apex of the cochlea. They are filled with perilymphatic fluid, whereas the cochlear duct contains endolymphatic fluid and is continuous with the semicircular canals and associated structures of the vestibular portion of the labyrinth.

The wall of the cochlear duct consists of several differentiated and highly developed areas, but I shall confine my description to the architecture of the organ of Corti, as it is the only part that is considered to be a neuroepithelium. The organ of Corti is a complex specialization appearing as a ridge of cells in the epithelium of the cochlear duct lying upon the basilar membrane (Fig. 4). It consists of rows of sensorireceptor cells, the hair cells, and a variety of supporting cells. All of the supporting cells extend from the basilar membrane to the free surface of the organ, but they are separated from one another by large intercellular spaces except at their apices. The largest of these spaces is the inner tunnel, which runs the length of the organ of Corti and is bounded by the inner and outer pillar cells (rod cells) leaning toward each other. The pillar cells contain dense arrays of microtubules collected together with 6 nm thick microfilaments into axial bundles that course vertically through the cells from base to apex. The intercellular spaces between the pillar cells communicate with the intercellular spaces between the other supporting cells, particularly the neighbouring space of Nuel and the outer tunnel. A row of inner phalangeal cells lies on the inner side of the inner pillar cells and completely surrounds the

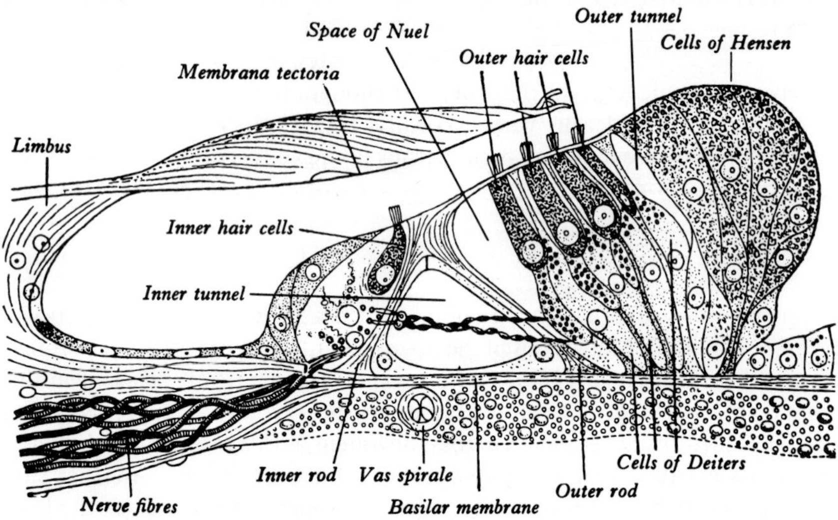

FIG. 4. Diagram of a transverse section through the organ of Corti. The separation between the tectorial membrane and the tips of the stereocilia on the inner and outer hair cells is an artifact. (From Warwick & Williams 1973, with permission. ©Longman Group UK Ltd. Published by Churchill Livingstone.)

single row of inner hair cells, except at their apical surfaces. Unlike the outer phalangeal cells, the inner ones are closely apposed and are not separated by enlarged intercellular spaces. The outer phalangeal cells (cells of Deiters) are columnar cells standing on the basilar membrane and enclosing the lower thirds of the outer hair cells in their cuplike apices. From each outer phalangeal cell a thin process reaches the surface of the organ where it expands into a flat plate attached to the adjacent hair cells. Both the outer phalangeal cell and its slender process contain an axial bundle of microtubules.

The sensorireceptor cells are hair cells arranged in regular rows on either side of the inner tunnel. The inner hair cells form a single row of columnar cells, each one lifted off the basilar membrane and enveloped in an inner phalangeal cell. The free apical surface of each inner hair cell is fitted with a cluster of microvilli arranged in the form of an arc with its concavity facing away from the inner pillars. The microvilli contain large numbers of axial filaments that extend deep into the thick terminal web at the apex of the cell. The outer hair cells line up in three parallel rows running the length of the organ of Corti. Each cell rests in the cuplike top of an outer phalangeal cell and a W-shaped cluster of microvilli extends from its apical surface. The concavities of the W face the outer pillar cells.

The microvilli of the hair cells are known as stereocilia, so named from the mistaken notion that they were non-motile cilia. But they do not have the internal fine structure of cilia, although there is a basal body and an associated centriole in the apical cytoplasm under the peak of the W or the arc of microvilli. The microvilli of the hair cells stretch out toward the tectorial membrane, into which their tips are embedded. The tectorial membrane is an extracellular proteinaceous material secreted by the cells covering the limbus at the inner angle of the scala media. Vibrations induced in the basilar membrane by sound waves transmitted to the fluid in the scala vestibuli and scala tympani distort the stiff stereocilia with a resultant depolarization or hyperpolarization of the hair cell, depending upon direction. The change in potential results in the release of transmitter from the base of the cell onto terminals of afferent nerve fibres, which are activated to carry impulses to the central nervous system.

The afferent fibres consist of two types originating from different types of cell bodies in the spiral ganglion (Simmons & Liberman 1988a). About 95% of the cells are Type I cells, which give rise to large, heavily myelinated fibres, the radial fibres, that contact individual inner hair cells at their bases and around their sides with large boutons containing few vesicles. The radial fibres ramify only sparsely and individual ones end on a single inner hair cell. Any one hair cell may receive afferent endings from several Type I cells. In contrast, Type II cells give rise to unmyelinated, thin nerve fibres, the spiral fibres, that enter the organ of Corti with the radial fibres but then cross the floor of the inner tunnel and turn to run along the rows of outer phalangeal (Deiters) cells. After ramifying, the spiral fibres give off short digitiform processes that end in small boutons on the basal surfaces of the outer hair cells. Each fibre contacts as many as a hundred outer hair cells, largely in a single row. Furthermore, each outer hair cell receives the endings of several different spiral fibres (Liberman 1980, Simmons & Liberman 1988a,b, Liberman et al 1990).

In addition to this afferent innervation, there is an efferent system, originating in the superior olivary complex. Olivocochlear fibres coming from the medial periolivary cells terminate as large vesicle-filled endings on the outer hair cells, whereas efferent fibres coming from the lateral periolivary cells terminate as small boutons mostly on the endings of radial fibres around inner hair cells (Brown et al 1988). This efferent system is thought to inhibit activation of the afferents.

Like the olfactory epithelium, the membranous labyrinth originates from a placode, a thickening of the ectoderm. This placode forms dorsal to the first branchial groove and, invaginating, sinks into the mesenchyme between the myelencephalon and the metencephalon of the brain, as the otic vesicle. The epithelium lining this vesicle enlarges and differentiates into the membranous labyrinth with its several vestibular and cochlear neuroepithelial specializations.

In mammals the cochlear epithelium is apparently not capable of regenerating after destruction of the organ of Corti. Hair cells that are killed by toxic substances or by excessive stimulation are permanently lost.

The retina

Of the neuroepithelia to be considered in this symposium, the retina is the most complicated. It is also different from all the others in its origin directly from the rostral end of the neural tube instead of from an ectodermal placode. It really is a peripherally situated part of the central nervous system. It starts out as an evagination from the prosencephalon attached to the brain by the optic stalk. Later local invagination transforms it into the optic cup, a bilaminar epithelium turned in on itself. The inner lamina differentiates into the neural retina, while the outer lamina remains fairly simple and becomes the pigment epithelium. The invagination practically obliterates the lumen of the vesicle and accounts for the fact that the photosensitive elements, the rods and cones, of the neural retina face into the pigment retina instead of toward the light. The optic stalk becomes transformed into the optic nerve by the growth of axons from the innermost layer of retinal nerve cells toward the brain through intercellular spaces in the ependymal lining of the stalk.

The complexity of the retina is such that I shall confine my remarks to its general character in relation to the topic of this symposium. Detailed descriptions are available in numerous textbooks and monographs, the most recent being the book by Dowling (1987). The retina is a conspicuously laminated structure. The outer layer of rods and cones and the alternating layers of nuclei and interwoven nerve cell processes form one of the images of the nervous system that is most familiar to students of biology (Fig. 5). Although the retina can be

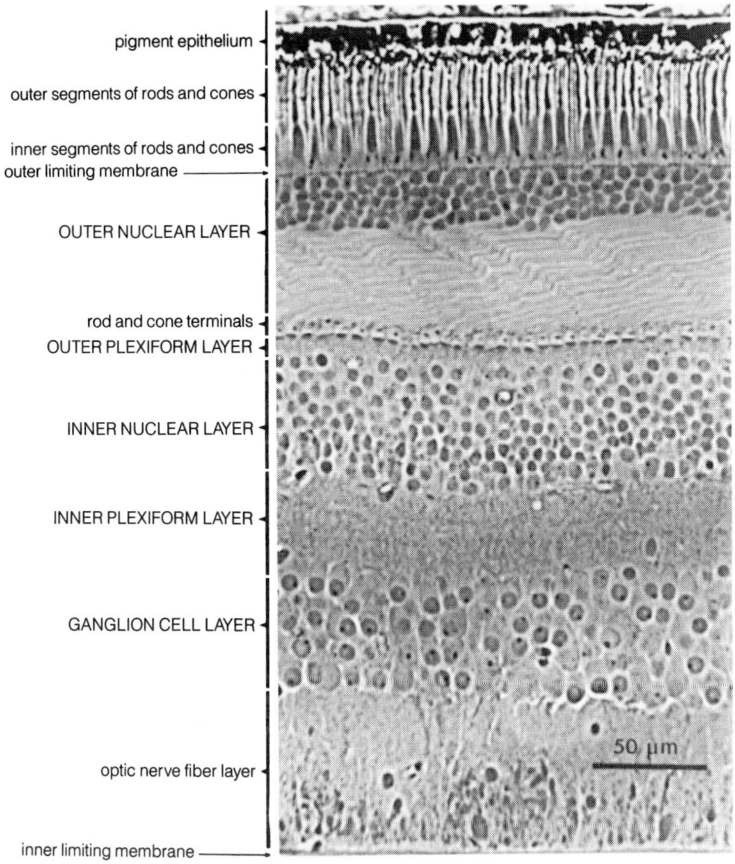

FIG. 5. Photomicrograph of a section through a human retina, showing the characteristic orderly lamination of its cells. (From Dowling 1987, with permission.)

described in the same terms as the other neuroepithelia—that is, as an assemblage of receptor cells and sustentacular cells resting on a basal lamina—its additional components, the vast numbers of nerve cells and their processes, give it an altogether different aspect of complexity. This is a structure that does not simply transduce external energy into nerve impulses, but also contributes to the analysis and refinement of those impulses so that the brain can construct a coherent and integrated visual image from them.

The outermost layer of the retina is the pigment epithelium, which derives from the outer lamina of the embryonic optic cup. This epithelium consists of a single layer of pigment-containing cuboidal cells connected to each other by typical junctional complexes. The apical surface of the cells is thrown up into folds that surround the outer segments of the receptor cells and into numerous slender microvilli. Although there are no structural connections between the photoreceptor layer and the pigment epithelium, the intimate contact between the two layers is necessary for the normal function of the photoreceptor cells. In addition to absorbing stray light, the pigment cells phagocytose the membranous tips of the photoreceptors, which are continually being cast off, and they also store vitamin A, which is a necessary precursor of the photosensitive pigments in the rods and cones.

The sustentacular cells of the neural retina, the Müller cells, correspond to the neuroglial cells of the central nervous system. They are radially oriented cells that extend through the whole thickness of the neural retina. Their processes are velate and filiform, insinuating themselves among the neural elements like the protoplasmic astrocytes of the brain. Müller cells contain vimentin filaments and form junctional complexes with the photoreceptor cells and one another at the outer limiting membrane, and again with one another at the inner limiting membrane, in effect defining the neural retina.

The photosensitive elements in the neural retina are the terminal projections of cell bodies densely packed into the outer nuclear layer. The apical process of each of these cells extends into the pigment epithelium and is divisible into two parts, an outer and an inner segment. The outer segment contains the photosensitive pigment. On the basis of the shape of the outer segment, the photoreceptor cells can be classified into two types, rods and cones. The difference in shape is, however, not always easy to discern, and the inner segments are often an easier criterion. The outer segment of the rod is a slender threadlike process of uniform calibre containing a stack of collapsed membranous discs. The inner segment, which is slightly thicker than the outer segment, contains numerous longitudinally oriented mitochondria in its outer part (the ellipsoid) and the other cytoplasmic organelles in the inner part (the myoid). The two segments are connected by a short, very narrow stalk containing nine pairs of longitudinal microtubules extending from a basal body in the tip of the inner segment. This connecting stalk is the vestige of the ciliary process out of which the outer segment derived.

The inner and outer segments of cones have an architecture similar to those of the rods, but there are certain differences of shape and detailed structure. Cone outer segments tend to taper toward their outer tips and the inner segment is usually broader and more obviously conical than that of the rods. In the fovea of primates, however, the cones are packed so closely together and are so attenuated that there is hardly any noticeable morphological difference between them and rods elsewhere.

The cell bodies of the photoreceptors give rise to axonal processes that extend inward to the outer plexiform layer, where they terminate in characteristic presynaptic expansions. Both the rods and the cones synapse with bipolar cells and with horizontal cells in complex synaptic formations. In addition, the terminal expansions of the rod and cone axons may themselves be interconnected by punctate or elongated gap junctions.

The perikarya of the bipolar cells make up the inner nuclear layer. Their dendrites reach into the outer plexiform layer to synapse with the terminals of the photoreceptor cells, whereas their axons descend into the inner plexiform layer to synapse with the dendrites of retinal ganglion cells and amacrine cells. The ganglion cells make up the innermost cellular layer of the retina, and they give rise to the optic nerve fibres, which course in a fibre layer on the innermost surface of the retina to the optic disc, where they penetrate through the thickness of the retina and, becoming myelinated, constitute the optic nerve.

The simple three-neuron chain described in the retina—photoreceptor to bipolar to ganglion cell—is a caricature of the circuitry in this neuroepithelium (Fig. 6). Not only are there many different kinds of each one of these primordial

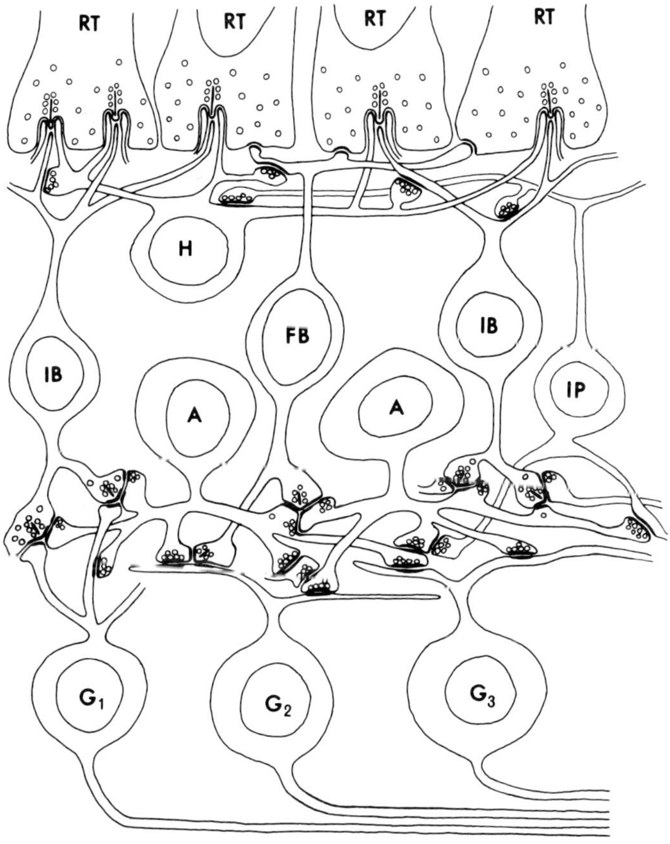

FIG. 6. Diagram of the principal synaptic interconnections in the vertebrate retina. Synapses are represented by thickened lines and associated synaptic vesicles, sometimes together with synaptic ribbons. RT, receptor terminal; H, horizontal cell, IB, invaginating bipolar cell; FB, flat bipolar cell; A, amacrine cell; G_{1-3}, ganglion cells with various kinds of input from bipolar and amacrine cells. (From Dowling 1987, with permission.)

types of cell, but there are also the horizontal cells lying in the outer rows of the inner nuclear layer and the amacrine cells lying in the innermost inner nuclear layer and even in the ganglion cell layer. And besides these elements, there are the interplexiform cells, which lie in the inner nuclear layer and send dendritic processes into the inner plexiform layer and axonal processes into the outer plexiform layer. Furthermore, one must remember the efferent fibres emanating from the brain (for example, the isthmo-optic nucleus in birds) and terminating on amacrine cells of the inner nuclear layer. All of these varieties of cells and their complex interactions, both morphological and physiological, lead to an intricate circuitry that is only now beginning to yield its secrets.

More to the point of this symposium, it is important to emphasize that this neuroepithelium has a distinctly different potential for plasticity and regeneration in different species and classes of vertebrates. In mammals and birds damage to the retina cannot be repaired by regeneration of the lost neural elements. In fishes and certain amphibians the retina continues to grow during the life of the animal; it can be restored after injury, and even, in some experimental situations, completely re-formed after being destroyed.

Conclusion

Neuroepithelia range in complexity from simple clusters of specialized cells to complex interconnected neural assemblies. In all cases their integrity depends upon the existence of connections with the nervous system, whether they give rise to the connections, as in the olfactory epithelium and the retina, or are merely innervated by peripheral sensory nerves. The capacity of neuroepithelia for regeneration varies not only according to the species of animal but also according to the age, structure, and natural history of the organs themselves.

References

Brown MC, Liberman MC, Benson TE, Ryugo DK 1988 Brainstem branches from olivocochlear axons in cat and rodents. J Comp Neurol 278:591–603

Costanzo RM, Graziadei PPC 1987 Development and plasticity of the olfactory system. In: Finger TE, Silver WL (eds) Neurobiology of taste and smell. Wiley, New York, p 233–250

Dowling JE 1987 The retina, an approachable part of the brain. Harvard University Press, Cambridge, MA

Graziadei PPC, Monti-Graziadei GA 1978 The olfactory system: a model for the study of neurogenesis and axon regeneration in mammals. In: Cotman CW (ed) Neuronal plasticity. Raven Press, New York, p 131–153

Graziadei PPC, Monti-Graziadei GA 1979 Neurogenesis and neuron regeneration in the olfactory system of mammals. I. Morphological aspects of differentiation and structural organization of the olfactory sensory neurons. J Neurocytol 8:1–18

Hinds JW, Hinds PL, McNelly NA 1984 An autoradiographic study of the mouse olfactory epithelium. Evidence of long lived receptors. Anat Rec 210:375–383

Kandel E, Schwartz J 1985 Principles of neural science, 2nd edn. Elsevier, New York
Kinnamon JC 1987 Organization and innervation of taste buds. In: Finger TE, Silver WL (eds) Neurobiology of taste and smell. Wiley, New York, p 277–297
Kinnamon JC, Taylor BJ, Delay RJ, Roper SD 1985 Ultrastructure of mouse vallate taste buds. I. Taste cells and their associated synapses. J Comp Neurol 235:48–60
Liberman MC 1980 Morphological differences among radial afferent fibers in the cat cochlea: an electron-microscopic study of serial sections. Hear Res 3:45–63
Liberman MC, Dodds LW, Pierce S 1990 Afferent and efferent innervation of the cat cochlea: quantitative analysis with light and electron microscopy. J Comp Neurol 301:443–460
Moran DT, Rowley JC, Jafek BW 1982 Electron microscopy of human olfactory epithelium reveals a new cell type: the microvillar cell. Brain Res 253:39–46
Morrison EE, Costanzo RM 1990 Morphology of the human olfactory epithelium. J Comp Neurol 297:1–13
Murray RG 1986 The mammalian taste bud type III cell: a critical analysis. J Ultrastruct Mol Struct Res 95:175–188
Reese TS 1965 Olfactory cilia in the frog. J Cell Biol 25:209–215
Royer SM, Kinnamon JC 1988 Ultrastructure of mouse foliate taste buds: synaptic and non-synaptic interactions between taste cells and nerve fibers. J Comp Neurol 270:11–24
Royer SM, Kinnamon JC 1991 HVEM serial section analysis of rabbit foliate taste buds: Type III cells and their synapses. J Comp Neurol 306:49–72
Simmons DD, Liberman MC 1988a Afferent innervation of outer hair cells in adult cats. I. Light microscopic analysis of fibers labeled with horseradish peroxidase. J Comp Neurol 270:132–144
Simmons DD, Liberman MC 1988b Afferent innervation of outer hair cells in adult cats. II. Electron microscopic analysis of fibers labeled with horseradish peroxidase. J Comp Neurol 270:145–154
Warwick R, Williams PL 1973 Gray's anatomy, 35th edn. Longman, London, p 1155

DISCUSSION

Raymond: Dr Palay, you mentioned that basal cells in taste buds of certain vertebrate species receive synapses. Could you elaborate on that?

Palay: In the mouse, Kinnamon and colleagues have looked in light and electron microscope preparations for synapses, because they wanted to define the gustatory receptor cell. They found in the mouse that all three of the intragemmal cell types, including basal cells, are innervated, whereas in the rabbit, this is not so (Kinnamon et al 1988, Royer & Kinnamon 1988).

Rubel: How sure are they that it's not true in the rabbit?

Palay: Kinnamon does say that it's possible that there are varieties of the innervated cell in the rabbit that may be closer to the basal cell in the mouse, so that as the intragemmal cell is developing from a basal cell to the definitive receptor cell, it may be innervated. This is not commonly found, however, in the rabbit.

Raymond: The basal cells are usually thought to be a mitotic population. Do mitotic cells receive synapses?

Palay: Apparently the basal cells in mitosis were not innervated, in rabbits.

Raymond: So the basal cell population could include not only a mitotic cell population, but also recently postmitotic cells which are beginning to differentiate?

Palay: Yes. With a cell that is in the process of differentiating, there is a threshold at which you can recognize when it has really begun to differentiate. So some of the basal cells may be further along in this process than others.

Rubel: As with psychophysics, as our judgement gets better, or as different markers become available, we see earlier signs of when a cell's fate is distinguishable.

Margolis: Are there adequate biochemical or immunohistochemical markers with which to distinguish the various stages of receptor cells in the taste system?

Palay: There are some markers, namely whether electron-dense granules are present or not, and whether the gustatory receptor cells have what appear to be monoamine granules.

Farbman: Studies on cell division in taste buds show that [^{3}H]thymidine is taken up mainly by epithelial cells just lateral to the taste bud. Epithelial cells, then, divide outside of the taste bud; then postmitotic cells move into the taste bud and become basal cells (Beidler & Smallman 1965). There have been occasional observations of mitotic figures within the taste bud proper, but the usual situation is that taste bud precursor cells are outside the bud and migrate in.

Raymond: Do they have to cross the basal lamina to reach the taste bud?

Farbman: No. They are the basal cells of the stratified squamous lingual epithelium lateral to the taste bud.

Raymond: What is the capsule around the taste bud?

Farbman: The 'capsule' consists of non-gustatory epithelial cells that are flattened against the periphery of the taste bud. They may be a special kind of glia-like cell, but so far as I know there are no markers that distinguish these cells from other non-gustatory epithelial cells nearby. The taste cell precursors would have to move laterally along the basement membrane to enter the taste bud. Whether they must cross a 'capsular' barrier or whether they must pass through a stage of differentiation as a peripheral or capsular cell is not known. In any event, it is apparent in electron micrographs where the non-gustatory epithelium ends and the taste bud epithelium begins.

Fernald: So are the basal cells intercalated among existing differentiated cells? And where do you find the other member of the pair of cells produced by mitosis?

Farbman: One daughter cell remains at the periphery, outside the bud, and the other enters the bud.

Rubel: Do the postmitotic cells slide across the stem cells?

Farbman: That isn't known. By Day 2 after [^{3}H]thymidine injection, the label is within the bud. That is all we can say.

Presson: On this question of the migration of the precursor cells in the taste bud, from the discussion I have a picture of active migration, where the precursor cells are coming from the outside and crossing some sort of boundary; but there is also the possibility that the migration may be passive, with only the precursor cells on the outside of the taste bud proliferating. Have timed [^{3}H]thymidine studies been done, to determine whether there is active migration or whether these precursor cells are just passively being pushed in, and maybe those cells then define the boundary of the taste bud?

Rubel: Are you really asking whether these are motile cells or not?

Presson: Yes; are they moving across other cells and then growing up, or are they just coming in laterally and the whole taste bud is being built by passive proliferation on the outside, which makes it look as if precursors are moving in?

Farbman: The picture hasn't really been analysed carefully in that way, yet.

Reh: The only way to address that is to see whether the degenerating cells are typically found in the centre of the structure; that's the prediction, if it is being built from the side. Do you see degenerating cells primarily in the interior of the taste bud?

Farbman: Not necessarily. But one doesn't see many degenerating cells in the taste bud. The disposal mechanism, whatever it is, seems to be very quick.

Lewis: Can one say fairly confidently that the taste bud cells and the keratinizing epithelial cells are derived from a common precursor, or is it possible that there's a separate lineage for the taste bud?

Oakley: That's still an open question. One way to characterize some of the cell types is to look at their keratin polypeptides, of which there are at least 19 different types. The basal cells in the gustatory epithelium express keratins different from those of the receptor cells, and the suprabasal cells have yet another keratin expression (see this volume: Oakley 1991). The differentiated elongated cells within the taste bud express keratin 19 and none of the surrounding cells does. These changes in keratin expression may help one to track cell lineage.

Lewis: But nobody has done clonal marking experiments, where a single cell is marked and its progeny are monitored, to see whether the clone includes both taste bud cells and keratinizing epithelial cells?

Oakley: No.

Farbman: We have done the following. Taste buds on the anterior part of the tongue are found within the epithelium of the mushroom-shaped fungiform papillae. In rats, there is a single taste bud at the apical surface of the papilla. We amputated the apical half of the fungiform papilla, including the taste bud. Within a few weeks, the papilla regenerated and contained a taste bud which was physiologically functional. However, when we amputated the entire fungiform papilla, including its base, there was epithelial wound healing but no regeneration of a papilla or taste bud (Hellekant et al 1987). We believe that the cytoarchitecture of the papilla must be related to the organization of the epithelial cells at the papillary base, and once that is destroyed it cannot regenerate. It is also conceivable that taste cell precursors or stem cells originate in the epithelium at the papillary base and slide along the basement membrane toward the epithelium, somewhat as intestinal epithelial cells slide from the crypts of Lieberkühn along the basement membrane up the microvilli, as shown by Chris Potten and others.

Oakley: Another interpretation is that the mesenchymal tissue beneath the taste bud is left intact and is able to influence the basal cells in the area surrounding the papilla and turn these basal cells into progenitors for gustatory cells.

Farbman: Yes. It is also interesting that in lower vertebrates the situation is different. In taste buds of catfish barbels, some cells don't have an epithelial character; that is, they don't have desmosomes. They may be Merkel-like cells. They may originate from a source other than the epithelium, and thus belong to a different lineage, perhaps the neural crest.

Rubel: You think that these cells have migrated out from the neural crest?

Farbman: It's possible.

Lewis: Earlier studies (e.g. Stone 1940, Speidel 1947) talk of the lateral line and the taste bud as if they are closely analogous systems—as though taste buds, like lateral line organs, might originate from a placode that gives rise to cells that migrate within the epithelium. Is that plausible?

Farbman: Nobody has shown it!

Calof: I have a related question. What is known about the embryonic origins of the taste bud, in contrast to the rest of the tongue epithelium, in mammals or in fishes? There is nothing like an oral 'placode', is there?

Farbman: One interpretation is that there are many placodes on the surface of the mammalian tongue and that each fungiform papilla develops from a placode. In the earliest stages of development of rat fungiform papillae one sees small epithelial bulges made up of elongated cells containing fewer surface microvilli and more mitochondria, compared to the interpapillary epithelial cells. These placode-like cells are mitotically silent; i.e., they do not incorporate [^{3}H]thymidine, for at least a period of time, perhaps 12–24 hours, while the epithelial cells between them continue to divide. These placode-like structures develop before the nerves reach the epithelium—that is to say, they are evident in the 15-day rat embryo; moreover, they develop in organotypic cultures of tongue taken from 14-day rat embryos (Farbman & Mbiene 1991).

Calof: Do you think those spots are already 'determined'? That is, if you could remove them from the tongue to a different tissue environment, would their fate be the same?

Farbman: Dr Lewis mentioned the lateral line organ. When these placode-like structures first appear on the rat tongue, they are arranged linearly, parallel to and on each side of the lingual midline. Later, a second, more laterally placed line of them appears. There seems to be a midline-to-lateral gradient, and a gradient from the tongue tip, where they first appear, to the tongue posterior. Interestingly, the innervation enters the fungiform papilla in the same sequence as the appearance of the placodes, but our studies with the electron microscope and DiI tracing have shown that innervation occurs after placode development (Farbman & Mbiene 1991).

Corwin: Is it possible that these collections of cells are not individual 'placodes' that originate and develop independently, but instead are derivatives of a single primordial placode that might have been present if you had looked earlier in development? The collections of taste bud cells that develop in lines on the tongue might have originated from such a placode through processes analogous to those that give rise to the lateral line organs which develop from a single, migratory placode in amphibian embryos.

Farbman: So far, we have not found any evidence for that, but we are continuing to look for clues that it may happen.

Pujol: To revert to the problem of the innervation of the basal cells (i.e., cells which have not differentiated into sensory cells), in the mammalian cochlea it is possible to observe cells which are not completely differentiated and are already innervated (Pujol & Lavigne-Rebillard 1985). One can see synapse-like structures in these cells. So a cell which will develop into a true sensory cell already has some nerve connections at its base. This perhaps helps the cell to complete its differentiation. A similar feature seems to occur in the taste bud, I think?

Farbman: Yes. Sensory cell differentiation will happen later, after the nerves enter these little placodes; then they will differentiate. If the nerve fibres don't come in, they do not differentiate. Bruce Oakley has shown this (Oakley 1985).

Pujol: But the fact that you find that the undifferentiated basal cell has nerve contacts is not surprising, because this also occurs within the inner ear, for instance (Pujol & Sans 1986).

Rubel: The question is this: if there are two basal-like cells sitting next to each other, and one has a nerve on it and the other doesn't, is that predictive of which one will develop into a hair cell, or a receptor cell in the taste bud?

Oakley: It may be a misnomer to call all of these cells basal cells. One thinks of a basal cell *layer*; some of the cells that we are talking about may be higher up in the epithelium, as part of the postmitotic suprabasal cells which have already been contacted by nerve fibres and may well be on their way to differentiation into receptor cells.

Rubel: But in the taste bud those may become receptor cells, whereas in cochlea, during regeneration (or during embryogenesis), these undifferentiated cells can either become hair cells or become 'support' cells. I am not sure that we can discriminate, at that time, on the basis of whether they are innervated or not, what type they will become.

Pujol: In the human embryonic cochlea it is very clear that only the cells that have nerve contact at the base will become sensory cells.

Corwin: Speidel (1947, 1948) and Wright (1947) demonstrated that normal hair cells could develop in completely denervated regenerating epithelia in the lateral line systems of *Rana* tadpoles. Jørgensen & Flock (1976) confirmed that by transmission electron microscopy.

In analogous experiments, Cotanche and I denervated the basilar papilla sensory epithelia of cochleas from embryonic chicks three days before the cell divisions that give rise to hair cells. Then the denervated ears were transplanted to the aneural chorioallantoic membranes of host embryos and allowed to develop without innervation for 12 days. The result was normal development of the cochlear hair cells, even to the extent of developing the location-specific gradients in stereociliary ultrastructure (Corwin & Cotanche 1989).

Farbman: This apparently does not happen in taste bud development. When we removed the 14-day fetal rat tongue and placed it in organotypic culture, we saw small fungiform-like papillae at two days, but they were not sustained when cultures were grown for a week. At that time, typical filiform papillae were present, but no fungiform papillae were sustained, nor were taste buds present in the absence of an innervation (Farbman 1973). On the other hand, when the fetal glossopharyngeal ganglion was included in the culture, some taste buds developed (Farbman 1972).

Corwin: In the lateral line system of amphibians, *maintenance* of neuromast sensory epithelia is also nerve dependent; if the sensory epithelia remain denervated for a month or longer, they begin to disaggregate. Surprisingly, even ectopic innervation by a spinal nerve is sufficient to prevent disaggregation and maintain the neuromasts (Speidel 1948, Corwin et al 1989). There appear to be different degrees of dependence on innervation in different sensory systems. Perhaps the undifferentiated sensory cells are all capable of starting to differentiate without the need for neural contracts, but in some sensory systems the receptor cells may require innervation to complete their development; in other systems, innervation may be required for the maintenance of normal sensory epithelium organization once the cells have differentiated.

Burd: Will these mini-papillae on the tongue disappear if there is no innervation?

Farbman: Yes. They become spiky, filiform papillae. That's all you see in a non-innervated tongue.

Calof: What would happen if you were able to explant the mini-placode into an alternative environment?

Farbman: This was attempted by Mistretta et al (1979). Grafts of tongue epithelium were made to an external cheek site in fetal sheep, a species in which taste bud development occurs prenatally. They found that gustatory papillae and taste buds did not develop under these conditions; only the filiform papillae were evident. Others have transplanted tongue fragments into the anterior chamber of the eye, both alone and in combination with either the vagus nerve sensory ganglion, or a sensory ganglion from the lumbar region. In both cases, taste buds developed (Zalewski 1972, 1973). However, the interpretation of these experiments is confounded by the fact that nerves ordinarily innervating the eye grow into these grafts as well (Farbman 1971), so it is not possible to say with certainty that the ganglion co-transplanted with the tongue was responsible for inducing taste bud formation, particularly when a few taste buds appear in the absence of transplanted ganglion (see also review by Oakley 1985).

References

Beidler LM, Smallman R 1965 Renewal of cells within taste buds. J Cell Biol 27:263–271

Corwin JT, Cotanche DA 1989 Development of location-specific hair cell stereocilia in denervated embryonic ears. J Comp Neurol 288:529–537

Corwin JT, Balak KJ, Borden PC 1989 Cellular events underlying the regenerative replacement of lateral line sensory epithelia in amphibians. In: Coombs S, Gorner P, Munz H (eds) The mechanosensory lateral line: neurobiology and evolution. Springer-Verlag, New York, p 161–183

Farbman AI 1971 Differentiation of foetal tongue homografts in the anterior chamber of the eye. Arch Oral Biol 16:51–57

Farbman AI 1972 Differentiation of taste buds in organ culture. J Cell Biol 52:489–493

Farbman AI 1973 Differentiation of lingual filiform papillae in organ culture. Arch Oral Biol 18:197–202

Farbman AI, Mbiene JP 1991 Early development and innervation of taste bud-bearing papillae on the rat tongue. J Comp Neurol 304:2–16

Hellekant G, Kasahara Y, Farbman AI, Harada S, Hård af Segerstad C 1987 Regeneration ability of fungiform papillae and taste buds in rats. Chem Sens 12:459–465

Jørgensen JM, Flock Å 1976 Non-innervated sense organs of the lateral line: development in the regenerating tail of the salamander *Ambystoma mexicanum*. J Neurocytol 5:33–41

Kinnamon JC, Sherman TA, Roper SD 1988 Ultrastructure of mouse vallate taste buds. III. Patterns of synaptic connectivity. J Comp Neurol 270:1–10

Mistretta CM, Bradley RM, Stedman HM 1979 Development of tongue epithelium grafts in fetal sheep. Soc Neurosci Abstr 5:130

Oakley B 1985 Trophic competence in mammalian gustation. In: Pfaff DW (ed) Taste, olfaction and the central nervous system. Rockefeller University Press, New York, p 92–103

Oakley B 1991 Neuronal–epithelial interactions in mammalian gustatory epithelium. In: Regeneration of vertebrate sensory receptor cells. Wiley, Chichester (Ciba Found Symp 160) p 277–293

Pujol R, Lavigne-Rebillard M 1985 Early stages of innervation and sensory cell differentiation in the human fetal organ of Corti. Acta Otolaryngol Suppl 423:43–50

Pujol R, Sans A 1986 Synaptogenesis in the mammalian inner ear. In: Aslin R (ed) Advances in neural and behavioural development. Ablex Press, Norwood, NJ, p 1–18

Royer SM, Kinnamon JC 1988 Ultrastructure of mouse foliate taste buds: synaptic and non-synaptic interactions between taste cells and nerve fibers. J Comp Neurol 270:11–24

Speidel CC 1947 Correlated studies of sense organs and nerves of the lateral line in living frog tadpoles. I. Regeneration of denervated organs. J Comp Neurol 87:29–55

Speidel CC 1948 Correlated studies of sense organs and nerves of the lateral line in living frog tadpoles. II. The trophic influence of specific nerve supply as revealed by prolonged observations of denervated and reinnervated organs. Am J Anat 82:277–320

Stone LS 1940 The origins and development of taste organs in salamanders observed in the living condition. J Exp Zool 83:481–506
Wright MR 1947 Regeneration and degeneration experiments on lateral line nerves and sense organs in anurans. J Exp Zool 105:221–257
Zalewski AA 1972 Regeneration of taste buds after transplantation of tongue and ganglia grafts to the anterior chamber of the eye. Exp Neurol 35:519–528
Zalewski AA 1973 Regeneration of taste buds in tongue grafts after reinnervation by neurons in transplanted lumbar sensory ganglia. Exp Neurol 40:161–169

Rules for the production of sensory cells

Julian Lewis

ICRF Developmental Biology Unit, Department of Zoology, University of Oxford, South Parks Road, Oxford OX1 3PS, UK

Abstract. During development and regeneration, cells divide, move, change their internal state, respond to environmental signals and die according to rules specified by their genome. These rules of cell behaviour are fundamental to any explanation of how sensory tissues are generated or repaired. An attempt is made to summarize and compare the rules underlying the development of four different sensory tissues: mammalian retina, avian auditory epithelium, mechanosensory bristle fields in *Drosophila* and, very briefly, *Drosophila* retina. There are parallels in cell lineage, in the timing of developmental decisions, in the postponement of the choice of differentiated state until the final cell division cycle, in the role of short-range lateral inhibition, and in the involvement of genes such as *Notch* and *achaete/scute*. These similarities may reflect the conservation of some basic mechanisms of sensory development that originated early in the evolution of multicellular animals. If so, studies in *Drosophila*, with its advantages for molecular genetics, may give important clues to the mechanisms of sensory development and regeneration in vertebrates.

1991 Regeneration of vertebrate sensory receptor cells. Wiley, Chichester (Ciba Foundation Symposium 160) p 25–53

In many sensory tissues, and in particular in the retina and auditory epithelium of higher vertebrates, cell division halts once the mature array of cell types has been generated. The possibility of regeneration after damage depends on the reasons for the halt during normal development. These reasons are generally different for the different cell types and the different sensory structures. Some cells, such as neurons, are intrinsically unable to divide; others, such as the hair cell progenitors in the avian cochlea, are halted by stop signals from their environment—by the presence or absence of diffusible growth factors, for example, or of contacts with specific types of extracellular matrix or with neighbouring cells of particular types. Such signals are important from a practical point of view, because they can, in principle, be artificially manipulated. They are, however, only a part of a larger story, and to make sense of them we need to understand their developmental context.

Development and regeneration are more than just a matter of stopping and starting cell division. Each cell follows a complex behavioural programme, stepping through a series of changes in its internal state, confronting developmental choices and responding to signals from its neighbours and surroundings. All this it does according to a set of rules that are defined by its genome and its developmental history. To discover those rules of cell behaviour, to show on the one hand how they generate the observed multicellular patterns, and on the other hand how they are dictated by the genome of the individual cell, is the central task of developmental biology.

This paper will try to summarize in a standardized format what is known of the rules of cell behaviour that generate some sensory tissues, in insects as well as in vertebrates, so as to identify common motifs and show how one may relate the events of development and regeneration to the functions of identified genes. The paper is intended as a preface to the more detailed accounts of specific systems to be given by other participants in this symposium. One of its chief purposes is to suggest a common language for stating conclusions and hypotheses in a well-defined way, in the hope that this will help us to make some useful comparisons. Comparative analysis has become all the more important with the advance of molecular genetics, which has revealed astonishing homologies between insects and vertebrates with respect to the patterning of the central body axis (Wilkinson & Krumlauf 1990). It is possible that the most direct route to an understanding of the molecular genetic controls of sensory system development in vertebrates may be through an understanding of the corresponding processes in *Drosophila*.

The rules of cell behaviour during development and regeneration can be summarized in a standard format

Cells have a position, an internal state, and an environment; they come into existence through cell division, they die, and occasionally they fuse with one another. Correspondingly, the range of things a cell can do can be classified under a few simple headings. In this paper, I shall concentrate on the *state transition rules* according to which cells change their state, divide, or die. To take stock of what we know and to get a sense of the complexity of the problem that a given system confronts us with, I summarize the rules in a formal, systematic way. This forces an explicit and unambiguous statement of conclusions and hypotheses, which can then be more easily scrutinized and amended. It also helps to clarify the parallels between one system and another. The formal summary can be set out in two parts: first, a list of the significantly different states that cells of the given system may adopt during normal development or regeneration; then a set of statements specifying, for each such state S, what a cell in that state will do next. Where there are several possible next steps, the account must state the environmental conditions that determine which one will be chosen. I shall use the simple notation shown below. Very often, the critical features of the environment for a given cell are defined by the states

S:	S′		Proceeds to state S′
S:	S		Makes no change
S:	÷	S′ + S′	Divides giving two daughters in state S′
S:	÷	S1 + S2	Divides asymmetrically, giving two daughters in different states S1 and S2
S:	X		Dies
S:	[E1] [E2]	S1 S2	Makes a choice between two or more options according to cues from the environment—in this example, adopting state S1 if in an environment [E1] and S2 if in an environment [E2]
S:	[prob p] [prob 1 − p]	S1 S2	Makes a stochastic choice, so that each option is associated with a probability rather than a determinative environmental condition

of its neighbours, which thereby govern the choices it makes. In this way, the rules give rise to coordinated multicellular patterns of cells in different states.

Lateral inhibition governs the production of hair cells and supporting cells in the ear

The auditory epithelium of a bird—the basilar papilla—consists of hair cells and supporting cells, arranged in a precise and regular pattern, and innervated by neurons of the adjacent cochlear ganglion (Fig. 1). The neurons are thought to derive ultimately from the same patch of embryonic ectoderm as the hair cells and supporting cells, like sensory neurons in insects, which likewise originate

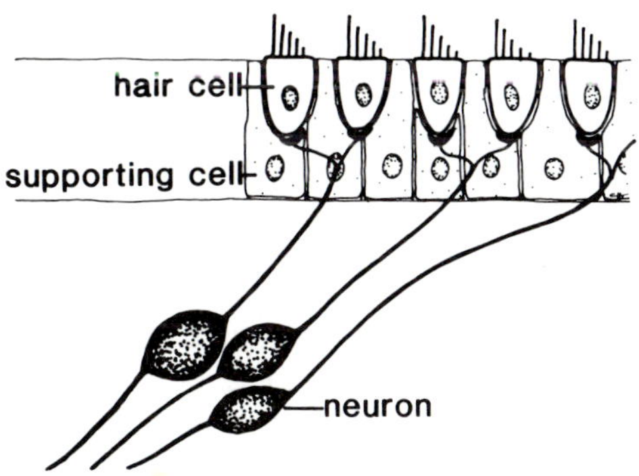

FIG. 1. Schematic diagram of auditory epithelium of a higher vertebrate such as a bird, showing the relationships of hair cells, supporting cells and neurons of the cochlear ganglion.

from the same portion of body surface that they will innervate (Knowlton 1967, Carney & Silver 1983, Ghysen & Dambly-Chaudière 1989). This sensory progenitor patch seems to be identifiable already in the very early otocyst by its expression of the *int-2* proto-oncogene (Wilkinson et al 1989). As in insect neurogenesis (Knust & Campos-Ortega 1989), a subset of cells in the avian embryonic sensory epithelium move basally and slip out into the underlying stroma to become neuroblasts (Carney & Couve 1989). In the vertebrate ear, we do not know how this neurogenesis is controlled so that not all the epithelial cells become neuroblasts; as we shall see below, studies in *Drosophila* may offer some clues. Neither do we know how the proliferation and final differentiation of the neuroblasts into neurons is controlled, or what determines when their cousins—the cells that remain epithelial—will stop proliferating and start differentiating (although there are hints that an autonomous maturation process is responsible, rather than a signal from the surroundings; Swanson et al 1990). It is only for the final phase of ear development, when the regular array of hair cells and supporting cells is generated, that we begin to have a clear picture of the underlying rules of cell behaviour. The key insights here have come from studies of regeneration.

The mature avian auditory epithelium has a rather simple structure. The supporting cells rest on the basal lamina, while the hair cells rest on the supporting cells; and thin projections from supporting cells, extending to the apical surface of the epithelium, separate each hair cell from neighbouring hair cells (Fig. 1). Apart from the nerve terminals synapsing with the hair cells, there appear to be only these two cell types in the epithelium, although it is possible that some other inconspicuous cell type(s)—some sort of stem cell, for example—might be present in the basal layer interspersed among the supporting cells. Under normal circumstances, the pattern of cell types is static: no cell division occurs.

If the hair cells in a patch of the epithelium are killed, by loud noise, for example (Corwin & Cotanche 1988, Ryals & Rubel 1988), cell division resumes in the surviving population of non-hair cells, and new hair cells are produced. From the pattern of tritiated thymidine labelling it can be seen that new supporting cells are generated along with the new hair cells. In the closely similar lateral line system, it has been shown directly that both the regenerated hair cells and the new supporting cells are daughters of supporting cells that have been roused to divide (Balak et al 1990). The simplest interpretation of these phenomena postulates that the epithelium does indeed consist of only two cell types—the supporting cell and the hair cell. Hair cells are unable to divide or change their character; supporting cells are able to divide and generate both hair cells and new supporting cells, but are inhibited from doing so by the presence of hair cells in contact with, or close to, them. In other words, hair cells exert a *lateral inhibition*, blocking production of further hair cells in their immediate neighbourhood and giving rise to a mosaic pattern of cells in alternate states. Lateral inhibition is a theme that recurs in the development of all the sensory systems I shall discuss here. In all of them, as in the avian ear, some crucial choices of cell state, regulated by lateral inhibition, are made at or after the final cell division cycle.

This is not quite the whole story, however. Girod et al (1989) and Rubel et al (1991) give evidence that new hair cells can be regenerated after acoustic damage in two ways: either from surviving supporting cells, as described above, or, less directly, from the adjacent population of so-called cuboidal or hyaline cells or border cells. These lie next to the inferior edge of the auditory epithelium, beside the region that suffers the most severe acoustic damage and loses most of its cells, and they appear to proliferate and migrate into that partially vacated region during regeneration, taking on the character of supporting cells and generating hair cells as progeny.

Rules for the genesis of auditory epithelium

The rules of cell behaviour emerging from all this can be summarized in the notation outlined earlier. Question marks and rows of dots stand for things we do not know.

Cell states

common sensory precursor, CSP; neuroblast, NB; neuron, N; restricted sensory precursor, RSP; unknown intermediate states, {?}; cuboidal/hyaline cell, C; supporting cell, S; hair cell, H

Transition rules

CSP:	?:....?: NB	
	?:....?: RSP	
NB:	?:....?: N	
N:	N	
RSP:	?:....?: C	
	?:....?: S	
	?:....?: H	
C:	[next to patch of S+H]	C
	[next to semi-vacant patch of S without H]	÷S+S
S:	[H contact]	S
	[no H contact, prob p]	H
	[no H contact, prob 1−p]	÷S+S
H:	H	

One might debate some details here. In particular, an alternative rule for S (the supporting cell) could be

S: [H contact] S
[no H contact] $\div S + H$

so that a supporting cell freed of inhibition always divides asymmetrically, throwing off one and only one hair cell. According to the rule proposed in the first place for S, a supporting cell in the absence of hair cells has a random choice: it may convert, with probability *p*, into a hair cell, or it may divide to generate two daughters similar to itself. Once a hair cell has been produced, it inhibits neighbouring supporting cells from becoming hair cells or dividing. Provided *p* is not too small, the result will be a static mosaic of hair cells and supporting cells, as observed. Such a rule would provide a straightforward explanation for the initially somewhat irregular pattern and asynchronous time course of hair cell production (Tilney et al 1986, Katayama & Corwin 1989).

Clonal analysis demonstrates the origin of all but one retinal cell type from a common precursor

The mature retina in higher vertebrates is a regular mosaic of non-dividing cell types, like the auditory epithelium, but much more complex. It contains at least four major classes of neurons—bipolars, horizontal cells, amacrine cells and ganglion cells—as well as rod and cone photoreceptors, Müller glial cells and astrocytes; most of these categories embrace a variety of subtypes. To determine the rules of cell behaviour that generate this system, we need to know the ancestry of the cells. A rich harvest of information on this score has been gathered by labelling random single cells in the developing retina with a heritable marker—an injected tracer molecule (Wetts & Fraser 1988, Holt et al 1988) or a replication-defective retrovirus (Turner et al 1990)—so as to produce clones of labelled progeny in the mature retina. If a single clone contains several differentiated cell types, these must all be derivable from a common precursor cell type.

It turns out that the marked clones in the retina are extraordinarily variable in their composition: provided that the marker has been applied early enough, one encounters almost every conceivable combination of cell types (other than astrocytes, which have a separate origin). This indicates, first and foremost, that all retinal cell types (other than astrocytes) are derivable from a common precursor cell type present in the embryonic retina. The variability of clone composition is most easily explained by supposing that the progeny of this precursor make developmental choices that are either intrinsically random or are governed by randomly variable environmental conditions—dependent, for instance, on the exact locations in which they come to lie and the states of the cells that they happen to have as neighbours.

The choice of differentiated retinal cell type is made late, under the influence of signals from neighbouring cells; but the timetable of competence to differentiate is dependent on cell-autonomous changes

Clonal analysis reveals a further important point about the timing of these developmental choices. Where clones consist of only two cells, these two sisters are often of dissimilar differentiated types. It follows that the choice of mode of differentiation is made very late: just as in the ear, either the final cell division is asymmetrical, or cells select their mode of differentiation after they have finished dividing. The latter possibility is favoured by the experiments of Adler & Hatlee (1989), who report that postmitotic cells from the retina of a bird embryo will differentiate differently according to whether they are left for some time *in situ* or explanted relatively promptly into culture.

Clonal analysis by itself is not enough to reveal the rules of cell behaviour, even though it provides some important information about them. Arbitrarily complex cell lineages may be generated by repeated application of a simple set of cell rules, and conversely many different sets of rules can generate the same cell lineage pattern. A crucial question that cannot be answered by clonal analysis alone concerns the control of timing: if differentiated cells begin to be produced at a certain time in development, is this because the precursor cells, in a cell-autonomous fashion, have progressed to a certain state of maturity, or is it because these precursors have received from their environment a signal for differentiation? Watanabe & Raff (1990) have tackled this question with respect to the production of rod cells in the rat, by culturing cells from developing rat retinas of different ages and observing when and how these cells differentiate, as a function of the culture conditions. They find that the cultured cell population generates rod cells at the same time as it would have done if it had been left *in vivo*, even if it is mingled with a 50-fold excess of retinal cells from a different stage of development. Evidently, the timing of differentiation of rods is not governed by environmental cues but by a cell-autonomous change in the state of the precursor cells—an internal 'maturation clock', in other words. The same appears to be true of the programme of cell division leading up to differentiation. The actual number of rods generated in these experiments, however, is strongly dependent on the nature of the other cells in the culture: whereas a cell-autonomous maturation clock governs the competence to divide or differentiate, environmental cues from neighbouring cells play a significant part in regulating the choice of pathway of differentiation.

Just as in the ear, there is good reason to think that at least some of these environmental cues mediate a lateral inhibition, such that differentiated cells of a given type suppress formation of more cells of the same type in their neighbourhood. This has been directly demonstrated for certain classes of retinal neurons (Reh & Tully 1986, Reh 1987). Thus ganglion cells, which differentiate early, may prevent adjacent precursor cells from giving rise to more ganglion cells, thereby forcing them instead to give rise to other cell types such as rods, which appear on the scene later (Reh & Kljavin 1989).

Rules for genesis of the mammalian retina

If we put all these observations on retinal development together and make the simplest assumptions consistent with them, we arrive at the following tentative, and no doubt grossly oversimplified, statement of the underlying rules of cell behaviour.

Cell states

common sensory precursors at different stages of maturation, CSP1, CSP2, ... CSP*n*; postmitotic precursor, PMP; ganglion cell, G; horizontal cell, H; cone, C; bipolar, B; amacrine cell, A; Müller cell, M; rod, R

Transition rules

CSP1: ÷ CSP2 + CSP2
CSP2: ÷ CSP3 + CSP3
etc., until, say,
CSP5: [prob p] ÷ CSP6 + CSP6
[prob 1 − p] PMP
CSP6: [prob p] ÷ CSP7 + CSP7
[prob 1 − p] PMP
etc., until
CSP*n*: PMP
PMP: [no G near] G
[G near, no H near] H
. . . .
[G near, H near,..., M near] R

Insect mechanosensory bristles: *achaete/scute* gene expression specifies competence to form them; lateral inhibition limits their numbers

The mechanosensory epithelia of the inner ear and lateral line in vertebrates have a counterpart in the system of mechanosensory bristles that are scattered over the body surface of a fly (Fig. 2). Each bristle, whether it be large (a macrochaete) or small (a microchaete), has the same basic structure (Fig. 3): it consists of a neuron, a glial sheath cell (the thecogen), a socket cell (the tormogen), and a shaft cell (the trichogen). These four cells all derive, via two division cycles, from a common precursor, the bristle mother cell, which itself

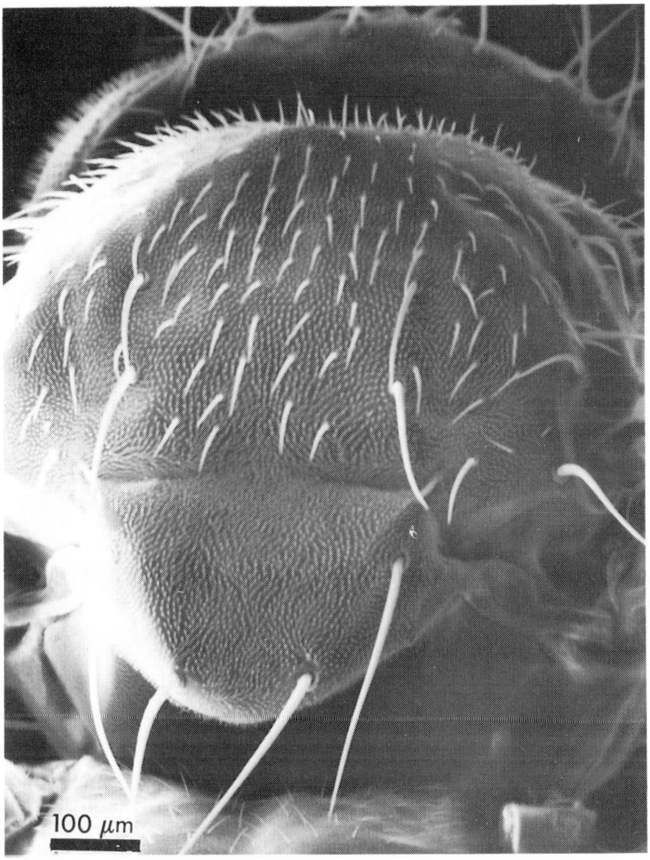

FIG. 2. Scanning electron micrograph showing the pattern of large and small bristles on the dorsal thorax of *Drosophila*. (Courtesy of Sheena Pinchin.)

shares a common ancestry with the surrounding ordinary epidermal cells. There is a parallel here with the development of the vertebrate ear, where again the neurons and the associated peripheral sensory cells arise from the same region of the embryonic body surface and remain close together; in both systems, the neurons or neuroblasts slip out of the surface epithelium into the underlying stroma as a first step in their differentiation.

The problem of explaining bristle production can be broken down into two questions: by what rules are some epidermal cells selected to become bristle mother cells? And by what rules does a bristle mother cell generate its differentiated progeny? The distribution of bristles varies over the adult body surface: some regions have none, others have many; at certain precisely predictable sites, large isolated bristles (macrochaetes) develop; elsewhere,

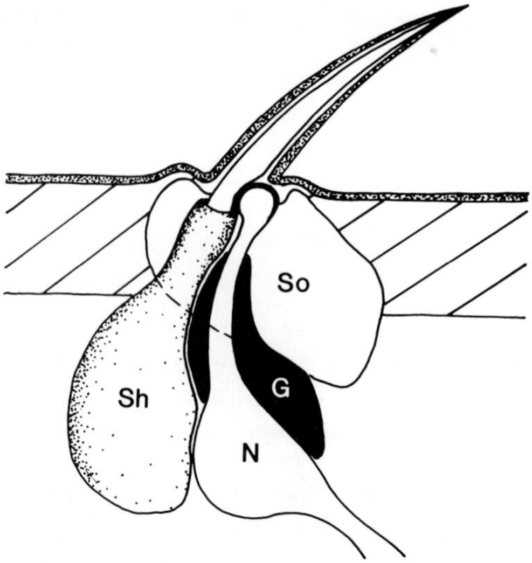

FIG. 3. Schematic diagram of a mechanosensory bristle of an insect, showing the four cells of which it is composed: neuron (N), glial sheath cell (G), shaft cell (Sh) and socket cell (So).

smaller bristles (microchaetes) may be densely scattered in a more random fashion. The distribution is thought to be governed by the pattern of expression of genes of the Achaete–Scute Complex (ASC), and specifically (for bristles of the adult fly) by the two related genes *achaete* and *scute*. One or other or both of these are expressed in the developing insect at just the sites where bristles will form—in discrete spots corresponding to future macrochaetes, and diffusely where fields of microchaetes are to be generated, such as the notum (the back of the thorax) (Romani et al 1989). If *achaete* and/or *scute* is deleted, the corresponding bristles fail to develop; and if these genes are ectopically expressed, bristles develop ectopically. The gene products must therefore be counted as determinants of the capacity to form bristles: they define what has been called a *proneural* state (Ghysen & Dambly-Chaudière 1988, 1989). However, not every cell that expresses *achaete* or *scute* will give rise to a bristle: in the case of macrochaetes, only one out of a cluster of perhaps half a dozen cells; in the case of a field of microchaetes such as that on the central notum, only one cell out of every 10 or 20 (at least, that is the ratio of bristles to epidermal cells in the adult structure). It appears that once a cell has adopted the character of bristle mother, it inhibits closely neighbouring cells from doing the same, and forces them instead to become ordinary epidermal cells. This is confirmed by experimental studies (reviewed by Ghysen & Dambly-Chaudière 1988, 1989, Simpson 1990). For example, a genetic mosaic can be created in which the cell at the normal site of a macrochaete is genetically incapable of forming a macrochaete, while some of the neighbouring cells are genetically normal. One of these normal neighbours, relieved of lateral inhibition, may then generate a macrochaete, at a slightly misplaced site, instead of undergoing its usual epidermal fate (Stern 1954).

The four cells of a bristle interact to select their appropriate differentiated states

The underlying rules of cell behaviour are simplest to state, and best analysed, for the medial microchaete field of the notum. Here there is apparently no further epidermal cell division after the division that gives birth to the bristle mothers, except on the part of the bristle mothers themselves (Hartenstein & Posakony 1989). Each of these (no matter where it lies on the body surface) divides twice to generate four initially indistinguishable granddaughters. One of these then moves to an internal position and begins to differentiate as a neuron, poking up a dendrite into its sister, which becomes the glial sheath cell; the glial sheath cell in turn is embraced by one of its two cousins, which becomes the shaft cell, and this in turn is embraced by its sister, which becomes the socket cell. Thus the four cells differentiate in a sequence corresponding to their final concentric arrangement, beginning with the centrally placed neuron. Given the regular lineage relationship between the four cells of the bristle, one might be tempted to guess that lineage, through asymmetrical cell divisions, would control their characters; but there is evidence against this. As in the other systems we have discussed, the choice of differentiated state seems to be made after the final cell division. For example, on the anterior wing margin, where bristles lie exceptionally close together, two sister cells that would usually remain associated in a single bristle occasionally become separated and incorporated into different bristles; here, instead of forming two different cell types, they may both differentiate in the same way, but always in accordance with the context, so that each bristle contains the canonical set of four cell types (Hartenstein & Posakony 1989). Another line of evidence, from genetic studies, will be described below. But first let us summarize the rules of cell behaviour corresponding to the account just given.

Rules for genesis of mechanosensory bristles

Possible cell states

proneural common sensory precursor, CSP; epidermal cell, E; bristle mother, BM; first generation descendant of bristle mother, BM1; second generation descendant, uncommitted, BM2; neuron, N; glial sheath cell, G; shaft cell, Sh; socket cell, So

Transition rules

CSP:	[no BM near, prob 1 − p]	CSP
	[no BM near, prob p]	BM
	[BM near]	E
E:	E	
BM:	÷ BM1 + BM1	
BM1:	÷ BM2 + BM2	
BM2:	[no N contact]	N
	[N contact, no G contact]	G
	[G contact, no Sh contact]	Sh
	[Sh contact]	So

The *Notch* gene mediates lateral inhibition

According to the above scheme, lateral inhibition operates twice, first to control the production of bristle mothers, then to control the differentiation of their granddaughters. This is beautifully confirmed by studies of *Notch* mutants in *Drosophila*. *Notch* is one of a small set of genes implicated in another lateral inhibition phenomenon, regulating the segregation of neuroblasts from epidermis to form the central nervous system in the embryo: when *Notch* is defective in the embryo, the inhibition fails and all the epidermal cells in the neurogenic region become neuroblasts, with lethal consequences (Knust & Campos-Ortega 1989, Hoppe & Greenspan 1990). The *Notch* product is a transmembrane protein with partial homology to epidermal growth factor and to various transmembrane receptors, and it has been shown to be capable of mediating cell–cell adhesion (Fehon et al 1990). Without it, cells seem unable to exchange the signals that bring about lateral inhibition.

Using a temperature-sensitive allele of *Notch*, one can switch the function of the gene on and off at will, and so test its role at much later stages, during bristle development. If *Notch* function is made defective in the period after the cell division that generates the bristle mother, but before the bristle mother herself begins to divide, the consequence (in the mid-notal microchaete field) is that *all* the epidermal cells become bristle mothers. If *Notch* function is made defective later, during or after the final divisions of the bristle lineage, there is no such disturbance of the numbers of bristle mothers and epidermal cells, but now the four granddaughters of each bristle mother all differentiate into neurons (Hartenstein & Posakony 1990). These observations, and similar findings with certain other genes (Simpson 1990), provide strong evidence (1) that lateral inhibition indeed operates repeatedly, (2) that it regulates the characters of cells that have finished dividing and yet are still uncommitted, and

(3) that at each step the neurogenic or neural option is the default adopted in the absence of a signal to the contrary. All these features are embodied in the cell rules for insect bristle formation stated formally above, and they all have an echo in the sets of rules suggested, more tentatively, for the vertebrate ear and eye.

Sequences of short-range cell–cell interactions regulate the choice of cell type postmitotically in the insect retina

The developing retina of *Drosophila* provides yet another illustration of the same principles. I shall discuss it very briefly here, since there have recently been several excellent reviews (Tomlinson 1990, Banerjee & Zipursky 1990). The insect eye consists of an astonishingly regular array of ommatidial units, each comprising eight photoreceptor neurons and fourteen accessory cells of various sorts, all occupying stereotyped positions (Ready et al 1976). Clonal analysis, exactly analogous to that performed on the mammalian retina, shows again (1) that all these cells derive from a common precursor; (2) that the choice of cell type is not dictated simply by lineage but by environmental influences; and (3) that these influences act at the last moment, regulating a decision a cell takes after it has completed its final mitosis (Lawrence & Green 1979). By studies of mutants and of genetic mosaics it has been possible to pinpoint the cells that emit some of the critical signals and to show that many of these act at very short range, probably via cell–cell contacts. Similarities to the bristle system are not merely abstract and formal. In fact, four of the fourteen accessory cells of each ommatidium go to form a mechanosensory bristle, exactly like those elsewhere on the body. Moreover, *Drosophila* eye development and bristle development are affected in a precisely analogous way by *Notch* mutations. *Notch* function can be switched off during ommatidium development, using a temperature-sensitive allele, as in the studies of bristle formation described earlier. Every one of the crucial cell–cell interactions is then liable to be disrupted, depending on the timing of the treatment; and a deficiency of *Notch* in the initial stages causes almost all the cells, relieved of signals from their neighbours, to develop similarly as photoreceptor neurons (Cagan & Ready 1989).

Conclusion

The retina of an insect and the mechanosensory bristles on its body surface seem at first sight to be radically different sensory structures—at least as different from one another as either is from the eye or the ear of a vertebrate. Yet, at a fundamental level, in terms of genes and rules of cell behaviour, they seem to have important features in common. Turning from comparisons between parts of the body to comparisons between species, one can see resemblances between insect retina and mammalian retina, with respect to both the molecular biology of the photoreceptors and the generative rules of cell behaviour. Recently genes have been identified in vertebrates that are similar, respectively, to *achaete/scute* and to *Notch*, both in sequence and in pattern of expression (Johnson et al 1990, Coffman et al 1990). The developmental parallels between different sensory systems in vertebrates and insects that we have encountered in this review may thus reflect a true homology—a system of genetic controls for the production of sensory cells that has been conserved in evolution as sense organs and species diverged. If so, students of vertebrate sensory development and regeneration may have much to learn from *Drosophila*.

Acknowledgements

I am indebted to Catherine Haddon for the drawings, to Sheena Pinchin for her *Drosophila* photograph, and to Paul Martin, David Ish-Horowicz, Ian McKay and Sherry Granum for comments on the text.

References

Adler R, Hatlee M 1989 Plasticity and differentiation of embryonic retinal cells after terminal mitosis. Science (Wash DC) 243:391–393

Balak KJ, Corwin JT, Jones JE 1990 Regenerated hair cells can originate from supporting cell progeny: evidence from phototoxicity and laser ablation experiments in the lateral line system. J Neurosci 10:2502–2512

Banerjee U, Zipursky SL 1990 The role of cell–cell interaction in the development of the *Drosophila* visual system. Neuron 4:177–187

Cagan RL, Ready DF 1989 *Notch* is required for successive cell decisions in the developing *Drosophila* retina. Genes Dev 3:1099–1112

Carney PR, Couve E 1989 Cell polarity changes and migration during early development of the avian peripheral auditory system. Anat Rec 225:156–164

Carney PR, Silver J 1983 Studies on cell migration and axon guidance in the developing distal auditory system of the mouse. J Comp Neurol 215:359–369

Coffman C, Harris W, Kintner C 1990 *Xotch*, the *Xenopus* homolog of *Drosophila Notch*. Science (Wash DC) 249:1438–1441

Corwin JT, Cotanche DA 1988 Regeneration of sensory hair cells after acoustic trauma. Science (Wash DC) 240:1772–1774

Fehon RG, Kooh PJ, Rebay I et al 1990 Molecular interactions between the protein products of the neurogenic loci Notch and Delta, two EGF-homologous genes in *Drosophila*. Cell 61:523–534

Ghysen A, Dambly-Chaudière C 1988 From DNA to form: the *achaete–scute* complex. Genes Dev 2:495–501

Ghysen A, Dambly-Chaudière C 1989 Genesis of the *Drosophila* peripheral nervous system. Trends Genet 5:251–255

Girod DA, Duckert LG, Rubel EW 1989 Possible precursors of regenerated hair cells in the avian cochlea following acoustic trauma. Hear Res 42:175–194

Hartenstein V, Posakony JW 1989 Development of adult sensilla on the wing and notum of *Drosophila melanogaster*. Development 107:389–405

Hartenstein V, Posakony JW 1990 A dual function of the *Notch* gene in *Drosophila* sensillum development. Dev Biol 142:13–30

Holt CE, Bertsch TW, Ellis HM, Harris WA 1988 Cellular determination in the *Xenopus* retina is independent of lineage and birth date. Neuron 1:15–26

Hoppe PE, Greenspan RJ 1990 The *Notch* locus of *Drosophila* is required in epidermal cells for epidermal development. Development 109:875–885

Johnson JE, Birren SJ, Anderson DJ 1990 Two rat homologues of *Drosophila achaete–scute* specifically expressed in neuronal precursors. Nature (Lond) 346:858–861

Katayama K, Corwin JT 1989 Cell production in the chicken cochlea. J Comp Neurol 281:129–135

Knowlton VY 1967 Correlation of the development of membranous and bony labyrinths, acoustic ganglia, nerves, and brain centers of the chick embryo. J Morphol 121:179–208

Knust E, Campos-Ortega JA 1989 The molecular genetics of early neurogenesis in *Drosophila melanogaster*. BioEssays 11:95–100

Lawrence PA, Green SM 1979 Cell lineage in the developing retina of *Drosophila*. Dev Biol 71:142–152

Ready DF, Hanson TE, Benzer S 1976 Development of the *Drosophila* retina, a neurocrystalline lattice. Dev Biol 53:217–240

Reh TA 1987 Cell-specific regulation of neuronal production in the larval frog retina. J Neurosci 1:3317–3324

Reh TA, Tully T 1986 Regulation of tyrosine-hydroxylase-containing amacrine cell number in larval frog retina. Dev Biol 114:463–469

Reh TA, Kljavin IJ 1989 Age of differentiation determines rat retinal germinal cell phenotype: induction of differentiation by dissociation. J Neurosci 9:4179–4189

Romani S, Campuzano S, Macagno ER, Modolell J 1989 Expression of *achaete* and *scute* genes in *Drosophila* imaginal discs and their function in sensory organ development. Genes Dev 3:997–1007

Rubel EW, Oesterle EC, Weisleder P 1991 Hair cell regeneration in the avian inner ear. In: Regeneration of vertebrate sensory receptor cells. Wiley, Chichester (Ciba Found Symp 160) p 77–102

Ryals BM, Rubel EW 1988 Hair cell regeneration after acoustic trauma in adult *Coturnix* quail. Science (Wash DC) 240:1774–1776

Simpson P 1990 Lateral inhibition and the development of the sensory bristles of the adult peripheral nervous system of *Drosophila*. Development 109:509–519

Stern C 1954 Two or three bristles. Am Sci 42:213–247

Swanson GJ, Howard M, Lewis J 1990 Epithelial autonomy in the development of the inner ear of a bird embryo. Dev Biol 137:243–257

Tilney LG, Tilney MS, Saunders JS, DeRosier DJ 1986 Actin filaments, stereocilia, and hair cells of the bird cochlea. III. The development and differentiation of hair cells and stereocilia. Dev Biol 116:100–118

Tomlinson A 1990 The molecular basis of pattern formation in the developing compound eye of *Drosophila*. Seminars Cell Biol 1:229–239

Turner DL, Snyder EY, Cepko CL 1990 Lineage-independent determination of cell type in the embryonic mouse retina. Neuron 4:833–845

Watanabe T, Raff MC 1990 Rod photoreceptor development in vitro: intrinsic properties of proliferating neuroepithelial cells change as development proceeds. Neuron 2:461–467

Wetts R, Fraser SE 1988 Multipotent precursors can give rise to all major cell types of the frog retina. Science (Wash DC) 239:1142–1145

Wilkinson DG, Krumlauf R 1990 Molecular approaches to the segmentation of the hindbrain. Trends Neurosci 13:335–339

Wilkinson DG, Bhatt S, McMahon AP 1989 Expression pattern of the FGF-related proto-oncogene *int*-2 suggests multiple roles in fetal development. Development 105:131–136

DISCUSSION

Raymond: Dr Lewis pointed out that the development of bristles in the *Drosophila* eye is a lineage dependent or clonal event, in that all four cells of the bristle are progeny of the same clone; whereas photoreceptor cells of the ommatidium are not clonally derived but differentiate according to specific cell–cell interactions. Does this imply a real difference in developmental mechanisms?

Lewis: No, not at a fundamental level. The argument is summarized in the text of my paper (p 35–37). The evidence comes from observations of an exceptional part of the *Drosophila* body where the bristles lie extremely close together, at the anterior margin of the wing. There, when you do the clonal analysis, you occasionally find that sister cells (descendants of a single bristle mother cell) are allocated to adjacent bristles instead of the same bristle. Then, just as in the retina, they may differentiate in the same way, rather than in different ways (Hartenstein & Posakony 1989). That is the non-genetic evidence that it's not a lineage mechanism that defines how the cells will differentiate. The normal course of differentiation is dependent on an 'accident of position': because all four descendants of the bristle mother cell happen to find themselves isolated in a single cluster, they interact to generate the four standard elements of the mechanosensory structure; it is not the case that each one is autonomously playing out a certain role in a lineage programme.

Burd: What is the expression of *Notch* in the adult fly? For instance, can *Notch* be operating to prevent differentiation at certain times in the adult?

Lewis: I don't know. In *Drosophila* there is no epidermal cell division once the adult state has been reached, so everything is static, and the maintenance of those controls may become unnecessary. But in holometabolous insects, which go through repeated moults, like *Rhodnius prolixus*, new bristles are generated at each moult in a way that reflects the rules I described (Wigglesworth 1940, Simpson 1990). This implies that the same mechanisms must continue in operation. But I don't know any direct evidence on whether *Notch* is expressed in adult insects.

Burd: I wonder whether a gene similar to *Notch* is present in vertebrates, in, say, olfactory neurons, and might be used for lateral inhibition.

Lewis: There are no answers to that yet.

Margolis: What is the biochemical or molecular nature of the vertebrate *Notch* counterpart?

Lewis: The *Notch* gene product, and its *Xenopus* counterpart (the *Xotch* gene product), are transmembrane proteins with homologies to epidermal growth factor (EGF), which itself is cleaved from a receptor-like transmembrane protein (Kidd et al 1989, Johansen et al 1989). There are experiments with transfected cells in culture showing that the *Notch* product can mediate cell–cell binding through an interaction with the product of another related gene called *Delta* (Fehon et al 1990).

Calof: The laboratories of Dr Michael Young and Dr Spiros Artavanis-Tsakonas have looked at the expression of the *Notch* protein during *Drosophila* development, and the question is more complicated (Kidd et al 1989, Johansen et al 1989). This relates to Gail Burd's point. The genetic evidence suggests that the absence of *Notch* results in a decision to 'go neural', but the papers from the Young and Artavanis-Tsakonas groups show that the *Notch* protein is expressed by both neural and epidermal precursors during the developmental period when neuroblasts are determined. In other words, the *Notch* protein is expressed by all the cells of the epidermis at this time; the presence or absence of this protein does not correlate with the developmental decision to become a neuroblast rather than an epidermal precursor. The null allele of *Notch*, resulting in the overproduction of neuroblasts, results from an absence of *Notch* function, but the absence of the *Notch* protein can't be responsible for the determination of neuroblasts during normal *Drosophila* development.

Lewis: Yes, *Notch* is widely expressed. It is thought that it is needed to mediate the transmission of a signal, not that it is itself the signal.

Corwin: In the *Drosophila* retina the *sevenless* gene product is the receptor that determines whether the presumptive R7 photoreceptor cell of an ommatidium responds to a signal from cell eight (R8). In ommatidia that are mosaic for wild-type and for the mutant called *bride of sevenless*, even the wild-type R7 precursors do not differentiate as R7s, if the neighbouring R8 cell carries the *bride of sevenless* mutation (Reinke & Zipursky 1988). So in that situation both sides of a cell–cell signalling interaction are known, the receiver and the sender. The molecular biology studies, together with the immunocytochemical demonstration that the *sevenless* product is localized to just the part of the R7 cell membrane that faces the cell R8, have shown that the determination of cell fate is contact mediated (Tomlinson 1988).

Reh: In respect to *Xotch* distribution in the vertebrate (*Xenopus*), this gene is present in the nervous system as well as in other areas of the developing frog. In the eye, the cells at the marginal part of the retina (which include the left-over proliferating cells that give rise to all the different types of retinal neurons and glia) have the greatest concentration of *Xotch* mRNA. It appears that fully differentiated retinal cells no longer express mRNA for this gene, but those that are still differentiating, at the margin, retain this expression. It's not clear, at the level of resolution, whether some of the early differentiating retinal neurons might also contain it. So one would not necessarily expect *Xotch* to be expressed in fully differentiated neurons, but it might be expressed in neurons that are beginning their differentiation process, and thereby exerting lateral inhibition on the cells that have yet to differentiate, or to decide whether they will differentiate. So maybe you will find a *Notch* homologue in the early differentiating basal cells, but not in fully differentiated cells, in a taste bud.

Fernald: Dr Lewis, could you describe more fully why you think *Notch* is involved both in adjusting the probabilities in your postulated transition phase, and also in the differentiation choice? You stress that *Notch* is a vehicle for information, not that it *is* the signal. Yet it is surprising that it both regulates a decision whether or not to divide, and determines the differentiated state.

Lewis: Notch controls cell division in the insect epidermis only in so far as it controls the decision whether the proneural precursor cell takes on the character of a bristle mother cell or not. In the case where you switch off *Notch* function shortly after determination of the bristle mother cell, but before it has embarked on its divisions, the standard set of two division cycles ensues. So the sequence of divisions of the bristle mother cell is independent of *Notch*.

Ryals: Thus the expression of this gene is not the signal for mitotic activity, but the signal for the type of differentiation that will occur?

Lewis: Yes, I am suggesting that role for the *Notch*-dependent signal.

Raymond: I like your attempt to simplify these problems with mathematical formulations, but with respect to *Notch*, the reader might get the wrong idea. This gene is only one of a number of neurogenic genes, all of which function together and interact with one another. One should not conclude that *Notch* is the only gene acting to regulate the decision of whether to be an epidermal cell or a neuron. For example, there is a whole family of neurogenic genes that function at this level.

Lewis: That's an important point that I neglected to make. There is for instance another gene, *shaggy*, mutation of which has exactly the same consequences for bristle production (Simpson 1990).

Steinberg: Do all the neurogenic genes involve lateral interaction?

Lewis: They are all believed to be involved in transmission or reception of signals between neighbouring cells (Knust & Campos-Ortega 1989).

Rubel: There is certainly a hypothesis about lateral interaction, but in many of these cases we can't yet say what mechanism is involved. These genes are involved in cell–cell interaction, but is it operating by contact-induced lateral interaction or can it be across two cells, or three cells, etc.? In many instances we cannot say where the signal is or where the receptor is.

Corwin: In some cases, such as the *sevenless* locus in *Drosophila*, we do appear to know where the signal and the receptor are located (see Tomlinson 1988 for a review). You can even say that when you have a single amino acid substitution in the tyrosine kinase domain of the *sevenless* product, which appears to be a transmembrane protein receptor with an intracellular tyrosine kinase domain, you deactivate the receiving cell's response to the signal molecule (Basler & Hafen 1988).

Fernald: Dr Lewis spoke about what a cell might 'read' from its environment as to what its fate will be. Sue McConnell (1991) argues that in the ferret cerebral cortex it may be as early as the S phase that cells are reading from their environment what their ultimate fate would be. Do you know any evidence from *Drosophila* about when in the cell cycle these extracellular signals are read and that final determination is made, and why it might occur at one time in the cycle rather than another?

Lewis: The signals can still operate after cycling is finished. In the *Notch* experiments with bristle development, one can switch off *Notch* function after the bristle mother cell has finished her two cycles of division, and still, at that late stage, alter the differentiation of cells so that all the cells become neuronal (Hartenstein & Posakony 1990). So the decision as to how cells differentiate is still plastic after cells have finished dividing. And the same is true in the retina.

Reh: For how long after cell division do the cells remain plastic with respect to their differentiation?

Lewis: I don't know exactly. The bristle mother cell goes through two cell divisions which are about eight hours apart. Hartenstein & Posakony (1990) get an effect if they switch off *Notch* function at the time of completion of the second of those divisions and keep it switched off for four hours.

Corwin: There is a programmed end to the functioning of these determining signals as well, after the specialization of cells has occurred (Bowtell et al 1989).

Lewis: Yes.

Raymond: You just mentioned 'plastic' differentiation, which at first seems a *non sequitur*! Could you expand on your idea about the common sensory progenitor cell (CSP), and the 'cascade' of transitions (CSP1, CSP2 . . . . CSPn), which suggests that the cell is changing as it goes along this pathway? How stable are those changes, or how plastic? Can CSP cells go in the inverse direction?

Lewis: According to the experiments of Watanabe & Raff (1990), both the timing of the production of differentiated rat retinal cells and the pattern of proliferation are determined by the age of the retina from which the cells are taken, and not by their environment. So there is 'cell autonomy' in the CSPs of the retina, in the sense that they carry a record of their developmental age and progress towards maturity, irrespective of their environment.

Can one alter this internal 'maturation clock'? In at least one system, it seems that one can. This is the optic nerve. Here, cells (in an analogous way to the retinal cells) appear to count their way through eight cell cycles, on their way to generating oligodendrocytes, and then autonomously differentiate (Lillien & Raff 1990); at least, this is what happens in a standard culture medium containing the growth factor PDGF, and the onset of differentiation coincides with a loss of responsiveness to the mitogenic influence of PDGF. You can, however, upset this programme of cell behaviour by adding basic fibroblast growth factor (bFGF) to the medium: this apparently stops the 'maturation clock' and enables the cells to remain responsive to PDGF and carry on dividing indefinitely without differentiating (Bogler et al 1990, McKinnon et al 1990). So one can manipulate the growth factors and upset the cell-autonomous process. But I can't think of any case where the maturation of sensory precursor cells is actually driven into reverse.

Watt: You discussed the auditory epithelium in birds and said that the idea of asymmetrical divisions of the supporting cells was less attractive than a probabilistic model. Are there lineage-marking experiments to support that?

Lewis: No! With that question in mind, I looked hard at the papers that I cited by Corwin & Cotanche (1988), Balak et al (1990), Ryals & Rubel (1988) and Girod et al (1989) to see whether their patterns of [^{3}H]thymidine labelling allow one to say whether a hair cell is always a sister of an associated supporting cell, and I couldn't see the answer. I must admit that part of my reason for finding the probabilistic model, involving postmitotic selection of cell fate, more attractive was that I wanted to emphasize the parallel with the other sensory systems.

Rubel: The evidence seems to be clear from Brenda Ryals' laboratory, where they looked at the difference between what happens if you destroy just hair cells, and if you destroy both hair cells and supporting cells.

Ryals: We have examined how many support cells and hair cells show [^{3}H]thymidine uptake after noise exposure (Ryals & Westbrook 1990). We find many more support cells showing label uptake than hair cells; but we probably damage both cell types mechanically. By using ototoxic agents we avoid this mechanical trauma. We are now examining the inner ears of birds given ototoxic doses of gentamicin. By determining the types of cells with [^{3}H]thymidine uptake in these animals, we may be able to better understand cell lineage.

Lewis: If you see more supporting cells produced than hair cells, then unless there is selective hair cell death, you must suppose that there can be symmetrical divisions of supporting cells.

Rubel: It's a mistake to think of hair cells and supporting cells as a binary system. It has been shown in taste buds and in the olfactory system that a variety of 'supporting' cells exist.

Lewis: Yes. I glossed over that for lack of time in my talk (but see p 29). I think the issue will be discussed in detail later in the symposium.

Zakon: With the model that you propose for the interaction of hair cells and support cells, would you predict that if you could destroy many of the support cells, every support cell would be next to a hair cell and there would be no division, to replace the hair cells? Or, since another aspect of your model is that every hair cell is next to a support cell, if hair cells were now next to each other, would that change them in some way that could be signalled to the support cells, so that they would start dividing to replace the lost support cells? This is a bit speculative, of course!

Lewis: I would have to predict that a system with a surplus of hair cells relative to supporting cells would be static, and would fail to make up its deficiency of supporting cells. But what Dr Rubel was driving at, in relation to his experiments (Girod et al 1989), is that adjacent to the auditory epithelium there is a population of cells which, after acoustic damage, can also be roused to divide, move into the damaged region, and there generate hair cells and supporting cells. So there are other sources of cells for regeneration. But I don't really know the answer to your speculative question.

Calof: Is this lateral, 'resting' population of hair cell precursors known to give rise to hair cells only in certain regenerative situations, or is it actually the source of hair cells during normal development, but it has been difficult to see? This is related to the question of whether the supporting cell is really a neural progenitor in the auditory system, because the evidence is slender at best in the olfactory system that the supporting cell lineage and the neuronal lineage are the same. I think they are probably separate.

Rubel: We can address that better in the vestibular system than the auditory system, to the extent that we almost always see pairs of [^{3}H]thymidine-labelled cells. But that doesn't really answer your question. It's a problem of following the [^{3}H]thymidine-labelled cells. We need to do a very closely timed pulse-labelling study.

Corwin: Not just that. Jay Jones has used time-lapse video microscopy to observe the divisions that supporting cells go through after laser microbeam ablation of all the pre-existing hair cells in a lateral line epithelium in the salamander, *Ambystoma mexicanum*. The observations were almost continuous, spanning the period from ablation until new hair cells had been formed by regeneration. By playing the recorded sequence backwards, he was able to observe that the regenerated hair cells were two cell divisions removed from differentiated supporting cells.

Calof: So there may be an intermediate stage, which is also what the evidence tends to support in the olfactory epithelium.

Rubel: We can't determine that yet unless we do very short duration pulse-labelling studies, and do grain counts, to see how many divisions have occurred.

Lewis: You need proper clonal analysis to settle the question.

Rubel: That's right; and you have to define very closely the original precursor that you are marking.

Corwin: Ken Balak (Balak & Corwin 1988, Corwin et al 1989) injected rhodamine-labelled dextran into individual supporting cells in lateral line neuromasts of axolotl salamanders (*A. mexicanum*) and waited to see what cell types would develop after divisions. In one case, both hair cells and supporting cells later contained rhodamine dextran, indicating that they arose from divisions of a labelled supporting cell. Unfortunately, there was only the single positive case from months of work, so we do not know that the result is representative. It does fit with the hypothesis that supporting cells may divide to produce both hair cells and supporting cells. It also agrees with our results from phototoxic and laser ablation experiments (Balak et al 1990) and with Jay Jones's results.

Farbman: In the olfactory system, we attempted in many experiments to grow disaggregated fetal rat olfactory epithelial cells in culture, in order to study axonal outgrowth. Few cells grew axons unless they were in direct contact with another cell. It was difficult to determine whether the other cell was olfactory or supporting, because we did not have good cell markers at the time. It appeared to us then that there was a contact-mediated phenomenon that played an important role in determining whether olfactory sensory cells would grow an axon (Gonzales et al 1985). We were not concerned, in these experiments, with a cell lineage issue, because we assumed that lineage had been determined by the time we removed the cells from the animals.

Calof: Our work shows that if you alter culture conditions—if, for example, you change the substratum on which olfactory receptor neurons are grown or on which explants are made—you can alter neuronal differentiation. It's a well-known property of extracellular matrix molecules like laminin that they promote neuronal differentiation and axon outgrowth in a wide variety of neuronal types, and that's true in our culture system as well (Calof & Lander 1990). Whereas another ECM molecule like fibronectin, which is a great supporter of adhesion in this culture system, is relatively non-permissive for neuronal differentiation (Calof & Lander 1990). When you talk about axon outgrowth in culture, the conditions are crucial. All these kinds of extracellular matrix molecules like laminin are made by cells, in culture and *in vivo*, and would be a component of the intercellular interactions observed. In my experiments the cells are always grown on a substratum, which is a rather complex mixture of fibronectin and laminin on polylysine (Calof & Chikaraishi 1989). We found that neuronal differentiation occurred on such a substratum at the level of axon outgrowth and of the expression of differentiative markers in the absence of exogenous growth factors, except what might be produced by basal cells and olfactory neuron precursor cells in the cultures themselves.

Rubel: But do you see axon outgrowth if the cell isn't contacting another cell?

Calof: If they are touching a laminin substratum, we see axon outgrowth by cultured olfactory receptor neurons. Often, but not always, these neurons contact other cells in the culture. We have never systematically investigated whether contact with other cells, at some point during the culture period, is a requirement for olfactory receptor neurons to extend axons *in vitro*.

Lopashov: I have a question for Dr Lewis. Have you artificially changed the succession in your cell generations, or not? I have done such experiments, using eye vesicles from the frog *Rana temporaria* (Lopashov & Hoperskaya 1970). I dissociated the eye vesicles into separate cells and added to them an equal volume of dissociated cells of early gastrula ectoderm, and mixed them. Retina developed after 5–6 days in culture, with the main structure similar to the structure of normal explanted eye rudiments, with photoreceptors on its surface. Unfortunately, the cells were not marked, as is usual now, but the unity of the structure and the absence of a number of dying cells make it probable that an assimilation of gastrula ectoderm cells into retina takes place, and that cells that have passed through different numbers of cell cycles can join to form a common structure.

Lewis: Gavin Swanson and his colleagues (1990) did experiments that are perhaps analogous, taking the early otic vesicle from the chick embryo, stripping it of all surrounding cells, and planting it in a foreign mesenchyme, in the limb bud. In that foreign environment the epithelium differentiated almost exactly in the same way and with the same timetable as if it had been left in its original site. So, whatever the interactions may be that govern proliferation and differentiation, they seem to occur only within the epithelium: the epithelium proceeds with its programme independently of signals from outside it.

Presson: My work deals with a system that is generating and not regenerating, namely the postembryonic statoacoustic system in fish. I am also concerned with how many different kinds of support cells there are and, even more fundamentally, what *are* supporting cells? We don't know very much about their function; we have no cellular markers for them. Therefore we need a good cellular analysis, so that we know at the cellular level what cell types we are dealing with and what is the cell that takes up tritiated thymidine and divides. We need more information about how to know whether a cell is a 'support cell', in terms of function.

Palay: The supporting, or sustentacular, cell has been defined as a neuroepithelial cell that is non-neural. Therefore the definitions are always going to be ambiguous, because you have to define the opposite cell (the neural cells) by some mechanism; then what is left over is the supporting cell.

Presson: This is clearly not sufficient as a definition.

Palay: No; it is not going to be good enough for the new kind of molecular analysis. That's why these interchanges between supporting and neural cells are partly semantic in nature.

Reh: It might be possible to find some commonality between the support cells of the different sensory systems, and between the neuronal cells that are generated in these systems as well, by the use of some of the genes homologous to those found in *Drosophila*. For example, *Xotch* (in *Xenopus*) is expressed at a particular phase during the genesis of retina. It may be that when one looks at a developing taste bud or olfactory epithelium, the cells that will give rise to these neurons transiently express, at some period in the progression from mitosis to differentiated neuron, one of these genes. It is possible that *achaete/scute* or its homologues, the two *mash* genes in the vertebrate, might turn out to be good markers that allow us to make certain distinctions regardless of morphology, based more on molecular grounds.

Rubel: But you are including the requirement that they have to be expressed transiently at the right time in development, and so forth?

Reh: Although these may be poor markers for neurons *per se*, because *Notch* is expressed in other cell classes they might be good markers in the sense that when you are dealing with an identified population of cells it would allow you to sort out where in the cycle those cells are—in terms of their signalling or in their differentiated pathways.

Corwin: In some of these epithelia we could add to our knowledge without getting into the molecular identification of cells. For example, by serial section EM we could find out whether the cells that some refer to as 'basal cells' actually have configurations that are different from those of the adjacent supporting cells. Although those cells are called basal cells, we may be dealing with cells whose nuclei are in a basal layer, but which might actually be part of a pseudo-stratified epithelium, not a stratified epithelium. We do not know the answers to such basic questions.

Rubel: Are you saying that we should be making cell distinctions on these grounds, in terms of cell types?

Corwin: People already have made cell distinctions, in effect, by using *names* such as 'basal cell'. In most cases it has been proposed that basal cells do not reach the luminal surface of the epithelium and, therefore, are distinct from the supporting cells that reach the lumen. Yet, the name 'basal cell' has been applied in the absence of the scientific work that would be required to rule out the possibility that those cells reach the lumen further along in the tissue, but outside the plane of the section examined. Serial section data would resolve this.

Ryals: So one morphological definition for a support cell is that it surrounds a nervous cell and also reaches the lumen of the epithelium? And a basal cell is different since it doesn't necessarily surround a nervous cell, nor does it reach the lumen?

Corwin: I think of 'support cells' as a *set* of cell types, which consists of subsets of cells such as mantle-type supporting cells, possibly basal cells, and other types of supporting cells. That would be consistent with the past use of the term, at least.

Farbman: We are hobbled by the term 'support cell' that was coined more than a century ago by the earlier histologists, which we are using without really knowing what it means.

Margolis: At the functional level, what do we mean by a 'support' cell?

Rubel: And to add to that, is there any 'rule' derived from cell biology, or from any other area, that cells that have become functional can't change what they do and what proteins they produce?

Lewis: Let me offer a definition which you could hope to apply rigorously (though you might find nothing that fits the definition). A supporting cell would be a cell which stands in a particular anatomical relation to hair cells, 'cupping' them and interposing between them, and which behaves according to the rules that I set out.

Rubel: The question, though, is whether one, two, or more types of 'support cells' exist.

Lewis: The relevant categories to distinguish are defined in terms like those I just suggested—that is, in terms of the rules of behaviour that they obey.

Presson: Does the supporting cell have to be a postmitotic cell?

Reasner: No. In olfactory systems, supporting cells can divide in their own more superficial layer, which is separate from the basal proliferative compartment.

Presson: And do they function in some postmitotic sense—some differentiated sense? Do they contribute something to the epithelial function?

Reasner: In standard anatomical terms, the new cells appear to be normal supporting cells; but again one needs cellular markers to suggest their exact class and possible function.

Calof: There is at least one antigenic marker for sustentacular (support) cells in the rat, called SUS-1. It is relatively specific for supporting cells of the rat olfactory epithelium and some cells in submucosal glands (Hempstead & Morgan 1983). Whether that antigenic similarity means a similarity of origin is not certain, but there are regeneration experiments which suggest that the submucosal glands may provide a source of sustentacular cells during regeneration of the olfactory epithelium (Matulionis 1976). So there is one class of support cell markers.

Margolis: This doesn't address function, though.

Calof: No, it simply says that there is this population of cells that can be distinguished on the basis of this molecular marker. The supporting cells are themselves a proliferating cell type, and after certain kinds of induced-degeneration paradigms these cells can be seen to remain as an intact layer above the degenerating olfactory receptor neurons. So they appear to be separate and presumably have some kind of function that is distinct from the neuronal transmission function of the epithelium.

Fernald: It sounds more and more as if the definition of the support cell is like the definition of obscenity: everyone knows it when they see it! But the real issue here is why, if we now know that incredible cell transitions are possible through destruction and repair and renewal, do we place such emphasis on morphology in order to get some kind of classical description, when maybe all these cells are interchangeable in some fundamental sense.

Presson: I think that *is* the fundamental question: are they interchangeable?

Rubel: We agree that some cell types are sufficiently differentiated that they don't re-enter the mitotic cycle. One thing that we want to know is which cell is entering the mitotic cycle and what it is doing before that time.

Cotanche: My bias has always been to think more about the sensory cells in development and regeneration, but in fact the supporting cells are functioning long before the hair cells differentiate, producing the tectorial membrane, both in the developing embryo and in regeneration. Within an hour after you cease blasting chicks with noise, the supporting cells are turning out new tectorial membrane, and in the embryo they are creating it before hair cells have differentiated, throughout the entire cochlea. So the supporting cells are not just bridesmaids waiting for the hair cells to develop and then playing a secondary role: they are already well under way and functioning. My question would be: do some supporting cells stop making tectorial membrane and become hair cells, or do nascent hair cells remain below the surface and only protrude when they begin to differentiate a stereociliary bundle, at a time when the supporting cells are already differentiated and functioning?

Rubel: Good question! That is the point. We don't have any idea.

Ryals: In other sensory systems, are support cells secretory?

Farbman: In frogs, the olfactory supporting cells do seem to be secretory, but they are apparently not so in mammals. Mammalian olfactory supporting cells resemble transport cells morphologically. They are thought to participate in modifying the ionic components of the mucus on the olfactory epithelial surface.

Steinberg: In the vertebrate retina we know more about supporting cells than in some of the other sensory epithelia. There are two kinds: one is the pigmented epithelial cell, which sits as a separate epithelium facing the photoreceptors; the other type is the glial cell (Müller cell) which fits more with Dr Palay's definition of a supporting cell. In both cases their functions in the normal maintenance of the retina are well known. In development, the retinal pigmented epithelium presents another example of a support cell that develops well before the sensory cells, and probably helps to create the ionic environment in which the outer segments of photoreceptors develop.

Reh: It's interesting that the supporting cell can maintain a very mature phenotype, for example of the pigmented cell, and still remain capable of undergoing quite a large phenotypic change.

Presson: The Müller glial cell can divide, but no one has suggested that it can give rise to neurons. It is clearly very well differentiated functionally, with multiple roles.

Rubel: What does the Müller cell do before it divides? Is it like skin? Does it dedifferentiate in a major way?

Raymond: Müller cells and pigmented epithelial cells can divide both *in vivo* and *in vitro*, with retention of differentiated characteristics that allow one to recognize the cell.

Steinberg: It depends whether you are talking about mammals or not, and the adult or the embryo. A pigmented epithelial cell in an adult mammal doesn't normally divide. It can dedifferentiate and then divide, and then leave the epithelial sheet, becoming a roving macrophage.

Raymond: It need not dedifferentiate entirely before it divides in culture. For example, the cultured cells still contain pigment granules.

Steinberg: I do not believe that it can manufacture any more pigment; the pigment already present becomes diluted as it divides.

Fernald: One thing we have learned from recent studies of glial cells is that they can be transformed from passive potassium pumps to amazing cells having multiple ionic conductances as well as other functions, including the production of growth factors. I don't know that anyone is looking into the ion conductances of the support cell. It would be interesting to see what is common to all so-called support cells and what is specialized, in terms of maintaining ion fluxes;

and how their ion conductances change as they differentiate, or as they receive different signals.

Rubel: Or as they may dedifferentiate and start becoming mitotic again!

Costanzo: Ironically, in the olfactory system it seems that one of the supporting cell's functions is support. Olfactory supporting cells have a foot process anchored to the basement membrane and extend to the surface of the epithelium. If you cause all the olfactory neurons to degenerate, by cutting their axon processes, the supporting cells remain intact and serve as the supporting structure of the olfactory epithelium. There is also recent morphological evidence that these support cells may have other roles. We now know that the olfactory dendrites, and the neuron itself, seem to hug the support cells, and in fact there are invaginations of supporting cell membrane around the olfactory neurons, and bridges (structural connections) between support cells and neurons (Morrison & Costanzo 1989).

Corwin: Supporting cells may also function in metabolic support. In auditory, vestibular and lateral line epithelia, the hair cells are far away from the basal lamina and the underlying vasculature. There are interdigitations of the lateral membranes where the hair cells and supporting cells are in contact in some of those epithelia, and prominent Golgi apparatus near those sites in the supporting cells; so there is a possibility of trophic interaction.

Costanzo: Yes; I would agree that the olfactory supporting cells are not purely physically supportive. They probably have other functions, including trophic ones.

References

Balak KJ, Corwin JT 1988 Hair cells originate from supporting cell progeny during regeneration in the lateral line system. Assoc Res Otolaryngol Abstr 11:107

Balak KJ, Corwin JT, Jones JE 1990 Regenerated hair cells can originate from supporting cell progeny: evidence from phototoxicity and laser ablation experiments in the lateral line system. J Neurosci 10:2502–2512

Basler K, Hafen E 1988 Control of photoreceptor cell fate by the *sevenless* protein requires a functional tyrosine kinase domain. Cell 54:299–311

Bogler O, Wren D, Barnett SC, Land H, Noble M 1990 Cooperation between two growth factors promotes self-renewal and inhibits differentiation of oligodendrocyte–type-2 astrocyte (O-2A) progenitor cells. Proc Natl Acad Sci USA 87:6368–6372

Bowtell DDL, Simon MA, Rubin GM 1989 Ommatidia in the developing *Drosophila* eye require and can respond to *sevenless* only for a restricted period. Cell 56:931–936

Corwin JT, Cotanche DA 1988 Regeneration of sensory hair cells after acoustic trauma. Science (Wash DC) 240:1772–1774

Corwin JT, Balak KJ, Borden PC 1989 Cellular events underlying the regenerative replacement of lateral line sensory epithelia in amphibians. In: Coombs S, Gorner P, Munz H (eds) The mechanosensory lateral line: neurobiology and evolution. Springer-Verlag, New York, p 161–183

Calof AL, Chikaraishi DM 1989 Analysis of neurogenesis in a mammalian neuroepithelium: proliferation and differentiation of an olfactory neuron precursor in vitro. Neuron 3:115–127

Calof AL, Lander AD 1990 *In vitro* analysis of neuronal precursor migration from mouse olfactory epithelium: evidence for regulation by extracellular matrix. J Cell Biol 111:491a

Fehon RG, Kooh PJ, Rebay I et al 1990 Molecular interactions between the protein products of the neurogenic loci *Notch* and *Delta*, two EGF-homologous genes in *Drosophila*. Cell 61:523–534

Girod DA, Duckert LG, Rubel EW 1989 Possible precursors of regenerated hair cells in the avian cochlea following acoustic trauma. Hear Res 42:175–194

Gonzales F, Farbman AI, Gesteland RC 1985 Cell and explant culture of olfactory chemoreceptor cells. J Neurosci Methods 14:77–90

Hartenstein V, Posakony JW 1989 Development of adult sensilla on the wing and notum of *Drosophila melanogaster*. Development 107:389–405

Hartenstein V, Posakony JW 1990 A dual function for the *Notch* gene in *Drosophila* sensillum development. Dev Biol 142:13–30

Hempstead JL, Morgan JI 1983 Monoclonal antibodies to the rat olfactory sustentacular cell. Brain Res 288:289–295

Johansen KM, Fehon RG, Artavanis-Tsakonas S 1989 The *Notch* gene product is a glycoprotein expressed on the cell surface of both epidermal and neuronal precursor cells during *Drosophila* development. J Cell Biol 109:2427–2440

Kidd S, Baylies MK, Gasic GP, Young MW 1989 Structure and distribution of the Notch protein in developing *Drosophila*. Genes Dev 3:1113–1129

Knust E, Campos-Ortega JA 1989 The molecular genetics of early neurogenesis in *Drosophila melanogaster*. BioEssays 11:95–100

Lillien LE, Raff MC 1990 Differentiation signals in the CNS: type-2 astrocyte development in vitro as a model system. Neuron 5:111–119

Lopashov GV, Hoperskaya OA 1970 Origin and modes of distribution of inducing agents in development. In: Lopashov GV, Rott NN, Tumanishvili GD (eds) Intercellular interactions in differentiation and growth. Nauka, Moscow, p 52–64

McConnell SK 1991 The generation of neuronal diversity in the central nervous system. Annu Rev Neurosci 14:269–300

McKinnon RD, Mtsui T, Dubois-Dalcq M, Aaronson SA 1990 FGF modulates the PDGF-driven pathway of oligodendrocyte development. Neuron 5:603–614

Matulionis DH 1976 Light and electron microscope study of the degeneration and early regeneration of olfactory epithelium in the mouse. Am J Anat 145:79–100

Morrison EE, Costanzo RM 1989 Scanning electron microscopic study of degeneration and regeneration in the olfactory epithelium after axotomy. J Neurocytol 18:393–405

Reinke R, Zipursky SL 1988 Cell–cell interaction in the *Drosophila* retina: the *bride of sevenless* gene is required in photoreceptor cell R8 for R7 cell development. Cell 55:321–330

Ryals BM, Rubel EW 1988 Hair cell regeneration after acoustic trauma in adult *Coturnix* quail. Science (Wash DC) 240:1774–1776

Ryals BM, Westbrook EW 1990 Hair cell regeneration in senescent quail. Hear Res 50:87–96

Simpson P 1990 Lateral inhibition and the development of the sensory bristles of the adult nervous system of *Drosophila*. Development 109:509–519

Swanson GJ, Howard M, Lewis J 1990 Epithelial autonomy in the development of the inner ear of a bird embryo. Dev Biol 137:243–257

Tomlinson A 1988 Cellular interactions in the developing *Drosophila* eye. Development 104:183–193

Wigglesworth VB 1940 Local and general factors in the development of 'pattern' in *Rhodnius prolixus* (Hemiptera). J Exp Biol 17:180–200
Watanabe T, Raff MC 1990 Rod photoreceptor development in vitro: intrinsic properties of proliferating neuroepithelial cells change as development proceeds. Neuron 2:461–467

Regeneration in epithelial proliferative units as exemplified by small intestinal crypts

Christopher S. Potten

CRC Department of Epithelial Biology, Paterson Institute for Cancer Research, Christie Hospital, Manchester M20 9BX, UK

Abstract. Surface epithelia are subdivided structurally into a series of partially autonomous proliferative units. Within each unit the proliferative cells are organized into a lineage or hierarchy each with relatively few ancestral (stem) cells. This structural organization is strikingly illustrated by the gastrointestinal mucosa where the proliferative cells are packaged into crypts. In the mouse small intestine there are 250 cells in total per crypt of which about 150–160 are proliferative. The latter cells are arranged as a series of about 10 rings each of about 16 cells, starting at about the 4th position from the bottom of the crypt and running up to the 13th–14th cell position. The stem cells are believed to be located in the lowermost ring. We have been studying the regenerative process in mouse small intestinal crypts after various levels of injury. Exposure to small doses of cytotoxic agents (e.g. γ-radiation) reveals that some cells in the stem cell zone are very sensitive but that apparently unaffected cells in the zone are easily disturbed in their cell cycle characteristics by this minor damage. Such observations suggest that some of the controls on stem cell proliferation are extremely local in their action. The stem cells can detect the death of one of their members and the remainder respond accordingly. After more severe injury (drugs or higher doses of radiation) the first detectable changes also appear in the stem cell zone. It is concluded that the crucial cells in the regeneration process in epithelia are the stem cells. The cell kinetic properties during regeneration have been studied experimentally and modelled mathematically.

1991 Regeneration of vertebrate sensory receptor cells. Wiley, Chichester (Ciba Foundation Symposium 160) p 54–76

In various regions of surface epithelia the tissue can be seen to be subdivided into a series of morphologically or structurally distinct units which, because of the inclusion of proliferative cells, suggests that the units are associated with, or generated by, cell proliferation. Such units of proliferation are evident in the epidermis from various sites in mouse and man (see reviews in Potten 1981, 1983a, Potten et al 1983a). These sites include the oral epithelia, but most notably the dorsal surface of the mouse tongue (Hume & Potten 1976, Hume 1983), and various regions of the gastrointestinal tract, where the units of proliferation tend to be relatively large and are called glands, or crypts, as in the stomach, small intestine and large bowel (see reviews in Potten et al 1983b, Potten & Hendry 1983, Wright & Alison 1984). It is now becoming increasingly evident, and generally accepted, that the cell replacement process within epithelial tissues and hence within the proliferative units is achieved by an hierarchical cell lineage with relatively few lineage originators (stem cells) and many proliferating transitory cells. In fact it appears that in any tissue where extensive cell replacement is required, such hierarchical schemes apply (for example, in the bone marrow and testis) (Potten 1983b, Wright & Alison 1984, Potten et al 1979, Potten & Lajtha 1982, Potten & Loeffler 1990).

Here I wish to consider one epithelial site, the small intestine, and its proliferative organization into crypts in order to examine how the crypts and their cellular constituents respond to injury and regenerate the crypt and the epithelium. The epithelium of the small intestine consists of a simple columnar epithelium folded into complex configurations of differentiated cells on the villi and proliferative cells in the 6–10 crypts that serve each villus. The architecture and cell kinetics have been extensively reviewed elsewhere (Potten & Hendry 1983, Potten et al 1983b, Wright & Alison 1984). The structure of the crypts of the small and large bowel is illustrated in Fig. 1, with some relevant cell kinetic parameters for mouse and man.

Briefly, the crypts of the mouse small intestine contain about 150–160 proliferative cells of which between about four and 16 (16 is a full circumferential annulus of cells in the mouse) function as *actual stem cells*. Up to a second full annulus of cells (i.e. a total of 32 cells per crypt) may be capable of acting as stem cells if the first annulus is damaged and thus are *potential stem cells* or regenerative (clonogenic) stem cells (Potten et al 1987, Potten & Loeffler 1990). The special feature of this tissue is that the hierarchical status of cells can be related to their topographical position; that is to say, stem cell characteristics can be investigated by studying the features and behaviour of cells at a particular cell position in the crypt (see Fig. 1). The data suggest that the 4th cell position from the base in the small intestinal crypts of the mouse is the most likely position of the stem cells. However, it should be remembered that the stem cells are thought to be located above the highest Paneth cell and this may be the 2nd cell position in some sections or the 7th in others. The 4th position is the average value of the first non-Paneth cell in crypt sections. It is believed that the stem cells have a cell cycle time of about 24 h, while the dividing transit cells (T1–T8 in Fig. 1) cycle every 12 h in the mouse small intestine. This results in about one cell dividing every five minutes in every crypt. I shall now consider the consequences of damaging this system in terms of recovery and regeneration.

Small Intestine

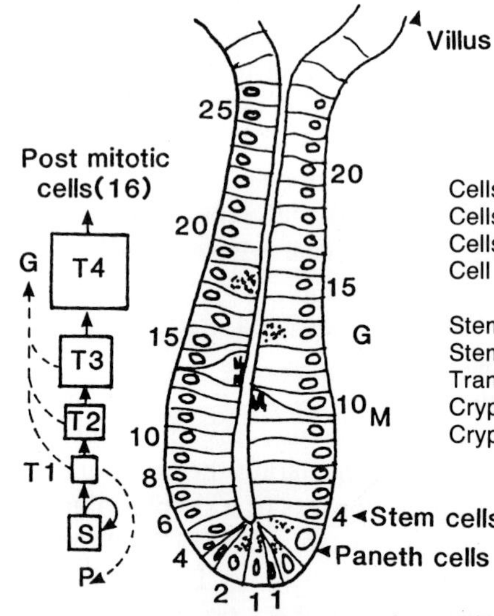

	Mouse	Man
Cells/column	25	34
Cells/circumference	16	22
Cells/crypt	250[+]	450*
Cell cycle	10–18 h[b]	27–73 h[b]
	12 h[d]	20–36 h
Stem cell cycle	≈24	>36 h[e]
Stem cells/crypt	4–16	?
Transit cell generations	4–6	>4–6
Crypts/villus	6–10	≈6
Crypts/intestine	$1–3\times10^{6}$	?

Large Intestine

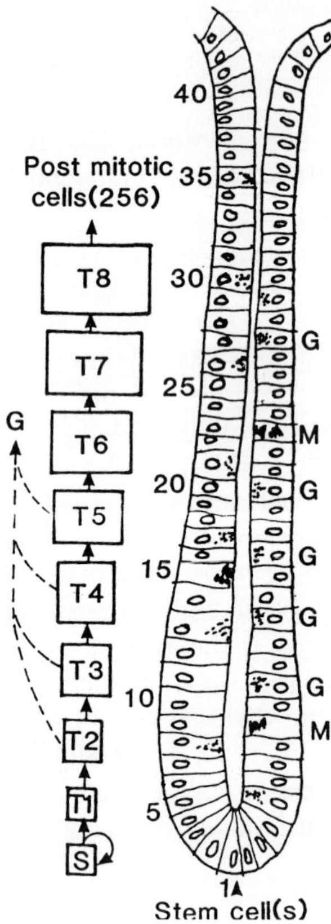

	Mouse	Man
Cells/column	42	82
Cells/circumference	18	46
Cells/crypt	300–450[‡]	2250*
Cell cycle	15–23 h[b]	72–96 h[b]
	24–36 h[a]	~36 h[c]
Stem cell cycle	>36 h	>36 h
Stem cells/crypt	1–8?	?
Transit generations	5-9	>5-9

Mild damage

I shall restrict consideration mainly to one cytotoxic agent, γ-radiation. However, the principles that will be deduced are probably general to many different types of cytotoxic insult. The first detectable change in the mouse small intestinal crypts after a small dose of γ-radiation (<1 Gy) is acute cell death at the histological level. The timing and morphological characteristics (at the light and electron microscope level), and the fact that the degree of cell death can be reduced or delayed by inhibitors of protein synthesis and by other metabolic inhibitors, all suggest that the cells die by apoptosis (programmed cell deletion)—i.e. by activation of a genetic programme (Kerr et al 1972, Potten 1977, 1990)—and so this term was adopted. The yield of apoptotic cells after low doses of radiation depends on the following facts:

1. The peak incidence tends to occur between three and six hours after exposure. New dying or apoptotic cells can first be observed about 1.5 h after exposure.
2. The yield of apoptotic cells at 3–6 h is strongly dose dependent up to about 1 Gy, above which a dose dependence is absent or greatly reduced.
3. The yield decreases at times beyond 6 h with a half-life of about 5 h for low doses of radiation and for some cytotoxic agents. After higher doses of radiation (5–10 Gy) the decay to control levels is slower and may be complicated by other subpopulations of cells dying according to different kinetics.
4. The apoptotic cells are not distributed at random throughout the proliferative compartment of the mouse small intestinal crypt, but have a strong proclivity for the cells at positions 3–5.

These observations are summarized in Fig. 2 (Potten 1977, 1990, Hendry et al 1982, Thakkar 1991). It is clear that each crypt contains, within the area suspected of containing the stem cells, a subpopulation of very sensitive cells which die through apoptosis. It is estimated that the crypt as a whole contains up to about six of these cells and their sensitivity can be expressed by the

---

FIG. 1. Diagrammatic representation of a vertical section through a mouse small intestinal crypt (*upper diagram*) and large intestinal crypt (*lower diagram*). Individual cells are shown with the cell numbering system used for scoring data related to the cell lineage or hierarchy that is shown on the left. The position of cells in the lineage can be related to topographical position. S, stem cells; T1–T8, transit cell generations; P, Paneth cells; G, goblet cells; M, mitosis. For the postmitotic cells, the stem cell amplification generated by the transit lineage is shown in parenthesis. Some numerical characteristics for mouse and man are shown on the right. *Assuming a 0.6 geometric correction factor (Potten et al 1988b). [+]Absolute count from whole crypts. [‡]Value depends on position: 300 for caecum, 450 for rectum. (Data from Potten et al 1983b. [a]From Kellett 1991. [b]From Wright & Alison 1984. [c]From Potten et al 1990b. [d]From Potten & Hendry 1983.)

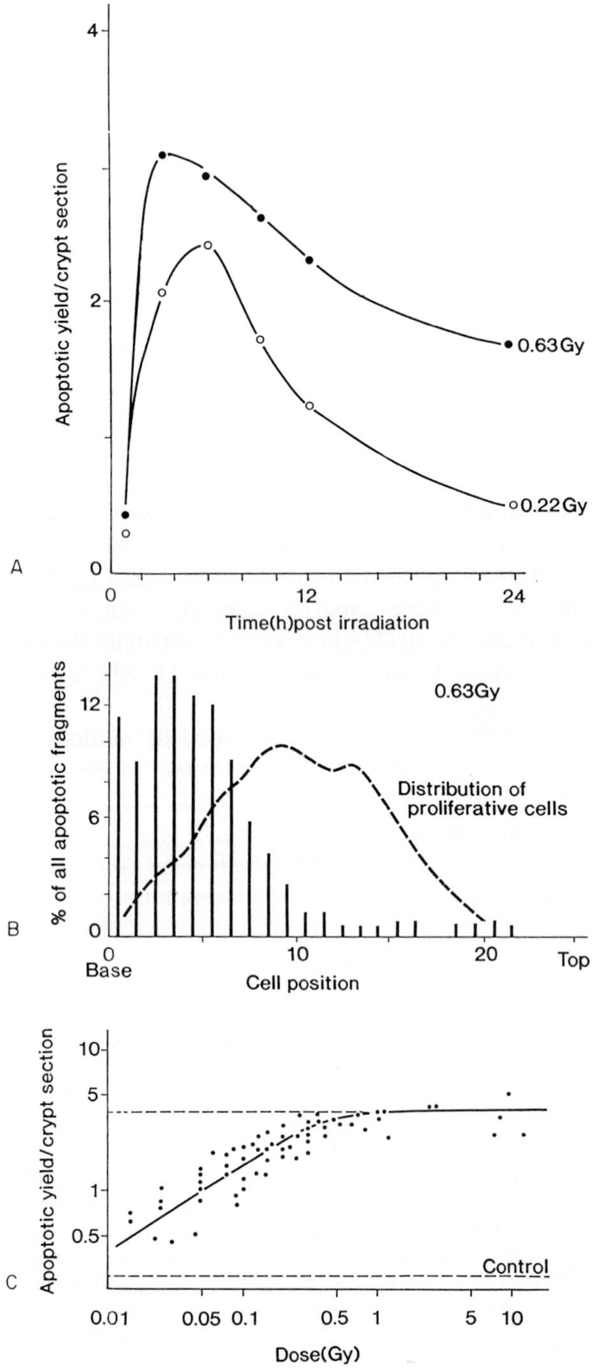
Apoptotic yield/crypt section
0.63Gy
0.22Gy
A
Time(h)post irradiation
% of all apoptotic fragments
0.63Gy
Distribution of proliferative cells
B
Base
Top
Cell position
Apoptotic yield/crypt section
Control
C
Dose(Gy)

mean lethal dose or D_0 extracted from a derived survival curve as the reciprocal of the slope (Potten 1977, Hendry et al 1982) of 0.1–0.3 Gy. These cells are so sensitive that some are killed by the tracer doses of tritiated thymidine often used in cell kinetic experiments. In fact, 0.925–1.85 MBq (25–50 μCi) administered to a mouse will cause roughly the same amount of cell death as 0.02–0.04 Gy (see Fig. 2) (unpublished data of K. Ijiri). Thus doses of the order of 0.05–0.1 Gy or 1.85–3.7 MBq will produce about one dead cell at cell position 4 in the small intestinal crypt. This can be regarded as a low level of damage.

Severe damage

After large doses of γ-radiation (8–20 Gy), several of the crypt cells appear to be dead, on morphological criteria, soon after irradiation (however, the apoptotic index rarely rises above 5–8 per crypt section at any time after any dose). Most of the regenerative (clonogenic) cells, or potential stem cells, are reproductively sterilized rather than histologically dead, because, above 8 Gy, crypts begin to be lost. Crypts disappear over a period of about two days if they possess no surviving clonogenic cells. Conversely, if one clonogenic cell survives, the crypt will be repopulated over the first three days after irradiation. At a dose of 8 Gy, on average most crypts contain 1–3 surviving clonogenic cells. By counting the number of surviving crypts on Day 3 we can obtain a crypt survival curve (Fig. 3B) which characteristically has a large broad shoulder, attributable to the product of the number of clonogens per crypt that have to be sterilized before the crypt is killed and the repair capacity of individual clonogenic cells (see reviews by Potten et al 1983b, Potten & Hendry 1985, Potten 1990). The repair capacity can be measured by appropriate split-dose experiments and is complete within about 4 h. If the maximum repair capacity is known, the number of clonogens per crypt can be estimated to be about 30–40 (Potten & Hendry 1975, 1983, Potten et al 1987). Using Poisson statistics and the number of surviving crypts, we can estimate the surviving fraction of clonogenic cells (Fig. 3C) (Withers & Elkind 1970). Using split doses of radiation with times sufficient for maximal repair between doses, the depreciation in surviving fraction attributed to small doses (of a few Gy) can be calculated and hence the full shape of the clonogenic survival curve can be deduced (Hendry 1979 and Fig. 3C).

FIG. 2. Data for apoptosis in the mouse small intestine. A. Yield of dead crypt cells against time after irradiation. B. Distribution with relation to cell position in the crypt (see Fig. 1 for cell position numbering) after γ-irradiation. The typical distribution of DNA-synthesizing (proliferating) cells is also shown for comparison (maximum labelling index about 45% over cell positions 6–14). C. Yield of apoptotic cells against radiation dose. The control background level of spontaneous cell death is shown by the lower dashed line. (A and B are taken, with permission of Taylor & Francis Ltd, from figures in Potten 1990, originally from Potten et al 1983b. C is from the same review but was originally from Hendry et al 1982.)

The split-dose approach allows the effects of smaller doses on fully repaired cells (equivalent to unirradiated cells) to be determined. Assuming that there is no heterogeneity in radiosensitivity among the clonogens, the derived curve can be taken as representative of the clonogen population at large. If heterogeneity exists, then the curve reflects the radiosensitivity of the most resistant component. The sensitivity of the clonogens is reflected by the D_0, which is 1.0–1.25 Gy generally for crypt survival curves, and is 0.9 Gy for the particular deduced clonogen survival curve shown in Fig. 3. These D_0 values are similar to those derived for the target cells for whole animal survival ($D_0 = 1.25$ Gy) when the $LD_{50/6-7}$ value is considered (Fig. 3A) (Hendry et al 1983). This suggests that crypt survival determines animal survival over the first 6–7 days, and clearly clonogen survival determines the crypt survival (see also review Potten 1990).

Similar split-dose approaches can be followed to measure the regrowth of clonogens (Potten et al 1988a); their doubling time has been determined to be 21 h. The cell cycle of the crypt-base cells during regeneration has been determined to be about 10 h (data of L. Geng in Potten 1990). Using weak β-particles to irradiate only the base of the crypts, Hendry et al (1989) showed that the clonogens are near the crypt base.

It is important here to draw a clear distinction between apoptotic death and reproductive death. The former is defined purely morphologically and the latter, purely functionally. Although it is possible that the cells that present as apoptotic cells at cell positions 3–5 are part of the clonogenic compartment, it is clear that not all sterilized clonogens become apoptotic. In fact it is evident from the length of time that the crypt and villus cellularities remain high that many

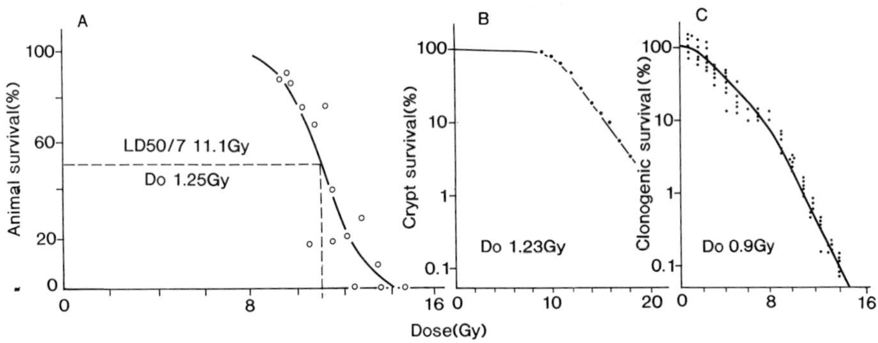

FIG. 3. Relationship between the survival of (A) mice ($LD_{50/7}$), (B) crypts and (C) clonogenic cells after acute whole-body γ-irradiation. The $LD_{50/7}$ is 11.1 Gy; the D_0 for the target cells for animal survival is deduced to be 1.25 Gy. The D_0 for the survival curve for crypts (B) varies between about 1.0 and 1.25 Gy for different experiments (1.23 here). The D_0 for the deduced clonogen survival curve (C) is 0.9 Gy. (Taken from Potten 1990 with permission of Taylor & Francis Ltd; the original data were from Hendry et al 1983 (graph A); B is from Potten et al 1983b; and C from Hendry 1979.)

TABLE 1 Summary of effects of whole-body acute doses of γ-radiation on small intestinal crypts in mice

Dose range	*Time*	*Effect*
0–1 Gy	3–6 h	Morphological cell death (apoptosis) Saturates at about 1 Gy Preferentially in stem cell region (cell position 4)
0–20 Gy	—	Clonogens sterilized but not many form apoptotic cells
9–20 Gy	2–3 d	Crypts destroyed
9–15 Gy	6–7 d	Animals die as a result of gut damage

sterilized clonogens end up as differentiated villus cells and may even divide once or twice before becoming villus cells; that is, they effectively become transit cells prematurely.

The consequences of whole-body or gastrointestinal exposure to γ-radiation in mice are summarized in Table 1. A few cGy will result in some acute cell death within the stem cell zone of the crypts. The number of dead cells increases with dose up to 4–6 per crypt at about 100 cGy (1 Gy), after which there is no clear dose dependence. The reproductive sterilizability of clonogenic cells increases with increasing dose up to about 8–9 Gy, when each crypt has an average of only one surviving clonogenic cell. Above 9 Gy these last surviving clonogenic cells are sterilized and crypts disappear. Above about 11 Gy, the crypt destruction is such that regenerative processes cannot prevent the symptoms of the gastrointestinal radiation syndrome, and death of about half the mice can be expected. After about 15 Gy, all mice will die within 6–7 days. These clearly represent severe levels of damage to the crypt.

I now wish to consider what we know about the regenerative processes at these various levels of damage in terms of the kinetic changes and the cells involved.

Cell kinetic changes after small doses of radiation

There are three quite distinct experiments, each of which clearly indicates that the cell kinetic behaviour of cells in the stem cell region of the crypt changes very rapidly after minor cytotoxic insults, such as the killing of a single cell in this region. These experiments will be briefly summarized.

1. Tsubouchi & Potten (1985) observed striking differences among curves of the percentage of labelled mitoses when 0.5 Gy of external γ-radiation with 92.5 kBq [^{3}H]thymidine, or 3.7 MBq (100 μCi) of [^{3}H]thymidine, was used, when compared with such curves generated using exceptionally small doses of [^{3}H]thymidine alone (92.5 kBq, 2.5 μCi) (Fig. 4). Both 0.5 Gy and 3.7 MBq of [^{3}H]thymidine are known to cause some apoptotic (acute) cell death in the stem cell region (crypt base). This cell death, it has been suggested, results in

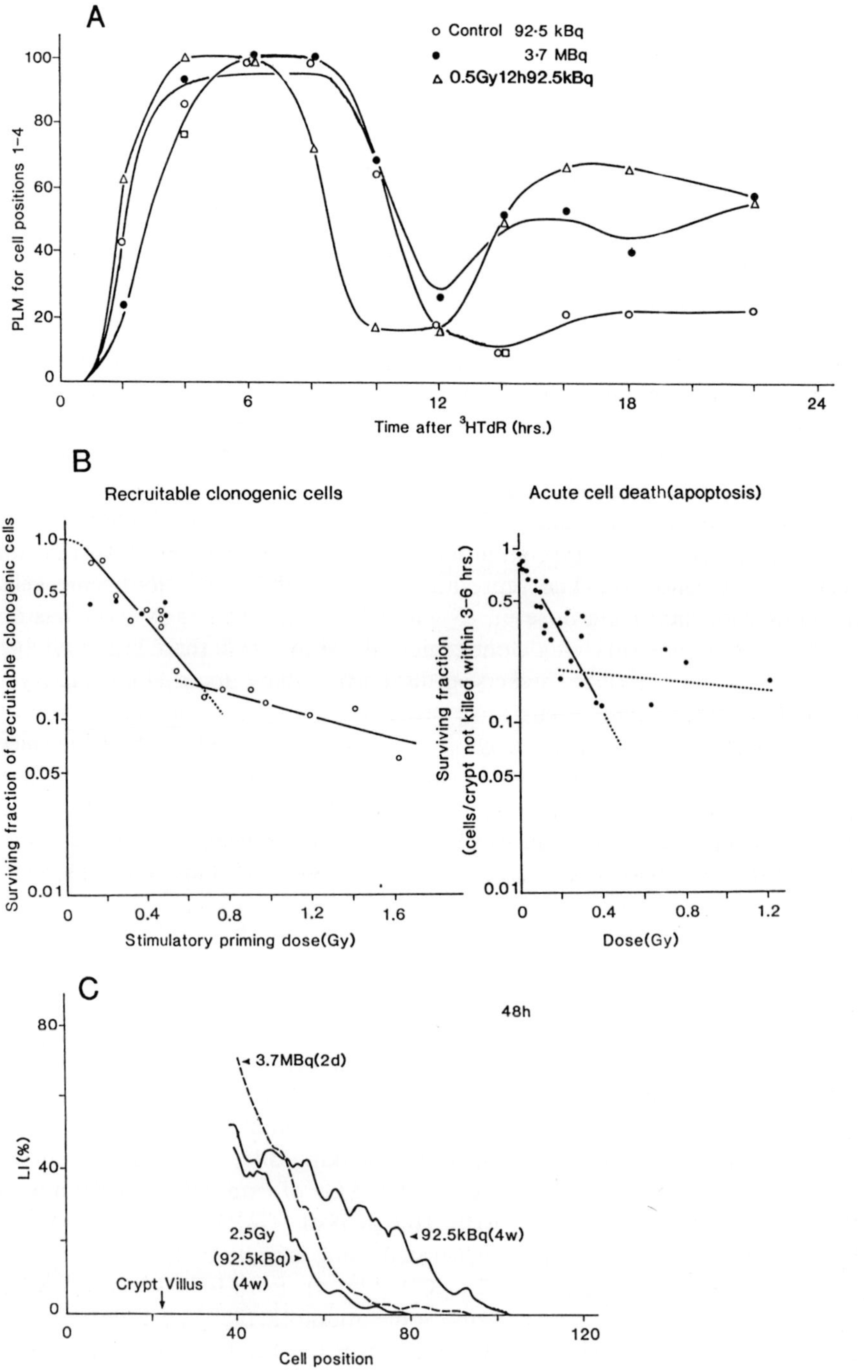
A
○ Control 92·5 kBq
● 3·7 MBq
△ 0.5Gy12h92.5kBq
PLM for cell positions 1–4
Time after ³HTdR (hrs.)
B
Recruitable clonogenic cells
Surviving fraction of recruitable clonogenic cells
Stimulatory priming dose(Gy)
Acute cell death(apoptosis)
Surviving fraction
(cells/crypt not killed within 3–6 hrs.)
Dose(Gy)
C
48h
3.7MBq(2d)
92.5kBq(4w)
2.5Gy
(92.5kBq)
(4w)
Crypt Villus
LI(%)
Cell position

the shortening of the cell cycle in the crypt base from about 25.5 h to about 13.1 h (after 0.5 Gy) (see also Potten 1986, 1990, 1991). There was also a significant burst in mitotic activity in the lower cell positions at about 16 h after 3.7 MBq of [^{3}H]thymidine. Many of these mitotic figures were labelled (Tsubouchi & Potten 1985).

2. Hanson and co-workers (Hanson et al 1980, Hanson & Boston 1983, Potten et al 1984), using a complex protocol in mice which combined a large acute dose of radiation (11 Gy) with a 12-hour cytotoxic metaphase accumulation regimen (colcemid or vincristine), noticed that small priming doses of radiation (0.13–0.5 Gy) resulted in a significantly reduced clonogenic survival over that expected from the metaphase arrest protocol combined with 11 Gy. The interpretation was that the small priming doses of radiation caused damage that triggered slowly cycling, or quiescent, clonogenic cells into rapid cell cycle, after which they were susceptible to the metaphase arrest regimen. The damage that triggered these changes was unknown, but these small doses of radiation are now known to result in rapid (apoptotic) cell death. It therefore seems reasonable to assume that it is the dead cells that trigger the cell cycle changes in the clonogenic cells. This is supported by the similarity in the shapes of the survival curve for the cells susceptible to acute cell death (apoptosis) and that for the recruitable clonogenic cells (Fig. 4B) (see also Potten et al 1984).

3. If [^{3}H]thymidine is given to animals it will label all the S (DNA synthetic) phase cells in the intestinal crypts. Shortly after labelling, these can be located autoradiographically in a broad band across the middle of the crypt, where the labelling index (LI) will be about 50%. If time is allowed between labelling and sampling, the labelled cells will divide and their daughters will migrate up the crypt and then onto the villus. By about 48 h they will be half to two-thirds of the distance up the villus. We have recently looked at the distribution of labelled cells on the villus in mice, 48 h after labelling using various doses of [^{3}H]thymidine and external γ-irradiation (Fig. 4C).

The results were somewhat surprising. With very small doses of [^{3}H]thymidine (92.5 kBq) that were assumed to be non-damaging, the leading edge (at approximately the half-peak height) was at about cell position 72. Little difference

FIG. 4. Proliferative changes after minor damage. A shows percentage labelled mitosis (PLM) curves obtained after a small dose (92.5 kBq) or a large dose (3.7 MBq) of [^{3}H]thymidine (^{3}HTdR). Also shown is a PLM curve which was begun 12 h after 0.5 Gy of γ-rays using the small [^{3}H]thymidine dose. (Redrawn from Tsubouchi & Potten 1985.) B. Survival curves for recruitable clonogenic cells (technique of Hanson et al 1980) and for apoptosis-susceptible cells. (Redrawn from Potten et al 1984.) C. Leading edge of labelling index (LI) versus cell position frequency plots obtained 48 h after a low or high dose of [^{3}H]thymidine or after 2.5 Gy of γ-irradiation (low [^{3}H]thymidine dose). Autoradiograph exposure times shown, four weeks and two days. A range of exposure times were analysed. (Unpublished data.)

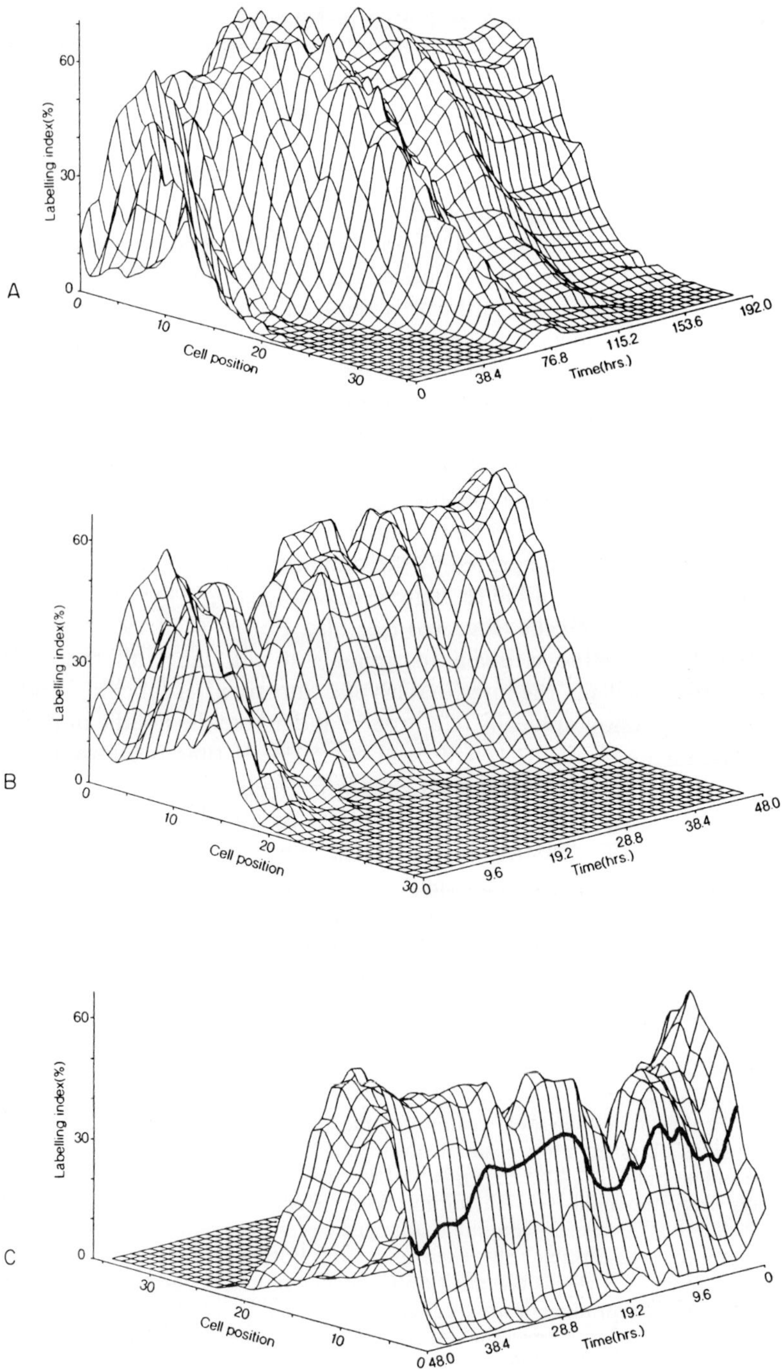

A
Labelling index(%)
0
30
60
0
10
20
30
Cell position
0
38.4
76.8
115.2
153.6
192.0
Time(hrs.)
B
Labelling index(%)
0
30
60
0
10
20
30
Cell position
0
9.6
19.2
28.8
38.4
48.0
Time(hrs.)
C
Labelling index(%)
0
30
60
30
20
10
0
Cell position
48.0
38.4
28.8
19.2
9.6
0
Time(hrs.)

was detected if 0.5 Gy was administered immediately after the labelled thymidine. However, if 3.7 MBq of [^{3}H]thymidine was given without additional irradiation, or 2.5 Gy was delivered immediately after the small dose of 92.5 kBq [^{3}H]thymidine, the labelled front lagged about 16–17 cell positions behind that for the 92.5 kBq dose (see also Potten 1990, 1991). The external irradiation and the high dose (3.7 MBq) of [^{3}H]thymidine are both known to cause cell death and, from the data above, are thought to stimulate rapid cell cycles in the stem cell zone of the crypt. Faster cell cycles might be expected to result in more rapid cell movement which would result in the leading edge being to the right of the control distribution. In fact the opposite is observed.

A possible explanation is that the induced rapid cycling is directed towards re-establishing the depleted stem cell population at the expense of cells entering the transit population. The data would be consistent with the equivalent of, on average, one complete stem cell cycle being directed towards stem renewal with no cells entering the transit population, spread over the 48 h period. This would result in a transit cell deficiency of 16 cells. These effects could be achieved if the stem cell self-maintenance probability was changed from 0.5 to 1.0 for one cell cycle or to some value between these figures over several cell cycles (48 h might be expected to cover a period of 3–5 stimulated stem cell cycles). It should be noted that this argument would only explain the results adequately if considerably more damage were sustained in the stem cell, and possibly also the early transit cell, populations than is suggested by the apoptotic yields alone (U. Paulus, personal communication). If the output from one entire stem cell cycle is lost to the transit compartment, this would imply that the stem cell population had been depleted to at least one-half of its original size.

Cell kinetic changes after large doses of γ-radiation

Various studies in the past have reported cell cycle changes in intestinal crypts after radiation or drug treatment in rodents. Lesher (1967) noted that the average cell cycle of all crypt cells was reduced from 13.1 h to about 10.4 h, one day

FIG. 5. Three-dimensional plots of labelling index, cell position and time after 8.0 Gy of whole-body γ-irradiation in mice. A. Full range of times analysed up to 192 h. There has been some smoothing of the data in relation to cell position and time. Altogether 34 different time points were analysed. The considerable expansion in the size of the crypt proliferative compartment can be seen at about 72–100 h. Some structure is evident throughout the time course of the experiment. B. The first 48 h expanded. The structure at the early times and lower cell positions is clearer. C. The reverse of the labelling index 'mountain' shown in B. Here the changes in the stem cell zone can be seen. The changes for cell position 4 have been highlighted. (Reproduced from Potten et al 1990 with permission of Taylor & Francis Ltd.)

after 3 Gy. Cairnie (1967) showed that the average crypt cell cycle time was reduced from 10.5 h to 9 h during continuous irradiation (3.5 Gy/day). However, the cells at the crypt base showed a larger reduction, from about 13–14 h down to 9–10 h. Al-Dewachi et al (1977) noted that hydroxyurea (1.84 g/kg) reduced the cell cycle at the crypt base from 15.4 h to 10.9 h. Further studies with cytosine arabinoside (440 mg/kg) reduced the crypt-base cell cycle from 15–17 h to 11–14 h (Al-Dewachi et al 1980). These studies suggested that one of the earlier events in crypt regeneration was a shortening of the cell cycle, particularly for the cells in the stem cell region. It could then be inferred that these crypt-base cells are important for regeneration. Similar conclusions have been reached in our own work using cytosine arabinoside (200 mg/kg) (Chwalinski & Potten 1989), where the first signs of regeneration after a total abolition of [^{3}H]thymidine labelling was at 12 h at cell position 4.

After irradiation (8 Gy), the clonogenic population is re-established with a doubling time of 21 h (Potten et al 1987) and cell cycle times for the crypt base of about 10 h (see data of L. Geng in Potten 1990). It would appear that in several situations where regeneration is required, the cell cycle of the cells in the stem cell zone is reduced to about 9–11 h. We have recently undertaken a detailed analysis of the spatial distribution of mitotic and labelled cells at 34 different times between 3 h and 192 h after a single acute dose of 8 Gy of γ-rays (Potten et al 1990a). The results are summarized for the labelling index by the graphs in Fig. 5. The results as a whole showed that:

1. The overall labelling index falls rapidly to reach a minimum value at 15 h, which is about one-third of the control value. The depressed labelling index persists until about 40 h post-irradiation.
2. The overall mitotic index does not fall until after 3 h and is near its minimum value at about 18 h. It persists at subnormal levels until about 30 h.
3. The overall labelling index shows a large overshoot, with its maximum at about 72 h. Normal values are regained at about 96 h. This overshoot is reflected by a spreading in the labelling index to higher positions in the crypt (see Fig. 5A).
4. The overall mitotic index shows a very large overshoot, with maximum values at about 72 h and 124 h. The index never regains control levels within 192 h. There are signs of peaks in the overall mitotic activity at about 10, 35, 70, and 120 h after γ-irradiation.
5. When the lower stem cell positions are analysed alone, more structure can be detected (Fig. 6). After an initial fall in both labelling and mitotic indexes, a series of peaks in stem cell proliferative activity are evident at about 10–12, 30–35, 60–68 and 120–132 h. The first peak begins to develop early, shortly after 3 h. This represents the first detectable signs of regenerative activity and it occurs in the stem cell region and at a time coinciding with the peak in apoptotic activity—that is, acute cell death.

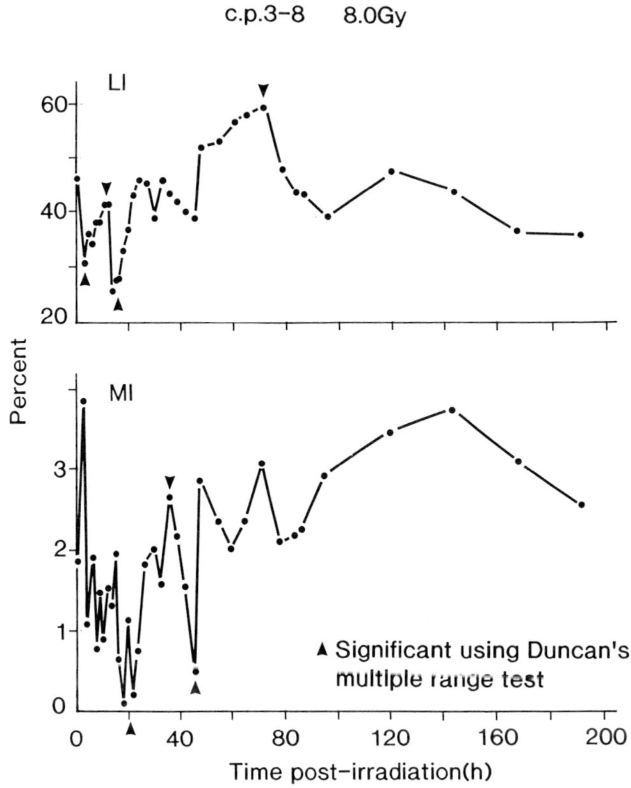

FIG. 6. Changes in cell proliferation (labelling index, LI; mitotic index, MI) for the lower cell positions of the small intestinal crypt in mice at various times after 8 Gy of γ-irradiation. The pooling of data from several positions allows the trends seen in Fig. 5C to be analysed statistically. (Taken from Potten 1990 with permission of Taylor & Francis Ltd.)

Discussion and conclusions

In summary, the indications are that the stem cells in the small intestinal proliferative unit, the crypt, are delicately balanced in terms of their proliferative controls. They are located in an annulus of variable height from the crypt base (the average is at the 4th position) and yet they appear to be able to detect the destruction of one of their members anywhere in the ring and respond rapidly by changing their rate of cell cycle progression. They must also be capable of detecting when the full complement is re-established. If more severe damage is incurred, their cycles remain shortened for longer periods of time. After more severe damage, some oscillations in crypt size may occur that can be largely attributed to changes in the transit compartment. It is interesting to speculate

TABLE 2 Summary of regenerative processes in the mouse small intestine after acute doses of whole-body γ-radiation

Dose range	*Time scale*	*Effect*
0–0.1 Gy	Very rapid	Changes in stem cell cycle parameters and mitosis
0–1 Gy	Rapid	Stem cell regenerative processes Cell cycle of stem cells reduced from 24 h to <16 h
1–8 Gy	Within a few hours	Crypt regenerative processes begin
8–20 Gy	0–3 d	Crypt regeneration, clonogenic doubling time ≈21 h, stem cell cycle ≈9–10 h
9–15 Gy	0–7 d	Mucosal regeneration, crypt fission and mucosal outgrowth

about the mechanisms involved in triggering the changes in cell cycle times for the stem cells. The only observable evidence of damage in most cases is the presence of apoptotic cells. The time of peak apoptotic yield coincides with the beginning of the first wave of proliferative activity for the stem cells. I would postulate that the apoptotic cells provide the signal for the changes in stem cell proliferation. The nature of this signal remains obscure at present.

TABLE 3 Effects of acute whole-body doses of γ-radiation on mouse small intestine

Effects seen strongly in stem cell region *All early effects (within a few hours)* Acute cell death Changes in cell cycle Mitotic delay (G_2/M arrest) Rapid stimulation into fast cell cycle Beginning of crypt regeneration	Clonogenic or stem cell effects
Effects seen on crypt *Intermediate time scale (more than a few hours)* Fall in proliferation Reduction in cell number (to 0 at high doses) Some repopulation	Transit cell effects
Effects seen on villus *Long time scale (Days 2–4)* Loss of epithelial cells Reduction in villus height Some crypts repopulated	Mature cell effects
Effects on GI system *Final effects (Days 4–7)* Loss of mucosal integrity leading to symptoms of GI syndrome Animal death (high doses) Some re-epithelialization (Crypt fission)	General effects on animal

In a broader context, the animals' well-being after irradiation is determined by the survival of intact crypts, and their survival in turn is determined by the survival and regeneration of the clonogenic (stem) cells (Tables 2 and 3). The primary determining element here is the interaction of any cytotoxic agent with the stem cells (see Potten 1990a).

Acknowledgement

This work has been supported by the Cancer Research Campaign.

References

Al-Dewachi HS, Wright NA, Appleton DR, Watson AJ 1977 The effect of a single injection of hydroxyurea on cell population kinetics in the small bowel mucosa of the rat. Cell Tissue Kinet 10:203–213

Al-Dewachi HS, Wright NA, Appleton DR, Watson AJ 1980 The effects of a single injection of cytosine arabinoside on cell population kinetics in the mouse jejunal crypt. Virchows Arch B Cell Pathol 34:299–309

Cairnie AB 1967 Cell proliferation studies in the intestinal epithelium of the rat: response to continuous irradiation. Radiat Res 32:240–264

Chwalinski S, Potten CS 1989 Crypt base columnar cells in ileum of BDF_1 male mice—their numbers and some features of their proliferation. Am J Anat 186:397–406

Hanson WR, Boston DL 1983 Cytosar-U (ARA-C) induced changes in mouse jejunal crypt epithelial kinetics and radiosensitivity to gamma rays and fast neutrons. Int J Radiat Oncol Biol Phys 9:515–521

Hanson WR, Henninger DL, Fry RJM, Sallese AR 1980 The response of small intestinal stem cells in the mouse to drug and irradiation treatment. In: Appleton DR, Sunter JP, Watson AJ (eds) Cell proliferation in the gastrointestinal tract. Pitman Medical, Tunbridge Wells, p 198–212

Hendry JH 1979 A new derivation, from split-dose data, of the complete survival curve for clonogenic normal cells in vivo. Radiat Res 78:404–414

Hendry JH, Potten CS, Chadwick C, Bianchi M 1982 Cell death (apoptosis) in the mouse small intestine after low doses: effects of dose-rate 14.7 MeV neutrons and 600 MeV (maximum energy) neutrons. Int J Radiat Biol 42:611–620

Hendry JH, Potten CS, Roberts NP 1983 The gastrointestinal syndrome and mucosal clonogenic cells: relationships between target cell sensitivity, LD_{50} and cell survival, and their modification by antibiotics. Radiat Res 96:110–112

Hendry JH, Potten CS, Ghaffor A, Moore JV, Roberts SA, Williams PC 1989 The response of murine intestinal crypts to short-range promethium-147 irradiation: deductions concerning clonogenic cell numbers and positions. Radiat Res 118:364–374

Hume WJ 1983 Stem cells in oral epithelia. In: Potten CS (ed) Stem cells: their identification and characterisation. Churchill Livingstone, Edinburgh, p 233–270

Hume WJ, Potten CS 1976 The ordered columnar structure of mouse filiform papillae. J Cell Sci 22:149–160

Kellett M 1991 PhD thesis, University of Manchester.

Kerr JFR, Whyllie AH, Currie AR 1972 Apoptosis: a basic biological phenomenon with wide-ranging implications in tissue kinetics. Br J Cancer 26:239–257

Lesher S 1967 Compensatory reactions in intestinal crypt cells after 300 roentgens of cobalt-60 gamma irradiation. Radiat Res 32:510–519

Potten CS 1977 Extreme sensitivity of some intestinal crypt cells to X and gamma irradiation. Nature (Lond) 269:518–521

Potten CS 1981 Cell replacement in epidermis (keratopoiesis) via discrete units of proliferation. Int Rev Cytol 69:271–318

Potten CS 1983a Stem cells in epidermis from the back of the mouse. In: Potten CS (ed) Stem cells: their identification and characterisation. Churchill Livingstone, Edinburgh, p 200–232

Potten CS 1983b (ed) Stem cells: their identification and characterisation. Churchill Livingstone, Edinburgh

Potten CS 1986 Cell cycles in cell hierarchies. Int J Radiat Biol 49:257–278

Potten CS 1990 A comprehensive study of the radiobiological response of the murine (BDF1) small intestine. Int J Radiat Biol 58:925–874

Potten CS 1991 The role of stem cells in the regeneration of intestinal crypts after cytotoxic exposure. In: Davis J (ed) Chemically induced cell proliferation (M. D. Anderson Cancer Center, Houston) Alan R Liss, New York, in press

Potten CS, Hendry JH 1975 Differential regeneration of intestinal proliferative cells and cryptogenic cells after irradiation. Int J Radiat Biol 27:413–424

Potten CS, Hendry JH 1983 Stem cells in murine small intestine. In: Potten CS (ed) Stem cells: their identification and characterisation. Churchill Livingstone, Edinburgh, p 155–159

Potten CS, Hendry JH 1985 The microcolony assay in mouse small intestine. In: Potten CS, Hendry JH (eds) Cell clones: manual of mammalian cell techniques. Churchill Livingstone, Edinburgh, p 50–60

Potten CS, Lajtha LG 1982 Stem cells versus stem lines. Ann NY Acad Sci 397:49–61

Potten CS, Loeffler M 1990 Stem cells: attributes, cycles, spirals, pitfalls and uncertainties. Development 110:1001–1019

Potten CS, Schofield R, Lajtha LG 1979 A comparison of cell replacement in bone marrow, testis and three regions of surface epithelium. Biochim Biophys Acta 560:281–299

Potten CS, Hendry JH, Al-Barwari SE 1983a A cellular analysis of radiation injury in epidermis. In: Potten CS, Hendry JH (eds) Cytotoxic insult to tissue: effects on cell lineages. Churchill Livingstone, Edinburgh, p 153–185

Potten CS, Hendry JH, Moore JV, Chwalinski S 1983b Cytotoxic effects in gastrointestinal epithelium (as exemplified by small intestine). In: Potten CS, Hendry JH (eds) Cytotoxic insult to tissue: effects on cell lineages. Churchill Livingstone, Edinburgh, p 105–152

Potten CS, Chadwick C, Ijiri K, Tsubouchi S, Hanson W 1984 The recruitability and cell cycle state of intestinal stem cells. Int J Cell Cloning 2:126–140

Potten CS, Hendry JH, Moore JV 1987 New estimates of the number of clonogenic cells in crypts of murine small intestine. Virchows Arch B Cell Pathol 53:227–234

Potten CS, Hendry JH, Taylor Y 1988a The doubling time of regenerating clonogenic cells in the crypts of the small intestine. Int J Radiat Biol 54:1041–1051

Potten CS, Roberts SA, Chwalinski S, Loeffler M, Paulus U 1988b Scoring mitotic activity in longitudinal sections of crypts of the small intestine. Cell Tissue Kinet 21:231–246

Potten CS, Owen G, Roberts SA 1990a The temporal and spatial changes in cell proliferation within the irradiated crypts of the murine small intestine. Int J Radiat Biol 57:185–199

Potten CS, Kellett M, Roberts SA, Rew DA, Wilson GD 1990b The measurement of in vivo proliferation in human colorectal mucosa using bromodeoxyuridine. Gut, in press

Thakkar N 1991 PhD thesis, University of Manchester
Tsubouchi S, Potten CS 1985 Recruitment of cells in the small intestines into rapid cell cycle by small doses of external or internal radiation. Int J Radiat Biol 48:361–369
Withers HR, Elkind MM 1970 Microcolony survival assay for cells of mouse intestinal mucosa exposed to radiation. Int J Radiat Biol 17:261–268
Wright NA, Alison M 1984 The biology of epithelial cell populations. Clarendon Press, Oxford, vol. 2:529–1247

DISCUSSION

Raymond: I am not clear what the evidence is that the stem cells in the crypts are different from the other cells of the so-called proliferative unit. You suggest that the stem cells have a different capacity or potential from the other proliferative cells in the unit.

Potten: We define 'stem cells' as undifferentiated cells with a large potential for division and a capability for sustaining a tissue for a long period of time—throughout adult life. On that definition, we can show that although cells in the upper regions of the crypt are proliferating rapidly (i.e. have a short cell cycle), they do not satisfy the criteria for a stem cell—namely, large division potential over a long time. We show this by mapping the movement of prelabelled dividing cells, by plotting labelling index distributions (a proliferative index) against cell position in the crypt, and following what happens with time.

As time passes, the labelled cells are constantly either moving, or being moved, to steadily higher positions in the crypt, at a velocity of one cell position per hour. You can also plot the cell 'velocity' against cell position (Kaur & Potten 1986). As cells move up the crypt to higher crypt cell positions, their migration velocity increases. Such data can be used to identify the cell position(s) where the movement does not occur—that is, the cell position(s) from which everything starts. This is around the 4th position in the crypt. So this is one way of tackling your question.

Another approach is to ask, from regeneration studies, how many cells in the crypt are capable of undergoing the large number of divisions required for regeneration. Here, we would like to equate the ability of cells to regenerate (in other words, divide a large number of times and produce many progeny, which is the same sort of criterion that we apply to the steady-state stem cells) with stem cell function. γ-Ray regeneration studies can be interpreted to provide information about the number of regenerative or clonogenic cells present in each unirradiated crypt (Potten et al 1987). In simple terms, this is part of the extrapolation back to the origin of a radiation survival curve when the dose is plotted against the number of surviving (regenerated) crypts. When we do this and take into account damage repair we arrive at values of around 32 clonogenic cells per crypt, which is considerably lower than the 150–160 proliferative cells. As I said, there are 16 cells in a ring, so this implies that two rings of cells possess this regenerative (stem cell) capability. If this is used as the way of identifying potential stem cells, it says that about 32 out of 150–160 proliferative cells possess regenerative (potential stem cell) capacity.

Raymond: I understand that as the cells are displaced towards the surface of the crypt, their fate is to undergo fewer and fewer divisions, and in that sense their proliferative potential has been restricted. I am still not clear whether there are any differences in the *potential* of those cells—not what they normally do, but their potential. In other words, if you took a cell out of the proliferative transit compartment and put it back down next to the base of the crypt, would it behave like a stem cell? Are there proliferating cells with different potentials, as there are in the haemopoietic system, where cells go through a sequence of discrete phases, for which there are distinct cell markers that can be correlated with different proliferative potentials?

Potten: We can't really do that transplantation experiment; all we can say is that the evidence from our radiation experiments suggests that the cells in one ring (probably the 4th from the crypt base) are performing the function of stem cells in steady-state conditions; they have the biggest division potential, and everything else can be traced back to them. If we ask how many of the other crypt cells can actually slot back into vacant stem cell positions and function as stem cells (i.e. potential stem cells), the radiation experiments suggest that only the cells in the next higher ring (the 5th cell position) can do so; those in the ring above that (i.e. in the 6th position) cannot. So one ring of cells, the actual stem cells, plus the immediate daughters of that ring in the next position up (the potential stem cells which can fall back and function as stem cells if we damage that ring), represent the full crypt stem cell complement.

But this whole question of how you define stem cells and then, having defined them, how many there are in a structure like the crypt, is still open. The estimates obtained range from one or two stem cells, depending on the type of experimental procedures that you look at, through to 16 or 32.

Farbman: You are suggesting that these cells have a way of communicating with each other, so that the 15 cells remaining after the killing of one cell know that one of their comrades is missing, and have to replace it. Do you have any idea of what the signals might be, or where they come from?

Potten: They are not 'holding hands' in a *circle* around the crypt, because they are in reality located over a range of cell positions: one cell may lie at cell position 2 while the next cell in the annulus may be at cell position 7.

Farbman: Might they be 'holding hands' in terms of gap junctions, or something like that?

Potten: Yes, but it looks as if they are too far apart to have this type of cell contact. Cells 'holding hands' in some way could communicate very easily laterally, but here we would have to envisage some chemical signal that diffuses across several cell diameters.

Farbman: There could be gap junctions on several surfaces of these cells.

Rubel: But cells several layers up will have difficulty in communicating with the lower layers.

Potten: Yes. Let me see if I can summarize the situation. We talk about the stem cells being at the 4th cell position but this is the mean of a range of possible positions for each stem cell in the annulus around the crypt. At one position in the ring the stem cell may be as low as the 2nd cell position from the bottom of the crypt while in the next position round the ring the stem cell may be at the 7th or 8th cell position. These two adjacent stem cells cannot have direct cell-to-cell contact (e.g. gap junctions). I can only imagine that there is a gradient of chemical message concentrated at the 4th cell position and the loss of one stem cell contributing to this message alters the overall balance of the message; and that this is detected by the other stem cells.

Fernald: You are treating cell damage as perfectly analogous to normal development, or normal use of the system, but it could be that in damage there is a circulating factor that is much more broadly communicated than under normal conditions.

Potten: That is absolutely right. I should have said that it is also possible that there is a positive signal released by a dying cell that is detected by the other stem cells. We must, however, remember that we are considering only minor levels of damage—an extension of the steady-state situation where an occasional cell dies without any known exposure to cytotoxic agents.

Watt: What you are saying seems to be consistent with the 'niche' hypothesis of Schofield (1978)—the idea that whether or not a cell is a stem cell depends on the microenvironment, which is, in turn, determined by other cells. To use this conference room as an analogy, the inner and outer rings of chairs constitute two different microenvironments. People sitting in the inner ring are equivalent to stem cells, and those in the outer ring are transit cells. If there are no free seats in the inner ring, someone entering the room will have to sit in the outer ring and, as a result, will lose stem cell properties and become equivalent to a transit cell.

Potten: That is an extension of the point raised by Dr Raymond. It is perfectly valid, but only up to a certain point. If there is a vacant chair in this ring of chairs, or cells, then a cell in the next ring above could occupy it, but not a cell from the ring above that. There appears to be a distinct limit to the distance up the crypt from which regenerative (clonogenic) cells can be recruited (Potten et al 1987).

Watt: That would be because differentiation is undirectional.

Potten: Yes; and that doesn't seem unreasonable to me, because the cells of the next layer (the 5th) above the stem cell layer (4th) are the immediate daughters (and twins) of each of the stem cells; that is, the stem cell progeny sit immediately above the stem cells. While they are still in that position, they are identical to their cell siblings (the stem cells) and can if needed fall back and function as stem cells. However, once they are displaced one cell position further up the crypt, they now have a new stem cell daughter cell between them and the stem cells, and then they lose the potential for regeneration. These cells have gone just that bit too far along the differentiation pathway, and cannot reverse this.

But all of this argument is dependent on how you interpret the numbers obtained from radiation experiments, and it all comes back to how you define the stem cell population and what set of experiments you are prepared to accept or consider, to answer the question of how many stem cells there are in the system. My view is that there are relatively few; I wouldn't like to put my reputation on 16 per crypt, but there is a small number that occupy part or all of a ring at cell position 4. It could be a few as four,and our views now are that it is more likely to be four than 16. However, up to 32 cells (i.e., two rings) are potential stem cells and can regenerate the crypt if damage is incurred in that area. The other point about this 'falling back' process is that it isn't always necessary, because cell kinetic experiments suggest that sometimes when there's an empty 'chair', or niche, the neighbouring stem cells divide and fill it.

Farbman: When you talk about the radiation-induced regulation of mitosis in the crypt, this is an up-regulation of mitosis (an increasing rate of mitosis). There is an implication that one can also down-regulate mitosis. Have you done any experiments to show that you can, in fact, down-regulate mitosis, from the basal level, using other signals, or using other means of invasion?

Potten: Not directly. But mitosis must be down-regulated, or switched off, when the 16 stem cells in a ring are re-established, because the crypt doesn't overshoot in size. We haven't succeeded in down-regulating mitosis from its steady-state level.

Fernald: Do these crypt cells show a circadian rhythm in their reproduction?

Potten: Crypt cells have a minor circadian rhythm. You have to look hard to find evidence for such rhythms in the gut; it's very damped.

Fernald: A circadian rhythm in cell proliferation would be one way in which you would have up- and down-regulation from the basal level. That could give you access to some causal factor that might underlie this basal rate.

Potten: That is possible; in fact, if we consider the circadian rhythm in gut, and other epithelial tissues, we believe there is evidence (Potten et al 1977) that circadian rhythms originate in the stem cells, and thus it is the stem cells that are actually being controlled by the factors that determine the circadian rhythms. They are therefore producing cells that enter the dividing transit population at different times of day, and this accounts for the different levels of 'dampening' of the rhythms that you see in different tissues. The cells desynchronize as they pass through the amplifying (transit) compartment.

Lewis: What are the effects of gastrin on gut cell proliferation?

Potten: A variety of gut trophic hormones have been studied over the years but most of these have relatively minor effects on cell kinetic parameters.

Rubel: Another question on regulation: if you make a small lesion, just disturbing one crypt, say, is there any hysteresis in the system, or do you see just the right amount of regulation? Is there any overshoot, in terms of the number of cells that enter the mitotic cycle?

Potten: The stem cell compartment seems to dampen itself out very quickly, with little overshoot (Potten et al 1990). The total proliferative compartment (which includes the transit cells), however, shows a considerable overshoot. This is partly, I think, because the stem cell population re-establishes itself quickly and then continues for a cycle or two at an elevated rate, and as a consequence produces more transit cells than the system requires, and it takes longer for the dampening-out to take effect through the entire lineage.

Rubel: So the stem cell compartment is not looking ahead at how many cells it is going to produce. It first produces the cells at a maximum rate, and then communicates to the proliferative compartment and says 'stop'?

Potten: The stem cell population seems to be regulating itself precisely; but the total population does not seem to do so.

Jørgensen: I am interested in the morphology of the dying cells. Do they always look like apoptotic cells, whether you use radiation, or a drug, to cause damage? Or do you find other types of dying cell?

Potten: For most of the cytotoxic agents that we have looked at, using low or moderate doses, the morphological appearance of the dying cells is consistent with the changes described for apoptosis. However, at very high doses of some of the cytotoxic agents you may precipitate a necrotic type of death.

Oakley: In the healthy, hairy skin of the mouse, is it your current view that there is a population of quiescent stem cells? If so, how large is that population, and is there any 'changing of the guard', or is it always the same cells which turn over?

Potten: We don't really understand what happens in damage or normal situations in skin. We believe that mammalian skin is divided up into proliferative units (Potten 1974), similar to those I described for gut, with small functional groups of cells, which in murine epidermis contain about 10 basal cells, within which there is a cell hierarchy. If that is the case, with only 10 cells per unit, there is probably only one stem cell for each group. How the epidermis manages with that type of arrangement is not clear. I don't know how it copes with the loss of an individual stem cell, and hence unit, and how new units are formed—that is, how the stem cells get from one proliferative unit to another unit, or indeed whether they are capable of doing this. It is possible that when regeneration is required in the epidermis, cells have to be recruited from other sites, for example from the hair follicles.

References

Kaur P, Potten CS 1986 Circadian variation in migration in small intestinal epithelium. Cell Tissue Kinet 19:591–600

Potten CS 1974 The epidermal proliferative unit: the possible role of the central basal cell. Cell Tissue Kinet 7:77–88

Potten CS, Al-Barwari SE, Hume WS, Searle J 1977 Circadian rhythms of presumptive stem cells in three different epithelia of the mouse. Cell Tissue Kinet 10:557–568

Potten CS, Hendry JH, Moore JV 1987 New estimates of the number of clonogenic cells in crypts of murine small intestine. Virchows Arch B Cell Pathol 53:227–234

Potten CS, Owen G, Roberts S 1990 The temporal and spatial changes in proliferation within the irradiated small intestinal crypts. Int J Rad Biol 57:185–199

Schofield R 1978 The relationship between the spleen colony-forming cell and the haemopoietic stem cell. A hypothesis. Blood Cells 4:7–25

Hair cell regeneration in the avian inner ear

Edwin W. Rubel, Elizabeth C. Oesterle and Pedro Weisleder

Hearing Development Laboratories, Department of Otolaryngology—Head and Neck Surgery, University of Washington, Seattle, WA 98195, USA

Abstract. The postembryonic production of hair cells in fish and reptiles has been known for several decades. Until recently it was assumed that this capacity was absent in the more highly specialized inner ears of birds and mammals. Recent research has shown, however, that birds have the capacity to rebuild a damaged inner ear. Summarized here are studies conducted in our laboratory which address the following questions: (1) Which are the precursors of the regenerated hair cells? (2) Are the new hair cells functional? (3) What are the ultrastructural properties of regenerated hair cells? and (4) Can the level of proliferation be regulated? Both the auditory and the vestibular systems of the avian inner ear were studied. Our results provide some answers to these questions. The implications of the results are discussed.

1991 Regeneration of vertebrate sensory receptor cells. Wiley, Chichester (Ciba Foundation Symposium 160) p 77–102

The potential for vertebrate hair cells to be proliferated and differentiate throughout life has been recognized for over 50 years (see Corwin et al 1989). For example, Stone (1933, 1937) studied regeneration of lateral line organs after tail amputation in amphibian embryos. He found that regenerated organs were numerically similar to the previous complement and that they were supplied by the migration of precursor cells from the last organ of the proximal tail stump. More recently, a number of groups have investigated both the continued production of hair cells in the saccule of rays and fishes as well as the regeneration of hair cells in the lateral line of amphibians (Balak et al 1990, Corwin 1983, Corwin et al 1989, Presson & Popper 1990a,b).

Until recently, however, it was generally accepted that the postembryonic production of hair cells is limited to cold-blooded vertebrates, as is the ability to restore damaged populations of hair cells. That is, birds and mammals have lost this ability during evolution. In both birds and mammals, hair cells that will normally populate the inner ear are produced relatively early in embryogenesis (Katayama & Corwin 1989, Ruben 1967). While some mitotic activity may continue after this time, the new cells have not been thought to differentiate into hair cells.

Two serendipitous findings suggested that birds were able to restore the population of hair cells after the elimination of embryonically produced hair cells. Cruz et al (1987) performed an experiment intended to examine the time course of destruction of hair cells in the chick cochlea after the administration of an aminoglycoside antibiotic. Neonatal chicks were given gentamicin injections for 10 days and allowed to survive for varying amounts of time ranging from one to 32 days. Cochleas were serially sectioned and the number of hair cells was counted at 100 μm intervals from the base to the apex. Hair cell counts revealed that after the 10-day aminoglycoside treatment there was a nearly total elimination of hair cells in the basal third of the cochlea. A week later the damage had spread to eliminate hair cells throughout the basal two-thirds of the cochlea, but counts at the basal pole revealed that the number of hair cells had been partially restored. After another two weeks the number of hair cells throughout the basal two-thirds of the cochlea appeared to be recovering toward normal.

While these results strongly suggested that new hair cells were being produced to replace those destroyed by the aminoglycoside treatment, other interpretations were considered. For example, it was possible that the aminoglycosides caused deterioration or dedifferentiation of hair cells to the extent that they were unrecognizable in the microscopic sections, and then recovery ensued. Even if new hair cells were taking the place of embryonically produced cells that had been destroyed by the drug, we could not conclude that these were newly produced cells. It was possible that the aminoglycosides or the resulting damage induced support cells to differentiate into new hair cells.

While that study was progressing, Cotanche (1987a) was examining the neonatal chick cochlea by scanning electron microscopy after acoustic trauma. Although the initial purpose of the study was to examine age differences in the position of damage produced by sound overstimulation (see Rubel & Ryals 1983, Ryals & Rubel 1985, Lippe & Rubel 1983, 1985), Cotanche also noted the repopulation of hair cells a few days after the noise damage. Again, this observation is open to several interpretations, including both cell regeneration and the recovery of stereocilia. Cotanche made two additional important observations; the apical surfaces of the cells that appeared to be repopulating the cochlea bore a striking resemblance to immature hair cells, and the sequence of differentiation paralleled the embryonic development of stereocilia.

Taken together, these two studies (Cruz et al 1987, Cotanche 1987a) suggested that the restoration of hair cells was not due to recovery, but represented newly created hair cells or transformed support cells. Three studies (Corwin & Cotanche 1988, Jørgensen & Mathiesen 1988, Ryals & Rubel 1988) then used [^{3}H]thymidine to label mitotically active cells in the inner ear of young and adult birds in order to prove that new hair cells were being produced. The radioactive tracer is incorporated into replicating DNA during the S phase of the cell cycle. It then remains in the nucleus throughout the life of the cell. If an abundance of radioactive thymidine is introduced into the environment of mitotically active cells during S phase, the progeny become radioactively labelled. Cells labelled by [^{3}H]thymidine can be identified at any subsequent time by autoradiography.

Using this method, Corwin & Cotanche (1988) and Ryals & Rubel (1988) demonstrated that damage to the avian cochlea causes a population of stem cells to re-enter the cell cycle and produce new cells, which subsequently differentiate into new hair cells as well as new support cells. Corwin and Cotanche used intense sound to destroy receptors in the cochlea of neonatal chicks. Tritiated thymidine was administered to 9–13-day-old chicks for 10 days after the sound exposure, and cochleas were processed for autoradiography. Both labelled hair cells and labelled support cells were observed. In a parallel study (Ryals & Rubel 1988) young, sexually mature quail (*Coturnix coturnix*) were exposed to an intense pure tone (1500 Hz) to destroy hair cells in the middle region of the cochlea. One group of birds was given [^{3}H]thymidine over the ensuing 10 days. The remaining birds were allowed to survive for 10, 30 or 60 days. Cochleas were sectioned; then either they were processed for autoradiography, or the numbers of hair cells were counted at 100 µm intervals from the basal to apical end. Ten days after the noise damage there is a massive reduction in the number of hair cells in the basal half of the cochlea. During the ensuing 50 days, the population of hair cells is restored to near-normal numbers. The birds treated with [^{3}H]thymidine provided convincing evidence that the repopulation of hair cells is due to the production and differentiation of a new generation of cells. Both support cells and hair cells were labelled in the lesioned area by [^{3}H]thymidine.

These two studies indicate that, in both neonatal and mature birds, damage to the cochlea induces a 'quiescent' population of precursor cells to re-enter the mitotic cycle. Newly produced cells then differentiate into either hair cells or support cells. Jørgensen & Mathiesen (1988) also reported on the production of new hair cells in the postembryonic inner ear of birds. In this case, normal adult budgerigars (*Melopsittacus undulatus*, Australian parrot) were given [^{3}H]thymidine, after which the vestibular epithelia were sectioned and processed for autoradiography. Scattered labelled hair cells and support cells were seen in each receptor epithelium. This result, which we have recently replicated in neonatal chicks (Roberson et al 1991), indicates that there is a low level of continual turnover of hair cells and support cells in the avian vestibular system.

Identity of the precursor population

We have begun to address a number of issues which may lead to further understanding of the process of hair cell regeneration in birds in the hope that the answers will suggest methods by which to induce this ability in mammals, including man. One of the foremost questions is the identification of the cellular

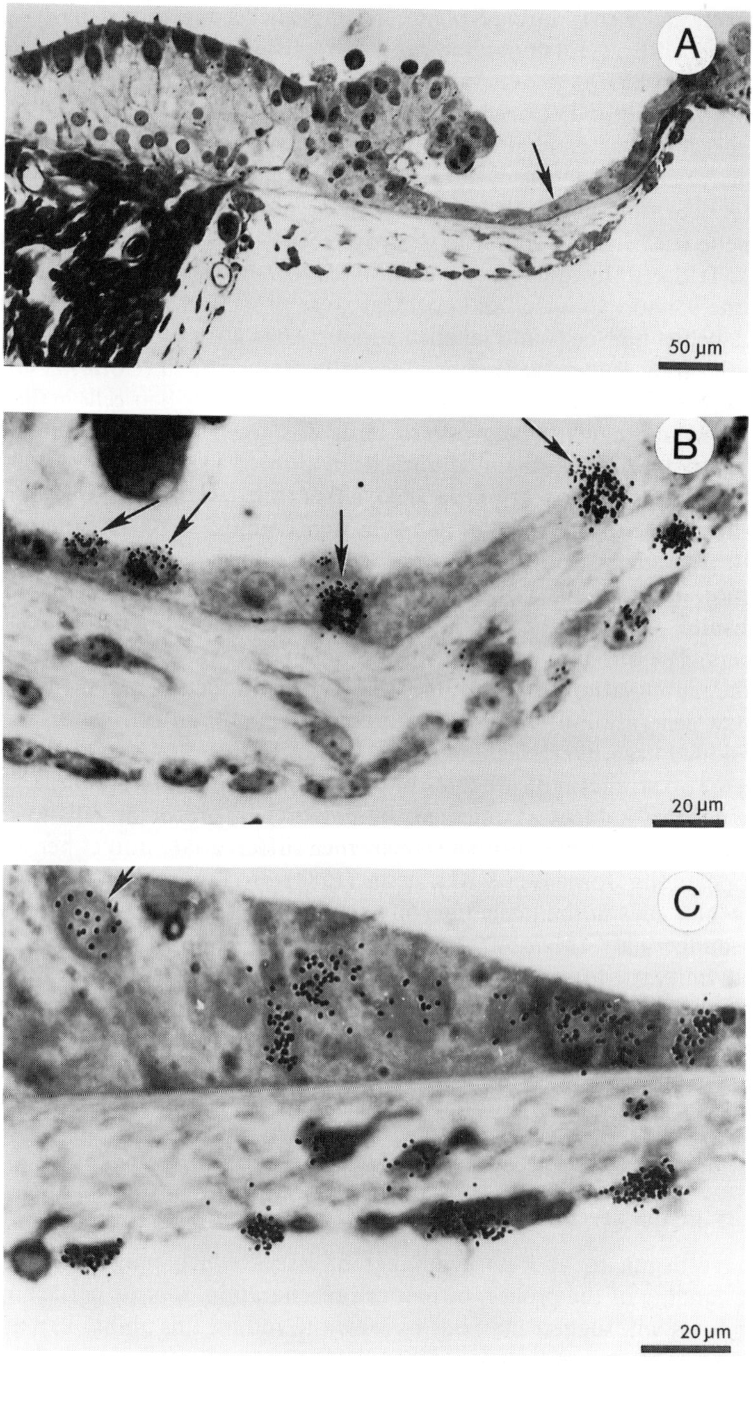
A
50 µm
B
20 µm
C
20 µm

population or populations which re-enter the mitotic cycle and produce new hair cells. This issue has been elegantly addressed by Balak et al (1990) in the lateral line organ of the axolotl, where it was confirmed that supporting cells in the centre of the receptor organ—the internal support cells—are progenitors of regenerated hair cells. In statoacoustic organs of fish, resident embryonic-like neuroepithelial cells give rise to new vestibular hair cells (Presson & Popper 1990b).

In our first experiment, designed to identify hair cell precursors in the avian cochlea (Girod et al 1989), we determined the cells which first become mitotically active after acoustic damage and then followed the fate of their progeny. Neonatal chicks were exposed to an intense pure tone (120 dB SPL, 1500 Hz) for 18 hours. Animals received injections of [^{3}H]thymidine over survival periods of 6, 15, 24 or 72 hours after exposure to the tone. One group of chicks was given an abundance of 'cold' thymidine after three days of [^{3}H]thymidine; these chicks were killed 30 days after noise exposure. Cochleas were processed for autoradiography or scanning electron microscopy (SEM). Figure 1A shows the appearance of the receptor epithelium six hours after the noise exposure. The epithelium is severely damaged and most of the short hair cells are extruded. At this time, no cells in the receptor epithelium or its vicinity are labelled. By 15 or 24 hours after noise exposure the undifferentiated epithelial cells lying at the inferior edge of the receptor epithelium are rapidly proliferating (Fig. 1B). As these cells continue to proliferate the single cell lamina is transformed to a layer 2–3 cells deep and labelled nuclei are abundant. Many labelled new hair cells as well as support cells can be observed toward the inferior side of the receptor epithelium three days after the noise exposure (Fig. 1C). New hair cells are easily recognized by the staining characteristics of their cytoplasm, by a thin apical process extending to the lumen already bearing a tuft of immature cilia, and by a basal process extending down toward the basilar membrane. In scanning electron microscopy, the tips of immature stereocilia are clearly recognized.

The ultrastructural characteristics of the potential hair cell precursors identified by Girod et al (1989), namely cells in the area of the hyaline or cuboidal

---

FIG. 1. Transverse light microscopic sections through the 1500 Hz region of chick cochleas after noise damage (1500 Hz tone at 120 dB SPL for 18 h). A. Sensory epithelium 6 h after completion of noise exposure. The inferior edge of the sensory epithelium is extensively damaged; hair cells and supporting cells are lost. Note the thin monolayer of cells spreading to cover the basilar membrane (arrow). B. Inferior region of the sensory epithelium 15 h after completion of noise exposure. Labelled nuclei (arrows) indicate mitosis within the cellular monolayer. C. Inferior region of the sensory epithelium three days after completion of the sound exposure. Labelled regenerated hair cell (arrow) with lightly staining cytoplasm and a large round nucleus adjacent to the region of active proliferation. Plane of focus is on the overlying silver grains, which puts the cells partially out of focus. (A–C from Girod et al 1989.)

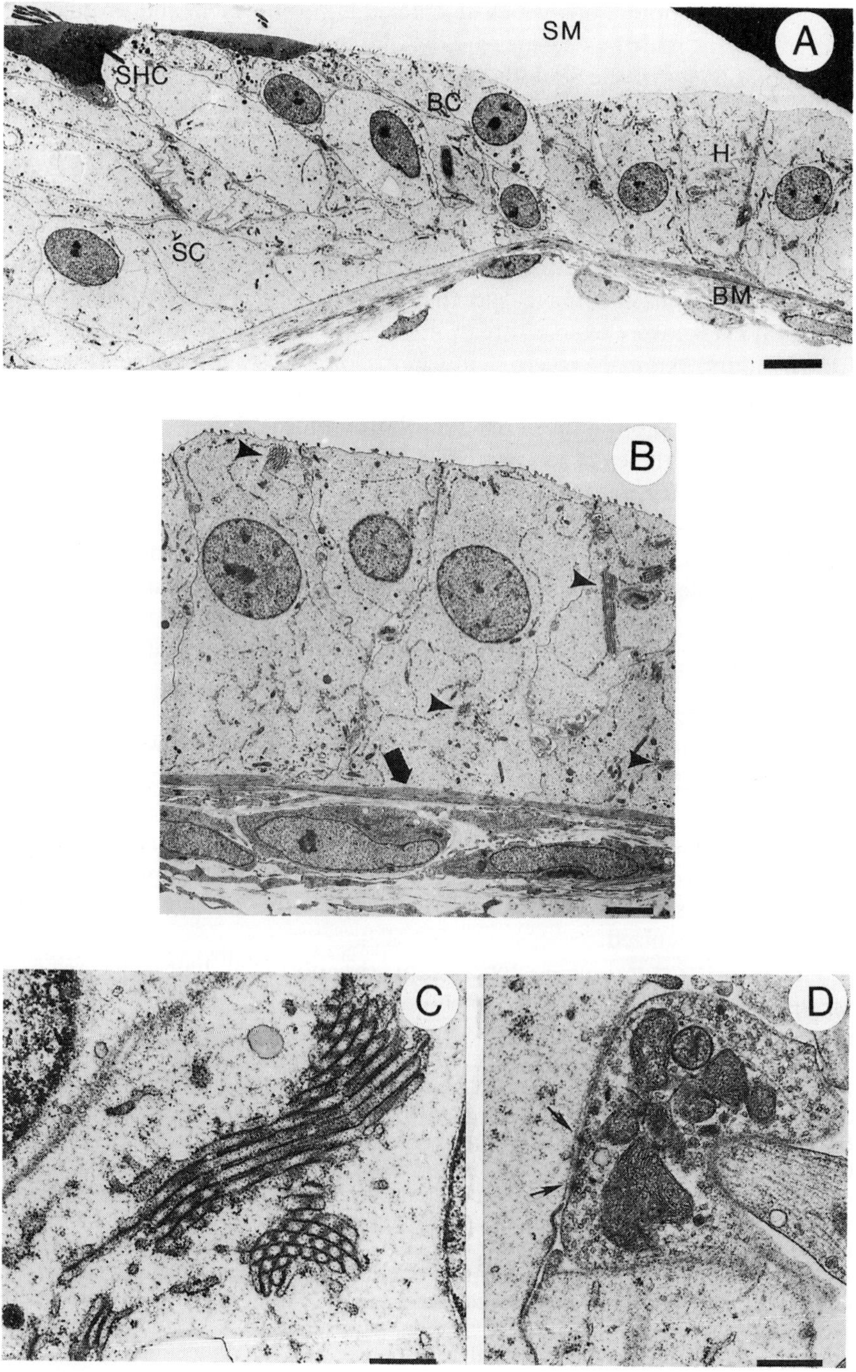
A
SM
SHC
BC
H
SC
BM
B
C
D

cells, have been studied (Oesterle et al 1990) in the normal chick cochlea. The inferior region of the basilar papilla in the chicken cochlea is shown in Fig. 2A. The inferior hair cell, a short hair cell, abuts an organ supporting cell. Several rows of cells, that we have termed border cells, separate this group of organ supporting cells from nearby hyaline cells. The hyaline cells are a row of cuboidal cells which lie inferior to the border cells and extend inferiorly towards the fibrocartilaginous plate. Border cells appear unspecialized, whereas hyaline cells are highly specialized. Dense bundles of filaments are present in the base of hyaline cells from the basal half of the papilla (Fig. 2B), and an unusual structure, a striated rough tubular aggregate, is present in hyaline cell cytoplasm (Fig. 2B and 2C). An intriguing new observation is that synaptic specializations are observed between neural elements and hyaline cells (Fig. 2D) and neural elements and border cells.

Our laboratory is now studying the ultrastructural characteristics of the regenerating receptor epithelium in the chick cochlea. Sound exposure parameters were identical to those of the earlier (Girod et al 1989) study, and chicks were killed 0, 12, 24 or 72 hours after the sound exposure. A consistent picture is emerging from the light and electron microscopic studies of the inferior region of the basilar membrane after noise

---

FIG. 2. Inferior region of the receptor epithelium and the neighbouring supporting cells in the normal chick cochlea. A. Electron micrograph showing the inferior region of the receptor epithelium in the chick inner ear and the neighbouring supporting cells. A short hair cell (SHC) is on the left. At the inferior edge of the receptor epithelium, a few cuboidal cells, that we have termed border cells (BC), separate the sensory region from the hyaline cells. Hyaline cells (H) are a group of cuboidal epithelial cells which rest on the basilar membrane (BM) and extend up to the scala media space (SM). Organ-supporting cells (SC). Bar, 5 μm. B. Hyaline cells from the high frequency region of the cochlea. Dense networks of fine filaments appear directly adjacent to the basal lamina (arrow). An unusual structure consisting of regular arrays of tubular membrane, a striated rough tubular aggregate, is often observed in apical and/or basal regions of hyaline cells (arrowheads). Bar, 2 μm. C. High magnification of a striated rough tubular aggregate. The tubular aggregate may reach up to 3–4 μm in length and 0.5 μm in diameter. Individual subunits are approximately 94 nm in diameter. Rough endoplasmic reticulum often lies in the immediate vicinity of the tubular aggregate and may be continuous with it. The rough striated tubular aggregate is observed in hyaline cells only; it does not occur in other cell types in and or adjacent to the receptor epithelium. Bar, 0.5 μm. D. Synaptic specializations are observed in nerve fibres juxtaposed to hyaline cells. A nerve fibre opposite a hyaline cell from the low frequency region of the cochlea is shown. Small presynaptic densities ringed by associated synaptic vesicles are present inside the nerve fibre (arrows) and are attached to the cytoplasmic face of the presynaptic membrane. Single densities may be visible or, as shown here, several densities may be clustered together. Analogous presynaptic specializations between neural elements and hyaline cells are observed in all regions of the cochlea; they are most numerous in the high-frequency regions. Bar, 0.5 μm. (D from Oesterle et al 1990.)

damage. Initially, support cells near the inferior margin of the receptor epithelium spread out to cover the area of the basilar membrane that has been damaged. This process involves major changes in the shape of the organ supporting cells and border cells. Minor shape changes may also occur in hyaline cells. The epithelial covering of the basilar membrane is seen as early as we have examined the tissue, at the end of 18 hours of sound exposure. Proliferation of epithelial cells does not appear to be involved in this process, because [^{3}H]thymidine labelling is not seen in this region six hours after the 18 hours of noise exposure.

The next phase involves rapid proliferation of this epithelial layer, which begins between six and 15 hours after exposure. The proliferating cells lie in the inferior region of the receptor and adjacent to the receptor cell epithelium. These cell types include cuboidal cells, hyaline cells, border cells and support cells under the hair cells. The specific cell type(s) which undergo mitotic activity are now under investigation. These new cells appear to form a pseudo-stratified epithelium and may show nuclear translocation from the basilar membrane to the lumen during mitosis (see Presson & Popper 1990a). Large multinucleated cells are seen 24 hours after exposure (Fig. 3). As cells leave the mitotic cycle, the epithelium thickens and some of the cells begin to differentiate into hair cells. The incipient hair cells can be recognized as they migrate toward the lumen by their relatively greater cytoplasmic staining density, which is associated more with organelles, than surrounding support cells. After the apical tip of the differentiating hair cell contacts the luminal surface, specializations begin to appear; these include small microvilli, immature stereocilia, and a cuticular plate. These and other surface modifications have been described in detail by others (Cotanche 1987a, Cotanche & Corwin 1991, Marsh et al 1990). During the next 30 days the new hair cells mature to the point where they are indistinguishable (except for [^{3}H]thymidine labelling) from normal hair cells in that part of the cochlea. As discussed later, even such individualized and specialized features as the number and height of stereocilia appear to be replicated in the postembryonically produced hair cells.

In a few cases we have observed hair cell loss limited to a strip of cells lying at the junction between short hair cells and tall hair cells (see Cotanche et al 1987). In these cases we have not observed proliferation of border or hyaline cells, which suggests that another cell population is generating the new hair cells. We have not yet identified this progenitor population; it could be the organ supporting cells, as suggested by Corwin & Cotanche (1988), or a population of stem cells that has heretofore remained unrecognized.

Since most of the damaged animals that have been studied with radioactive thymidine have sustained short hair cell damage, it is not surprising that regeneration of short hair cells has predominated. We have, however,

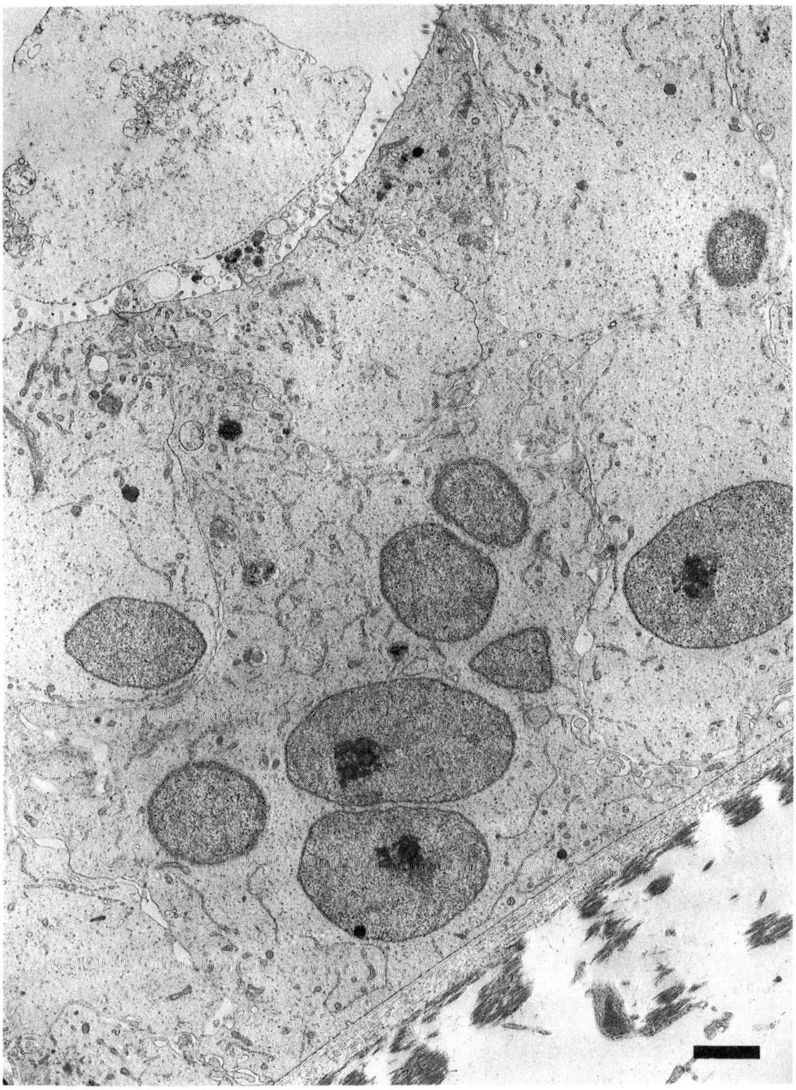

FIG. 3. Electron micrograph of a large multinucleated cell from the 1500 Hz region of a chick cochlea 24 hours after the completion of noise exposure (1500 Hz tone at 120 dB SPL for 18 hours). Bar, 2 μm.

occasionally seen labelled tall hair cells, suggesting that both types can be produced postembryonically. The few tall hair cells observed were always at an appropriate location, superior to the short hair cells. Early in the regeneration process after noise damage, the tectorial membrane—also destroyed by

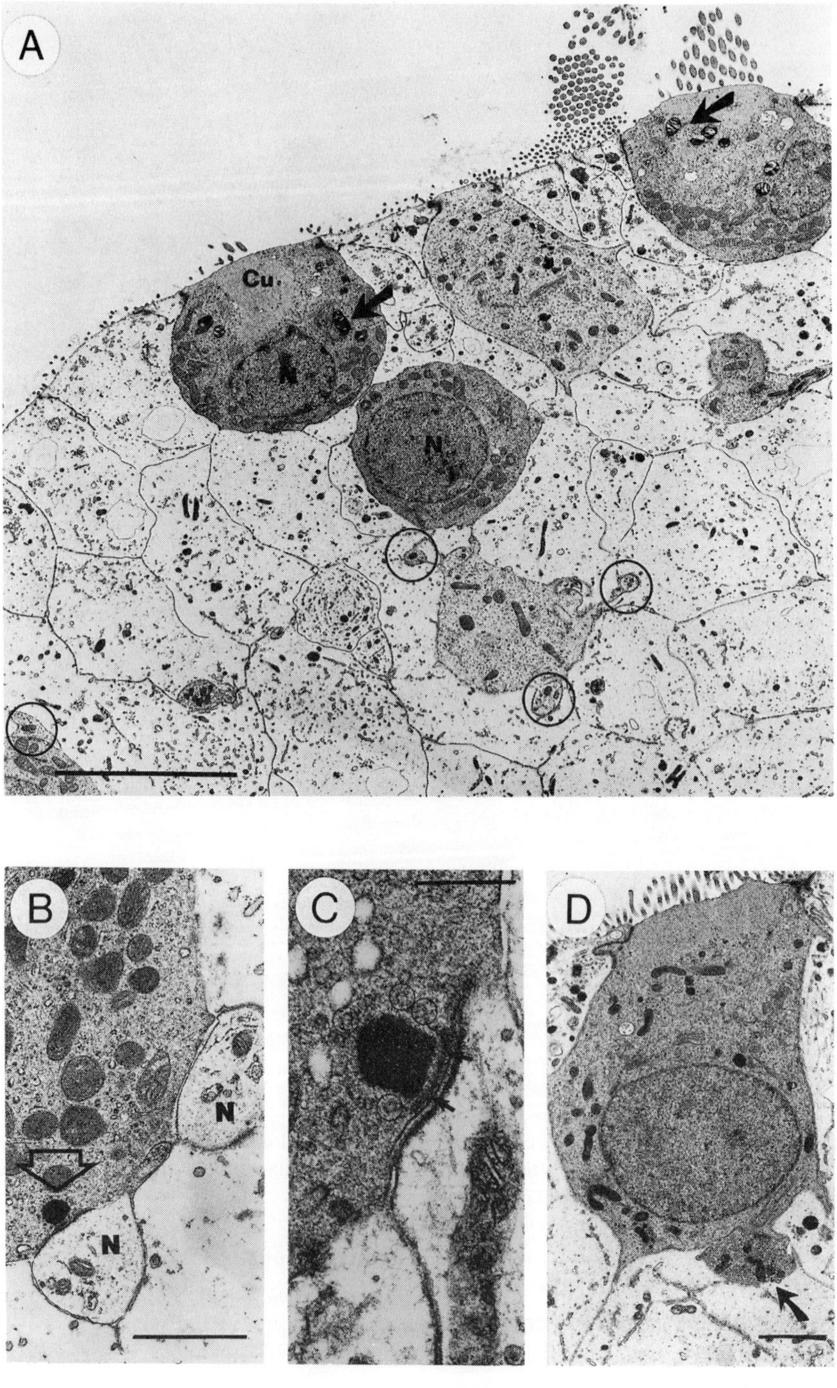
A
Cu
N
N
B
N
N
C
D

overstimulation—is re-formed (Cotanche 1987b). Its reconstitution may be important for the alignment of stereocilia bundles (Cotanche & Corwin 1991).

Do regenerated hair cells restore hearing?

The physiological status of the newly regenerated cells has been studied by Tucci & Rubel (1990) and Norton et al (1990). In the first experiment, tone-burst (250–4000 Hz) auditory evoked potentials (AEPs) were recorded from gentamicin-treated chicks (10 days of subcutaneous gentamicin injections, 50 mg/kg). AEP thresholds deteriorated by 30 to 40 dB until five weeks after treatment. Significant threshold improvements were noted between 16 and 20 weeks at all frequencies tested. In the second study, Norton and collaborators were interested in investigating whether the protracted time course of functional recovery was due to hair cell or neural dysfunction. Evoked otoacoustic emissions (EOEs) and AEPs were recorded from neonatal chickens that had received gentamicin injections for ten days. Five weeks after this treatment, hair cell function (EOEs) and evoked potentials were still abnormal. However, by 8–14 weeks the EOEs of the treated group were not different from those of controls, whereas the AEPs of the injected birds were still abnormal and did not approach normal levels until 22 weeks. The difference suggests that hair cell recovery precedes neural recovery by several weeks. In short, these experiments strongly suggest that the regenerated hair cells are functional and relay information to the central nervous system.

Ultrastructural properties of regenerated hair cells

Our group and other groups (e.g. Cotanche 1987a, Marsh et al 1990, Cotanche & Corwin 1991) have begun to study the ultrastructural properties of the

FIG. 4. Ultrastructural characteristics of regenerated hair cells in drug- and sound-damaged chickens. A. Transmission electron micrograph showing erupted and unerupted regenerated hair cells 28 days after gentamicin treatment. The cells at the surface are rotund with a basally located nucleus (N). Note increasing electron density as cells migrate toward surface. Cu, cuticular plate; arrows, myelin figures. Nerve fibres are circled. Bar, 10 μm. (From Duckert & Rubel 1990.) B. Small bouton afferent nerve endings (N) opposed to basal portion of regenerated sensory hair cell one day after termination of gentamicin treatment. Arrow, synaptic complex consisting of synaptic ball and vesicles. Bar, 1 μm. C. High magnification of synaptic complex in a regenerated hair cell one day after the termination of gentamicin treatment. Note row of vesicles surrounding synaptic ball (circles). A granular synaptic membrane (arrows) is found between the ball and the cell membrane. Bar, 0.5 μm. (From Duckert & Rubel 1990.) D. Regenerated hair cell three days after completion of sound exposure (1500 Hz tone at 120 dB SPL for 18 h). Note the efferent terminal at the base of the cell (arrow) and the immature stereociliary bundle and cuticular plate at the apical surface. Bar, 2 μm.

regenerated sensory epithelium after noise-induced or drug-induced damage. In this section we briefly describe ultrastructural observations on regenerated hair cells after aminoglycoside treatment. More detailed descriptions can be found in Duckert & Rubel (1990). Preliminary observations on the ultrastructural characteristics of regenerated hair cells after sound damage will also be discussed.

The two issues to be addressed here are related to the conclusions we have drawn from our functional analyses. First, it is of interest to determine whether regenerated avian hair cells make synaptic connections with the central nervous system and, if so, the time course over which these connections mature. Second, since in normal animals the characteristics of the stereocilia bundle vary precisely as a function of cochlear location, it is of interest to examine the number and length of the stereocilia and the orientation of stereocilia bundles on regenerated hair cells. These stereocilia properties are thought to be related to functional properties of mature hair cells and their coupling to the tectorial membrane.

Synaptic connections with regenerated hair cells

The principal questions we wish to address here are: (1) When can afferent and efferent terminals be recognized on the regenerating hair cells? and (2) What is their relative maturity? More detailed analyses involving quantitative studies of synaptology are in progress.

In both drug- and sound-damaged birds, immature cells destined to become hair cells can be recognized soon after they begin migrating from the basilar membrane toward the luminal surface. By comparison with the supporting cells, the cytoplasm of regenerating hair cells is more electron dense, as a result of an increase in the number of organelles. As the cells approach the luminal surface this difference increases. A good example is shown in Fig. 4A. The three immature hair cells lined up under the luminal surface show progressively decreasing electron density, but all are more dense than the surrounding support cells. Circled elements in Fig. 4A are nerve fibres in close approximation to an immature hair cell, which has yet to reach the luminal surface or to produce stereocilia.

Figures 4B and C show afferent synaptic terminals on regenerated hair cells. Afferent synaptic complexes on immature cells are seen as early as we have looked in drug-induced lesions in chicks. They are seen one day after the termination of gentamicin treatment but are more numerous at 1–4 weeks. They are found both on cells that have reached the luminal surface (erupted) which are beginning to produce stereocilia, and on unerupted migrating cells. The synaptic complexes appear immature but have the full complement of specializations including synaptic ball, vesicles, presynaptic density and postsynaptic density. The terminals are usually boutons and are less densely packed than on mature cells. Afferent synaptic complexes have not been observed in sound-induced lesions three days after sound exposure.

Vesiculated terminals, which are presumably efferents from the central nervous system, are also seen on both erupted and unerupted immature hair cells, in both sound- and drug-induced lesions. These have not been observed earlier than one week after aminoglycoside treatment, but have been observed three days after sound exposure (Fig. 4D). In both normal and regenerating cochleas they tend to occur most often on short hair cells. Typically they are packed with synaptic vesicles. Often, but not always, subsynaptic cisternae are seen. At early times these endings tend to be boutons, while at later survival times and in normal animals they form flattened cups around the base of the short hair cells.

Thus, after drug insult, both afferent and efferent terminals can be identified early in the regeneration process. We have observed both types of terminals on erupted and on non-erupted cells and all the normal components are present. Further analyses will be required to describe the maturation of synapses, which may be important for understanding the lag between the recovery of hair cell function (emissions) and the recovery of evoked potential thresholds. After noise insult, efferent but not afferent terminals can be identified early in the regeneration process, by three days following sound exposure.

Stereocilia

The maturation of stereocilia in regenerating hair cells has been described by a number of investigators (Cotanche 1987a, Henry et al 1988, Girod et al 1989, Duckert & Rubel 1990). Recently Cotanche & Corwin (1991) showed that the orientation of stereocilia bundles during regeneration after noise trauma in chicks initially varies over approximately 100 degrees. Then, over four days, the bundles became aligned with those of other regenerating cells and with the orientation of surviving hair cells surrounding the lesion. The change occurred 6–10 days after the sound exposure. Although the reorientation of stereocilia bundles after the induction of aminoglycoside ototoxicity in chicks has not been quantified, the same general phenomenon has been seen, but with a much longer time course (Duckert & Rubel 1990). Scanning electron microscopy observations indicate that five weeks after gentamicin treatment the stereocilia bundles throughout the basal two-thirds of the cochlea are disoriented. By 10–12 weeks some reorientation is apparent in the middle region but in the basal and mid-basal regions high variability still predominates. By 22 weeks all but the basal region appears to show consistent bundle orientation.

In the chick cochlea, stereocilia bundles vary systematically in the length and number of elements (Tilney & Saunders 1983). At the basal end the bundle consists of many (200–300) individual stereocilia and the tallest row extends roughly 5 μm from the hair cell surface; toward the apical end the number decreases and the height increases. Development of this gradient has been studied in detail by Tilney et al (1986, 1988). It is of considerable interest to understand

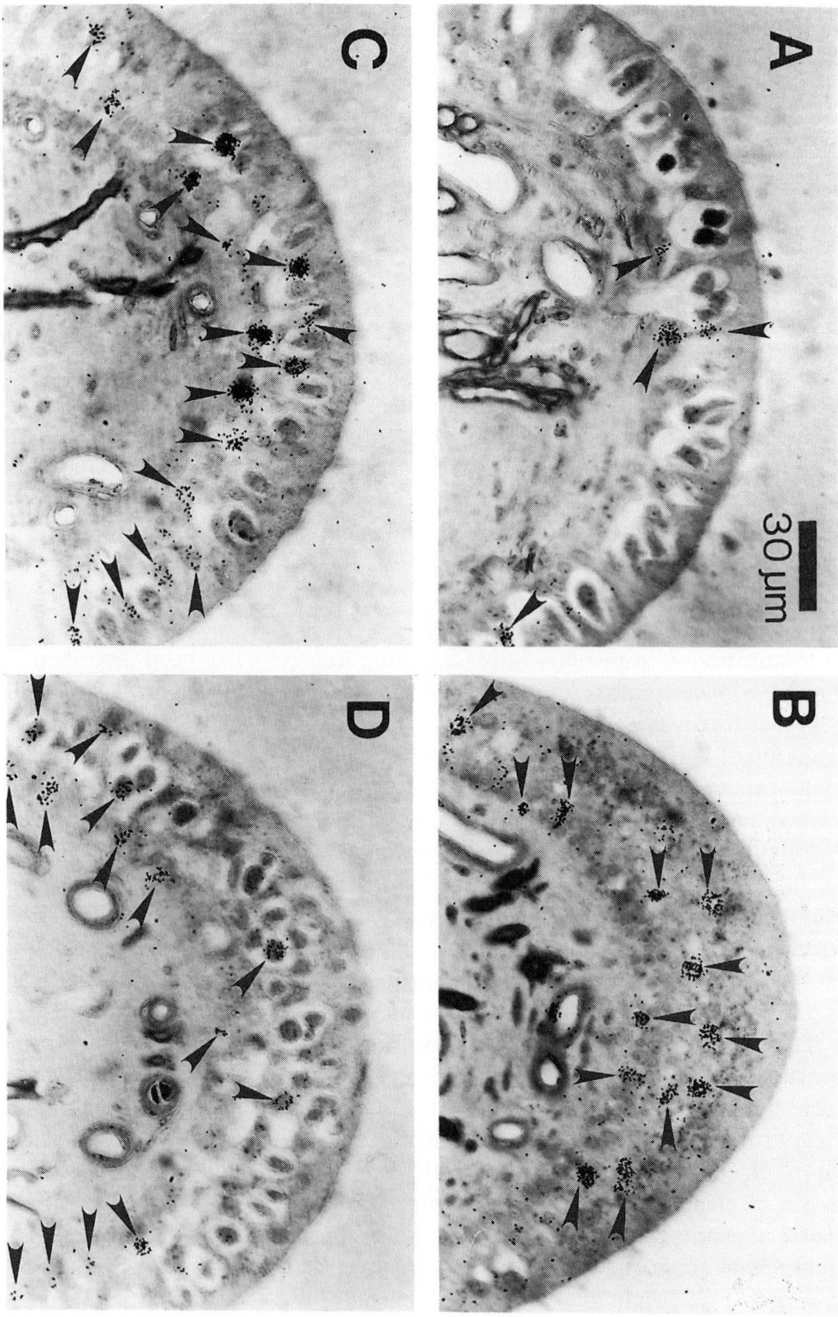
A
B
C
D
30 μm

FIG. 5. Up-regulation of hair cell regeneration in the avian vestibular system. Ampullary tissue from chicks killed after a three-day course of 20 μCi/g per day [^{3}H]thymidine. Arrowheads point to cells in the sensory epithelium which have incorporated this label. A. Ampulla from a control chick allowed to survive 60 days after [^{3}H]thymidine injections. Note the presence and normal complement of nerve calyces, Type I and Type II hair cells, and the single row of supporting cells. Only four labelled cells are seen. B. Ampulla from a chick treated with 600 mg/kg per day streptomycin sulphate for seven days and then given [^{3}H]thymidine. Note the absence of all nerve calyces, all Type I and most type II hair cells, and the presence of several rows of supporting cells. Numerous labelled cells are seen. C. Ampulla from a streptomycin-treated chick allowed to survive 20 days after [^{3}H]thymidine injections. The nerve calyces, which usually contain several Type I hair cells, can now be observed at the top of the organ. Most, however, contain only one cell. Labelled Type I and Type II hair cells as well as supporting cells are seen. D. Ampulla from a drug-treated chick allowed to survive for 60 days. The tissue has recovered its normal appearance. Numerous labelled cells can still be observed.

the signals underlying the ontogeny of this pattern and it might be expected that these signals are restricted to the embryonic period. We have begun to examine the number and height of stereocilia in regenerated hair cells. Few measurements have been obtained so far, but even cursory observation reveals the striking finding that after 20–25 weeks the normal patterns have been restored. The height of the stereocilia of regenerated hair cells appears identical to that of the remaining hair cells at any given position along the cochlea, and at the basal end the entire complement of new hair cells have short stereocilia. The number of stereocilia has been counted on several mature-appearing regenerated hair cells from the basal region in a 20-week survivor after gentamicin treatment. The number was between 150 and 170 per cell, which is comparable to the number reported by Tilney & Saunders (1983). Without more extensive measurements of stereocilia height and number throughout the length of the cochlea, the precision of this pattern cannot be evaluated, but our observations to date strongly suggest that the signals regulating this pattern persist in the mature avian cochlea. The site of these signals remains to be determined. They may be expressed by the genome of the stem cell population or be obtained from the local environment of the differentiating hair cells.

Regeneration of hair cells in the vestibular epithelium

Jørgensen & Mathiesen (1988) and Roberson et al (1991) demonstrated the ongoing production of hair cells in the postnatal avian vestibular system. Using DNA autoradiography, Jørgensen and Mathiesen were able to label supporting cells and hair cells in the ampullary and otolithic organs of sexually mature budgerigars. Roberson and his collaborators utilized immunocytochemical techniques in addition to autoradiographic methods, and corroborated Jørgensen and Mathiesen's results in chicks.

In a more recent study, we evaluated the effects of aminoglycoside toxicity on the capacity of the avian vestibular system to produce new hair cells (Weisleder & Rubel 1991). Newly hatched chicks received daily injections of streptomycin sulphate (600 mg/kg per day) for seven days. Starting on the fifth day, these animals, and age-matched control chicks, received either injections of [^{3}H]thymidine for three days or injections of 5-bromo-2′-deoxyuridine (BrdU) for five days. BrdU is a thymidine analogue which is incorporated into the DNA of dividing cells. BrdU is antigenically distinct from normal thymidine and can be recognized by a monoclonal antibody and labelled immunocytochemically. Streptomycin-treated and control chicks were killed one, 20 or 60 days after the last injection of proliferation marker. Vestibular organs were sectioned and processed for immunocytochemistry or autoradiography. The tissue of [^{3}H]thymidine-injected chicks was used to study morphological characteristics of the regenerated cells; that of BrdU-injected birds was used for purposes of quantification.

One day after the last proliferation marker injection, the tissue from streptomycin-treated chicks shows clear signs of vestibulotoxicity (Fig. 5B). As previously described by Wersäll & Hawkins (1962) and Duvall & Wersäll (1964), these signs include the loss of most Type I hair cells; a drastic reduction in the number of Type II hair cells; and a reduction in the number of myelinated fibres. At this time, BrdU immunostaining and [^{3}H]thymidine autoradiography reveal mitotic activity in the supporting cell layer of all organs in both treated and untreated animals (Fig. 5A and B). In the drug-treated group, however, the number of labelled nuclei is four-fold greater than the labelling seen in control chicks. Just as the damage to the sensory epithelium is generalized, so is the proliferative activity. Labelled nuclei are observed throughout the organs. There appears to be no specific area of proliferation, as described in other species (Corwin 1981, 1983, 1985). Instead, the pattern of generalized proliferation reported by Jørgensen & Mathiesen (1988) is seen. In both control and drug-treated animals positive nuclei in the supporting cell layer usually occur in pairs. In some instances they are stacked one on top of the other; in other instances they appear side by side.

Twenty days after the last cell proliferation marker injection, the aminoglycoside's damage can still be detected. Few Type I hair cells are present in the sensory epithelium. The number of Type II hair cells, however, has increased. Many supporting cells and several Type II hair cells are labelled (Fig. 5C). Labelled Type II hair cells usually occur on top of labelled support cells. In some cases two labelled hair cells occur side by side on the sensory epithelium. Interestingly, Type II hair cells are found throughout the organs' epithelia, including areas usually occupied only by Type I hair cells.

Finally, in the 60 days survival group the sensory epithelium has recovered its normal appearance. Type I hair cells can now be identified occupying their normal location at the top of the organ. Type II hair cells occupy the skirts of the organ as they do in the untreated animals. Labelled nuclei are found in Type I and Type II hair cells as well as in some supporting cells (Fig. 5D). The normal complement of each cell type appears to have been restored.

Two main conclusions can be drawn from these experiments. First, the avian vestibular system is able to recover from aminoglycoside toxicity. All cell types normally present in the vestibular sensory epithelium can be identified within 60 days of streptomycin treatment. Second, the level of proliferation in the avian vestibular system appears to be dynamically regulated. The level of proliferation in the sensory epithelium increased significantly in response to the drug damage. The observed increase in the number of cells with labelled nuclei is, apparently, an attempt by the organism to compensate for the massive loss of hair cells.

According to Ramprashad et al (1986), the gross and microscopic anatomies of the avian, reptilian and mammalian vestibular systems are essentially similar. For this reason we believe that understanding the process of regeneration in the avian vestibular system may provide important pointers toward the development of strategies for eliciting hair cell regeneration in the mammalian sensory epithelium.

Acknowledgements

The authors are keenly aware that the studies reported in this chapter were carried out by an outstanding group of collaborators, composed of students, postdoctoral fellows, otolaryngology residents and colleagues. They include Pamela Bohrer, Raul Cruz, Larry Duckert, Douglas Girod, Paul Lambert, David Roberson and Brenda Ryals. In addition, outstanding technical assistance and advice was provided by Dale Cunningham and Glen MacDonald. Expert secretarial assistance was provided by Terry Hogley. The research was supported by Public Health Service grant DC00395, NS grant 07097, funds from the Deafness Research Foundation and the Cora M. Poncin Foundation, and funds from the Departments of Otolaryngology—Head and Neck Surgery at the University of Virginia and the University of Washington.

References

Balak KJ, Corwin JT, Jones JE 1990 Regenerated hair cells can originate from supporting cell progeny: evidence from phototoxicity and laser ablation experiments in the lateral line system. J Neurosci 8:2502–2512

Corwin JT 1981 Postembryonic production and aging of inner ear hair cells in sharks. J Comp Neurol 201:541–553

Corwin JT 1983 Postembryonic growth of the macula neglecta auditory detector in the ray, *Raja clavata*: continual increases in hair cell number, neural convergence, and physiological sensitivity. J Comp Neurol 217:345–356

Corwin JT 1985 Perpetual production of hair cells and maturational changes in hair cell ultrastructure accompany postembryonic growth in an amphibian ear. Proc Natl Acad Sci USA 82:3911–3915

Corwin JT, Balak KJ, Borden PC 1989 Cellular events underlying the regenerative replacement of lateral line sensory epithelia in amphibians. In: Coombs S, Görner P, Münz H (eds) The mechanosensory lateral line: neurobiology and evolution. Springer-Verlag, New York, p 161–183

Corwin JT, Cotanche DA 1988 Regeneration of sensory hair cells after acoustic trauma. Science (Wash DC) 240:1772–1774

Cotanche DA 1987a Regeneration of hair cell stereociliary bundles in the chick cochlea following severe acoustic trauma. Hear Res 30:181–194

Cotanche DA 1987b Regeneration of the tectorial membrane in the chick cochlea following severe acoustic trauma. Hear Res 30:197–206

Cotanche DA, Corwin JT 1991 Stereociliary bundles reorient during hair cell development and regeneration in the chick cochlea. Hear Res 52:379–402

Cotanche DA, Saunders JC, Tilney LG 1987 Hair cell damage produced by acoustic trauma in the chick cochlea. Hear Res 25:267–278

Cruz RM, Lambert PR, Rubel EW 1987 Light microscopic evidence of hair cell regeneration after gentamicin toxicity in chick cochlea. Arch Otolaryngol Head & Neck Surg 113:1058–1062

Duckert LG, Rubel EW 1990 Ultrastructural observation on regenerating hair cells in the chick basilar papilla. Hear Res 48:161–182

Duvall AJ, Wersäll J 1964 Site of action of streptomycin upon inner ear sensory cells. Acta Oto-laryngol 57:581–598

Girod DA, Duckert LG, Rubel EW 1989 Possible precursors of regenerated hair cells in the avian cochlea following acoustic trauma. Hear Res 42:175–194

Henry WJ, Makaretz M, Saunders JC, Schneider ME, Vrettakos P 1988 Hair cell loss and regeneration after exposure to intense sound in neonatal chicks. Otolaryngol Head Neck Surg 98:607–611

Jørgensen JM, Mathiesen C 1988 The avian inner ear. Continuous production of hair cells in vestibular organs, but not in the auditory papilla. Naturwissenschaften 75:319–320

Katayama A, Corwin JT 1989 Cell production in the chicken cochlea. J Comp Neurol 281:129–135

Lippe WR, Rubel EW 1983 Development of the place principle: tonotopic organization. Science (Wash DC) 219:514–516

Lippe WR, Rubel EW 1985 Ontogeny of tonotopic organization of brain stem auditory nuclei in the chicken: implications for development of the place principle. J Comp Neurol 237:273–289

Marsh RR, Xu L, Moy JP, Saunders JC 1990 Recovery of the basilar papilla following intense sound exposure in the chick. Hear Res 46:229–238

Norton S, Tucci D, Rubel EW 1990 Comparison of acoustic and neural responses from avian ears following gentamicin. Assoc Res Otolaryngol Abstr 13:62

Oesterle EC, Cunningham DE, Rubel EW 1990 Ultrastructural characteristics of hyaline cells in chick inner ear. In: Peachey L, Williams D (eds) Proc XIIth Int Congr for Electron Microscopy, vol 3, p 452–453. San Francisco Press, San Francisco

Presson JC, Popper AN 1990a A ganglionic source of new eighth nerve neurons in a post-embryonic fish. Hear Res 46:23–28

Presson JC, Popper AN 1990b Possible precursors to new hair cells, support cells, and Schwann cells in the ear of a post-embryonic fish. Hear Res 46:9–22

Ramprashad F, Landolt JP, Money KE, Laufer J 1986 Comparative morphometric study of the vestibular system of the vertebrata: reptilia, aves, amphibia, and pisces. Acta Oto-laryngol Suppl 427:1–42

Roberson DW, Weisleder P, Bohrer PS, Rubel EW 1991 Ongoing production of sensory cells in the vestibular epithelium of the chick. Hear Res, submitted

Rubel EW, Ryals BM 1983 Development of the place principle: acoustic trauma. Science (Wash DC) 219:512–514

Ruben RJ 1967 Development of the inner ear of the mouse: a radioautographic study of terminal mitoses. Acta Oto-laryngol 220:1–44

Ryals BM, Rubel EW 1985 Ontogenic changes in the position of hair cell loss after acoustic overstimulation in avian basiliar papilla. Hear Res 19:135–142

Ryals BM, Rubel EW 1988 Hair cell regeneration after acoustic trauma in adult *Coturnix* quail. Science (Wash DC) 240:1774–1776

Stone LS 1933 The development of lateral-line sense organs in amphibians observed in living and vital-stained preparations. J Comp Neurol 57:507–540

Stone LS 1937 Further experimental studies of the development of lateral-line sense organs in amphibians observed in living preparations. J Comp Neurol 68:83–115

Tilney LG, Saunders JC 1983 Actin filaments, stereocilia and hair cells of the bird cochlea. I. The length, number, width and distribution of stereocilia of each hair cell are related to the position of the hair cell on the cochlea. J Cell Biol 96:807–821

Tilney LG, Tilney MS, Saunders JC, DeRosier DJ 1986 Actin filaments, stereocilia, and hair cells of the bird cochlea. III. The development and differentiation of hair cells and stereocilia. Dev Biol 116:100–118

Tilney LG, Tilney MS, Cotanche DA 1988 Actin filaments, stereocilia and hair cells of the bird cochlea. V. How the staircase pattern of stereociliary lengths is generated. J Cell Biol 106:335–365

Tucci DL, Rubel EW 1990 Physiological status of regenerated hair cells in the avian inner ear following aminoglycoside ototoxicity. Otolaryngol Head Neck Surg 103:443–450

Weisleder P, Rubel EW 1991 Born-again hair cells: the vestibular system story. Assoc Res Otolaryngol Abstr 14:154

Wersäll J, Hawkins JE 1962 The vestibular sensory epithelia in the cat labyrinth and their reactions in chronic streptomycin intoxication. Acta Otolaryngol 54:1–23

DISCUSSION

Fernald: I am a little concerned by your use of the term 'regulation' when you find an increased number of mitoses in response to what is clearly an overdose of some drug. I wouldn't consider that to be regulation. I think this is indicative of the problems that we may have, of distinguishing between the inferences you want to draw from damage experiments, as opposed to tracing natural developmental sequences. The real issue is the regulation of cell division in the normal course of events, as cells turn over slowly. It's not surprising that you get a massive increase in mitoses in response to widespread drug damage, but I am not sure you would learn from that how cell division is regulated normally.

Rubel: I agree with you, and maybe we don't want to say 'regulation' in the fine sense of the term. I can say, however, from these studies, that either the rate of the cycle, or the number of cells kicked into the mitotic cycle, is increased by causing cochlear damage. Hair cells are lost; we don't know if it's the actual loss of hair cells, or some other stimulus, which produces this increase. We think it's the number of cells that are entering the mitotic cycle that is increased.

Fernald: Let's turn it around, then, and ask what, in the normal course of events, would be a reasonable regulator of the generation of new hair cells in a normal ear, given normal hair cell loss.

Calof: Is there a turnover of hair cells in the normal avian ear?

Rubel: In the normal avian vestibular system, in all the epithelial surfaces (of which there are six) there is some ongoing, normal turnover. It is, in fact, turnover rather than production of permanent new cells, because there's not a large increase in the size of the organs. There is no proliferative zone; the proliferation seems to happen all over the receptor epithelium; and there's not a large increase in size.

Fernald: But with a slow rate of cell division and a slight increase in size, it might be the new generation of cells.

Rubel: It might be, but I would doubt it! We don't think there's any significant turnover, or increase in size, or ongoing proliferation, in the avian cochlea.

In many hundreds of sections we have occasionally seen a labelled cell, in normal birds, or in the non-damaged zone of the cochlea in noise-damaged animals; but we don't normally see proliferation. Once in a while Brenda Ryals has seen it, and so have we, but we attribute it to small amounts of damage due to noise in the colony, and things like that.

Ryals: I agree that it seems unlikely that there is hair cell proliferation in the cochlear partition of the normal avian ear. If such proliferation were occurring, we would predict that the number of cells within the partition would increase with age. We have counted cells in serial sections from the basilar papilla in young (three month) and old (3–6 years) quail and do not find any increase in the number of cells. In fact we find a slight decrease (6%) in the number of hair cells in older quail (Ryals & Westbrook 1988). So, although we occasionally see some cells with [^{3}H]thymidine uptake in normal, non-noise-exposed quail, we feel that these are probably replaced cells rather than added cells (Ryals & Westbrook 1990).

Reasner: Dr Rubel, could you compare the noise levels that you use to induce hair cell damage with the typical sound levels present in the colony room?

Rubel: No, because colony sound levels fluctuate so much. We have used between 110 dB and 125 dB SPL of an intense pure tone (the former level for a longer duration). Our standard level now is 1500 Hz at 120 dB for 18 hours.

Reasner: I think it is relevant to ask to what extent might typical stimuli act on a normal sensory system and be a causal factor in receptor cell turnover.

Rubel: Yes, but I don't know how the normal stimuli in our colony would compare to a farmyard.

Calof: Dr Rubel, I thought your view was that there is a separate precursor population in the avian cochlea, and that it is these cells that are the precursors of both the supporting cells and the hair cells? Therefore, in any of these paradigms where you induce hair cell loss, do you also see a loss of supporting cells? And do you then see the normal level of proliferation in the supporting cell population, or is that increased as well, by sound damage?

Rubel: In this field so far, there are two ways by which cochlear hair cell damage is produced experimentally: either with noise or with aminoglycosides. The changes are very different. Noise induces a loss of hair cells, and also loss of supporting cells. You also see loss of tectorial membrane. With this type of damage, induced by pure tone noise, we find proliferation both of hair cells and supporting cells. We have no evidence suggesting that they derive from different precursor cell populations. So the hair cells and supporting cells could be coming from the same or different precursors; we don't know.

With aminoglycosides, you kill all the hair cells, but there's very little loss of supporting cells. You get hair cells labelled; supporting cells are also labelled. We don't yet know whether equivalent numbers are produced in these two situations. Brenda Ryals has some figures.

Ryals: We find two to three times as many labelled support cells than hair cells, after noise damage. We are looking at the gentamicin-treated birds now.

Burd: You have not done short survival periods, then?

Rubel: We have done continuous labelling and then fixation. We have not done pulse-fix experiments yet.

Corwin: The last time I heard about this work, you distinguished hyaline and cuboidal cells as the sources of regenerated hair cells at the inferior edge of the papilla (Girod et al 1989).

Rubel: That was clearly partially wrong! We can now distinguish a subclass of cells that we call border cells based on their position, inferior to the most inferior short hair cell.

Corwin: That helps, because I wondered what the evidence was for border cells as a distinct subclass.

Rubel: In our first study (Girod et al 1989) we gave thymidine to chicks for various periods after noise damage and then fixed the cochlea. All we can determine from this study is the area of the papilla where we first see labelling. From that, we said that cells in the inferior region were usually the first to be labelled after noise trauma. Then this area becomes several cells in depth, and then we can recognize labelled immature hair cells in the inferior, damaged area.

In these cases, it appeared that the inferior portion of the sensory epithelium was destroyed and all that was left were the hyaline and cuboidal cells. Thus we concluded, perhaps prematurely, that they must be dividing and be the precursors for the new hair cells. At that time we had no cytological markers for the different cell types. We can now recognize the hyaline cells on the basis of the cytological characteristics I've discussed in the chapter. When we looked closer, right after damage, we realized that the marker for hyaline cells (namely, basal actin filaments and cytoplasmic tubular structure) is seldom seen in cells which have recently divided. We have looked for a variety of cytological and antigenic markers of various cell types but have not found any except for those mentioned for hyaline cells.

We have distinguished a group of cells we call border cells because they are positioned between the last short hair cell and the hyaline cells. There are about three cells adjacent to each other and they look like the 'support' cells which surround hair cells, except they do not have microvilli or vesicles at their apical pole. Immediately after damage these cells seem to remain, flatten out and then divide. We never see support cells that are directly under the remaining short hair cells dividing. However, at the margin between the tall and short hair cells, we definitely see support cells dividing. Thus we now feel that there may be specialized subpopulations of cells lying in particular regions of the epithelium which are capable of becoming mitotically active. This is not to say that they are cytologically recognizably different from the rest of the support cells, at this time.

Corwin: Part of my concern is over using the last short hair cell as the criterion for deciding where support cells end and border cells begin. This was the problem with the proposal that hyaline and cuboidal cells were the precursors. It may have looked as if it was these cells because the epithelium had become flattened due to the loss of hair cells. That is, it lost its stratified appearance and therefore now appeared like the unstratified region of the hyaline and cuboidal cells.

Rubel: When the damage is at the inferior edge, support cells are lost as well. Clearly something is spreading out or migrating in, to form the cellular layer covering the basilar membrane.

Cotanche: You still see the hyaline cells very clearly after the noise trauma; there's a clear border where they are, and your labelled cells are some way in from the hyaline cells. They are at least halfway across the basilar membrane from the hyaline cells.

Rubel: However, during the time between the noise exposure and fixation of the tissue, the cells can flatten and spread out. At the light microscope level we could not recognize that the epithelial layer of cells were not hyaline cells, because we had no markers. Now, at the EM level, we can.

Corwin: But how do you define the border cell?

Rubel: We define them as the cells inferior to the last short hair cell, and they reside directly over a structural change in the composition in the basilar membrane that can be clearly identified.

Corwin: Defining the border cell on the basis of that last short hair cell is not valid, because when damage occurs the most inferior hair cells are often lost.

Rubel: You are correct! But location on the basilar membrane is valid. You can also make measurements from the superior edge of the basilar membrane in normal and damaged animals and show that proliferation is usually seen first at about the distance from the superior edge that is consistent with the normal location of the border cells. All of this is indirect, of course; we cannot positively identify the progenitor cells until we have distinct antigenic or cytological markers for all of the cell types and perform closely timed pulse-labelling studies. We are trying to find clear antigenic markers, but have not been successful so far.

Steinberg: In terms of mechanism, have you obtained, or can you obtain, a dose–response curve with noise, and see if the level of noise which produces cell death is the same as the level that produces an increase in mitosis?

Rubel: We haven't done that. There's a significant variability in the amount of hair cell loss that you get with any amount of exposure. Another approach that we are taking now, in collaboration with Doug Cotanche, is to look at various times after aminoglycosides are given, to see whether we start getting proliferation before there's actual hair cell death, or whether proliferation starts while the hair cells are still present.

Calof: What is the mechanism of toxicity of the aminoglycosides?

Rubel: This is not known. There are some good theories.

Fernald: In the noise paradigm, since you are using 1500 Hz, can't you use that to localize along the basilar membrane where you are destroying the hair cells? This might allow a more subtle destructive force, so that you leave part of the membrane intact and destroy part by highly focused sound frequencies.

Rubel: Yes, that is exactly what we do. But that gives us a measure of where *along* the basilar membrane we are destroying cells, not where *across* it. The greatest variability is in where the damage is across (from superior to inferior) the basilar membrane and in the longitudinal extent of damage—not in the central point of damage.

Corwin: Dr Rubel, how do you know that the migrating cells that became innervated were not contacting the luminal surface?

Rubel: Dale Cunningham and Larry Duckert have done serial EM sections on two such cells. They show no processes reaching the lumen. In addition, we have looked at many, many sections; if there was an apical process, I think we would have seen some. We can see cells that are two cell diameters down in from the lumen with no indication of small thin processes going up to the surface, but which are innervated.

Corwin: At what stage is that, and how do you know that those are new cells?

Rubel: After aminoglycoside treatment, all the hair cells that are going to occur in the basal tip of the papilla are new cells. We have destroyed all the original hair cells at the basilar end.

Corwin: You know that 100% were destroyed?

Rubel: We know that about 99% were! And we have seen this innervation in many cells; so I don't doubt that they are new hair cells. We see a progression of the density of cytoplasmic staining as cells migrate up to become hair cells. We haven't followed cells as elegantly as you have in the lateral line, with time-lapse cinematography, so we can't say on that level; but this is a continuous progression, so I have little doubt that these are going to become hair cells.

Cotanche: How do you know they are not going the other way? You see this continual dying of cells after gentamicin; could they be cells that have died and are being resorbed?

Rubel: I can't tell the direction of movement, obviously, but I am taking a best guess on the basis of looking at a lot of electron micrographs.

Lewis: Is there any evidence for or against the notion that some hair cells are produced without a cell division, simply by a direct change in the character of supporting cells?

Rubel: We can't say whether an existing supporting cell turns into a hair cell after damage. The only way to do that would be to inject the animal with aminoglycoside and maintain the [^{3}H]thymidine label at a high level for a long time. We think that aminoglycosides induce delayed hair cell death. So we would need to see if, after a long period, all the hair cells were labelled. We haven't done that yet.

Burd: Using a fluorescent dye you could label all mature supporting cells that have surface contact on the epithelium, and combine that with tritiated thymidine labelling. You could then determine whether a hair cell was new, because it would be labelled with thymidine, or whether it was pre-existing, because it would be labelled with the dye.

Rubel: Or you could do the reverse experiment—label supporting cells with thymidine when they are really young, and proliferating, and then induce damage when the animals reach adulthood. You would have to keep a [^{3}H]thymidine-labelled animal for a long time, which is expensive.

Margolis: That wouldn't work; the question is whether a support cell will change into a hair cell.

Burd: A mature support cell, from what Ed Rubel said, should have a process to the surface of the epithelium and you could label it with the dye.

Rubel: Some 'support' cells reach the luminal surface; we have no idea whether all do.

Lewis: I am not so much concerned with the definition of a supporting cell, but just whether you can produce a hair cell without prior cell division.

Rubel: I don't know!

Pujol: Another possibility is that a dedifferentiation of the sensory cell occurs, as recently shown in denervated vestibular epithelium (Favre et al 1989); then a new differentiation follows.

Rubel: We have never seen anything suggesting that a cell with any of the properties of the hair cell was proliferating. It would have to be complete dedifferentiation.

Powers: You discussed the restoration of hearing in your paper. There are a couple of things we have to think about here, though I don't expect there to be answers to these questions. First is the quality of the restored hearing. You showed simply thresholds; when considering regenerating systems we have to ask whether there would be sound discrimination equivalent to or even approaching the level of the normal hearing system, and whether intensity–response functions are normal.

Rubel: That's right; all I have shown are the results of evoked potential studies using tone bursts. Nobody, to my knowledge, has examined frequency selectivity, yet.

Powers: The second point I want to bring up is that evoked potentials, while valuable as far as they go, don't tell us that the animal is hearing. Only its behaviour would tell us that.

Cotanche: I think these noise-exposed birds do hear again. When I was in Charleston I used to keep them in my lab, caged with birds that did hear. When they first came out of the noise they were placed in a cage with unexposed birds. The normal birds would always jump when someone came into the room, whereas the birds that had been exposed to noise would not. After 4–5 days, the damaged birds ended up responding to sounds. Perhaps they had learned to jump, but I think they were regaining their hearing!

Rubel: There are elegant psychophysical techniques that have been worked out for testing birds; in a couple of years we may see good studies on auditory perception.

References

Favre D, Bordure P, Sans A 1989 Morphological changes in the adult vestibular sensory epithelium after denervation: a new model of plasticity? 26th Workshop on Inner Ear Biology, Paris, September 3–6, 1989

Girod DA, Duckert LG, Rubel EW 1989 Possible precursors of regenerated hair cells in the avian cochlea following acoustic trauma. Hear Res 42:175–194

Ryals BM, Westbrook EW 1988 Ganglion cell and hair cell loss in coturnix quail associated with aging. Hear Res 36:1–8

Ryals BM, Westbrook EW 1990 Hair cell regeneration in senescent quail. Hear Res 50:87–96

Hair cell regeneration: the identities of progenitor cells, potential triggers and instructive cues

Jeffrey T. Corwin, Jay E. Jones, Akiko Katayama, Matthew W. Kelley and Mark E. Warchol

Department of Otolaryngology—Head & Neck Surgery and Department of Neuroscience, University of Virginia School of Medicine, Charlottesville, VA 22908, USA

Abstract. Hair cells are produced and accumulate in the ears of fish and amphibians as they grow during postembryonic life; hair cell regeneration occurs in lateral line organs in those groups and in the cochlea in birds. Continuous time-lapse microscopy has directly demonstrated that supporting cells divide to give rise to hair cells during regeneration in lateral line neuromasts. Supporting cells also appear to give rise to hair cells during regeneration in the avian ear, but additional cell types have been proposed as hair cell progenitors. Alternative interpretations of current evidence are discussed in relation to the possibility that supporting cells may be the common progenitor in all cases of hair cell regeneration. The regenerative proliferation of hair cells in birds occurs in populations of cells that are mitotically quiescent in undamaged ears. Evidence suggests that the extrusion of damaged hair cells and the breaking of intercellular junctional adhesions may be a trigger for regenerative proliferation. The potential triggering influence of phagocytes is also discussed. The differentiation of replacement cells during regeneration in the cochlea may be regulated by surface interactions between cells. A model that could account for the reconstitution of the mosaic pattern of hair cells and supporting cells is proposed.

1991 Regeneration of vertebrate sensory receptor cells. Wiley, Chichester (Ciba Foundation Symposium 160) p 103–130

Sensorineural hearing loss is a debilitating condition that affects roughly one in fifteen individuals in developed countries and approximately 80% of patients who have significant hearing impairment. It can be caused by exposure to loud sound, by treatment with aminoglycoside antibiotics, and by unspecified events in ageing. This condition is often called 'nerve deafness', but the primary insult is usually damage to hair cells in the cochlear epithelium. In the absence of evidence that mammalian hair cells can be replaced by mitotic proliferation, sensorineural hearing loss in humans has been considered irreversible. In mammals, hair cells appear to be produced only during embryonic development (Ruben 1967); the full complement is present at birth (Bredberg 1968).

In contrast, the ears of fish and amphibians add hair cells continually as the body grows (Corwin, 1981, 1983, 1985a, Jørgensen 1981, Popper & Hoxter 1984). In sharks, mitotic divisions in an individual epithelium can generate over a million hair cells during the course of postembryonic life (Corwin 1985b).

Hair cells can also be produced as replacements during regeneration. In amphibians, removal of a segment of skin that contains neuromast hair cell epithelia of the lateral line evokes mitotic divisions that produce pluripotent cells in a regenerative placode. These cells migrate into the newly regrown skin and give rise to approximately the same number of replacement neuromasts and hair cells as were removed (Stone 1937, Corwin 1986, Corwin et al 1989a). In birds, regeneration can replace hair cell losses in the cochlea (Cotanche et al 1986, Cotanche 1987, Cruz et al 1987). This regeneration is notable because of its dependence on trauma-evoked mitosis in populations of cells that are mitotically quiescent in undamaged ears (Corwin & Cotanche 1988, Ryals & Rubel 1988).

At the present time, several laboratories are completing descriptions of stages that occur during the continuous addition of hair cells in the ears of fish and amphibians, and stages in the regenerative replacement of hair cells that occurs in lateral line neuromasts and avian ears. The ultimate objective of understanding the mechanisms that trigger proliferation and control hair cell differentiation during regeneration hopefully will be realized in the future, in part on the basis of these studies. Here we provide an update that summarizes recent findings and outlines hypotheses in the framework of answers to three key questions: (1) What are the progenitors of new hair cells? (2) What are the potential triggers for the proliferation that results in the regenerative replacement of hair cells? And (3) How is the differentiation of newly produced hair cells guided so that appropriate, functional replacement cells develop in damaged epithelia?

Progenitor cells

Hair cells of the ear and the mechanoreceptive lateral line organs of vertebrates are highly specialized. They express a shared constellation of derived and distinctive characteristics (Table 1) (Corwin & Warchol 1991). The bulk of the evidence suggests that mechanoreceptive hair cells are a single cell type, that arises through the expression of a set of homologous genes in the same way that different skeletal muscle cells compose a single cell type. Yet, three different types of cell have been proposed to serve as postembryonic progenitors of hair cells. These include supporting cells (Corwin 1981, 1983, 1985a, 1986, Corwin & Cotanche 1988, Corwin et al 1989b, Balak & Corwin 1988, Jones & Corwin 1988, Girod et al 1989, Balak et al 1990, Duckert & Rubel 1990); hyaline and cuboidal cells (Girod et al 1989); and neuroepithelial cells (Presson & Popper 1990). Even given the taxonomic distances between the fish, amphibians, and birds investigated, it would be surprising if there were, in fact, three very different

TABLE 1 Lists of the many shared characteristics and the few differences among hair cells in the mechanoreceptive lateral line and the inner ear epithelia of vertebrates

Similarities	*Differences*
Placodal origin (adjacent to medulla)	Location
Afferent innervation (projects to lateral medulla)	Presence in amniotes
Efferent innervation (the same neurons send collaterals to hair cells in both)	
Stereocilia (staircase array, actin-filled with constricted bases)	
Eccentric kinocilium (with 9 + 2 arrangement of microtubules)	
Cuticular plate (beneath stereocilia)	
Synaptic bodies (at afferent sites)	
Subsynaptic cisternae (at efferent sites)	
Structural relationships (similar hair cell:supporting cell relations)	
Shared antigenicity (as shown by the Zn-1 monoclonal antibody)	
Shared electrophysiology	
Aminoglycoside sensitivity	
Pharmacology (indications of shared transmitter)	

cytogenetic origins for this one specialized type of receptor cell. Specialized cells usually arise through stereotyped cytogenetic lineages. The one notable exception occurs in the retina, where three different progenitors of photoreceptors have been identified (see Raymond 1991, Reh et al 1991: this volume). The occurrence of multiple progenitors in the retina appears to be the result of regression to less differentiated states in a single lineage. It is possible that lineage re-setting could occur in postembryonic production of hair cells as well, and may explain the diversity of suggested progenitors, but it is incumbent upon researchers to establish the existence of the multiple proposed precursors with rigorous documentation.

Supporting cells

Evidence consistent with the proposed role of supporting cells as postembryonic progenitors for both supporting cells and hair cells has accumulated as the number of studies has increased. Indirect evidence for a close lineage relationship between these cell types includes the coincidental autoradiographic labelling of adjacent hair cells and supporting cells in epithelia when tritiated thymidine was administered some days before fixation in postembryonically growing sharks

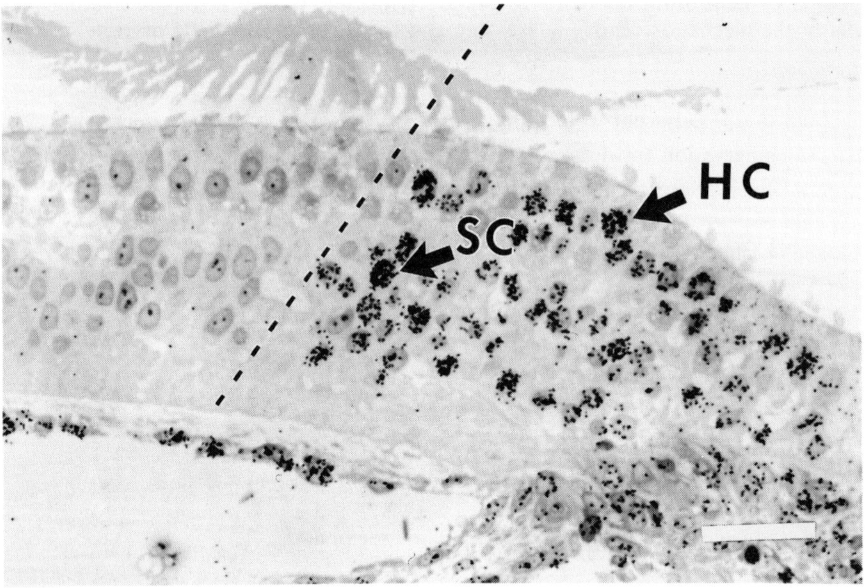

FIG. 1. A transverse section through part of the basilar papilla sensory epithelium from the cochlea of a chicken at embryonic day 15 (E15). On E6, a radioactive precursor to DNA, [^{3}H]thymidine, was injected into the yolk sac of this embryo, so that all cells that originated from divisions after E6 would be autoradiographically labelled. On E15, hair cells (HC) with nuclei near the lumen and supporting cells (SC) with nuclei near the base of the epithelium are recognizable. The results suggest a close lineage relationship between hair cells and supporting cells, such as the potential production of both types of cells from divisions of a single type of progenitor, because the cells segregate in the order of production and not as separate cell types. Labelled cells of both types are present to the right of the broken line and unlabelled cells to the left. Bar, 25 μm.

and toads (Corwin 1981, 1985a), in chickens during hair cell regeneration (Corwin & Cotanche 1988, Corwin et al 1989b), and in embryonic chickens during cochlear development (Fig. 1) (Katayama & Corwin 1989). Other indirect evidence includes the exclusive presence of supporting cells prior to the regenerative replacement of hair cells in experimentally damaged lateral line neuromasts of salamanders (*Ambystoma mexicanum*) (Corwin 1986, Balak & Corwin 1988), and in regions of the chicken cochlea damaged by overstimulation (Cotanche 1987) or by aminoglycoside poisoning (Duckert & Rubel 1990).

Direct evidence that supporting cells can be the progenitors of regenerated hair cells includes the results of intracellular injections of rhodamine dextran for lineage tracing in the lateral line (Corwin et al 1989b). Additional direct evidence includes continuous time-lapse microscopic observations of regeneration in lateral line neuromasts that were completely depleted of hair cells by laser microbeam irradiation (Balak et al 1990, Jones & Corwin 1991), and continuous time-lapse microscopic observations of neuromast regeneration after amputation of the tip of the tail in salamanders (Fig. 2) (Jones & Corwin 1988, 1991). The continuous records that have been made in lateral line neuromasts included observations of all cell movements, all cell divisions, and the progression of cell differentiation; therefore, they leave almost no room for doubt that supporting cell divisions can give rise to progeny that can differentiate as hair cells.

Hyaline and cuboidal cells

Hyaline and cuboidal cells that are located beyond the inferior (abneural) edge of the cochlear epithelium in chickens have been proposed as possible progenitors of regenerated hair cells by Girod et al (1989). This proposal was based on the observation that cells outside the edge of the sensory epithelium were heavily labelled by tritiated thymidine in chickens that were given the tracer during acoustic overstimulation or soon after and then sacrificed after a short survival (6–15 hours). With longer survival times (30 days), labelled cells were not found outside the edge of the epithelium, but labelled hair cells and supporting cells were observed. It was proposed that the disappearance of cell labelling outside the epithelium resulted from two processes: from repeated mitosis resulting in the dilution of the initially administered label, and from the migration of the labelled progeny into the epithelium, where they differentiated as hair cells and supporting cells.

We have conducted similar pulse injection studies in chickens, but have interpreted the results differently. In our experiments supporting cells are labelled early in regeneration, during the first 8 h after a 40 h period of overstimulation, both near the inferior edge of the epithelium and far from that edge. We conclude that here, as in the lateral line neuromasts, supporting cells may be the progenitors of regenerated hair cells, independent of where that regeneration occurs in the epithelium (Corwin et al 1989b). The different interpretations of the labelling results may reflect differences in the criteria used to decide whether a labelled cell is inside or outside the edge of the epithelium. When the outermost hair cells have been lost from the inferior edge of the avian cochlear epithelium, the stratified tissue changes to a simple cuboidal form, causing the damaged region of the sensory epithelium to appear indistinguishable from the cuboidal cell region that is external to the sensory epithelium in an undamaged ear. After some days the inferior edge regains its original structure as regenerated hair cells appear and form a new stratum that borders the lumen. The thickness of the epithelium can be an ambiguous indicator of the original location of the sensory epithelium's edge. If instead the tapered inferior edge of the basilar membrane is used as a landmark, then the correct identification of the original

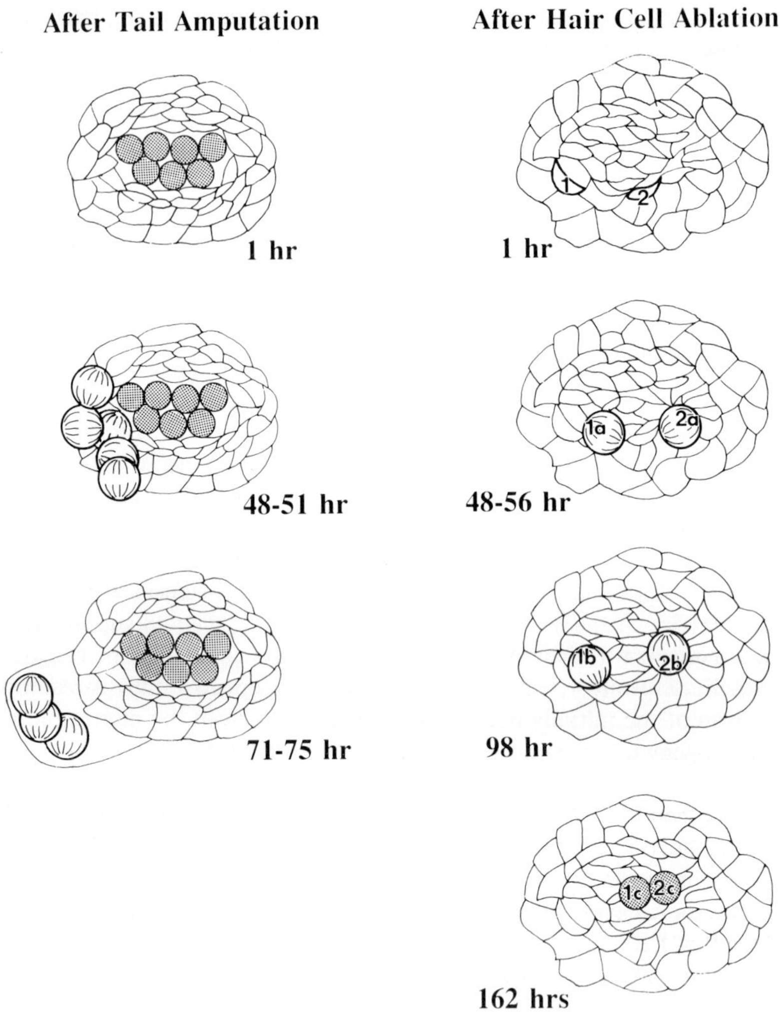

FIG. 2. Nearly continuous time-lapse video microscopic recordings have documented the cell divisions that give rise to replacement hair cells during two forms of regeneration that occur in the lateral line of the axolotl salamander. *Panels on the left:* Sequential views of the posteriormost neuromast on the tail stump after amputation of the tip of the tail and several neuromasts at time = 0. One hour after the amputation, the hair cells (stippled) and supporting cells of the neuromast remained mitotically quiescent. During the interval from 48 to 51 h after amputation, there was a sudden onset of mitotic activity in the posteroventral quadrant of the neuromast. Most of the cell divisions occurred in the mantle-type supporting cells at the outer edge. Cell divisions continued in the progeny of those cells and seeded the forming regenerative placode that extended posteriorly from the original boundary of that neuromast (e.g. 71–75 h). The cells in the placode continued to divide and migrated posteriorly into the regenerated portion

boundary between the sensory epithelium and the true cuboidal epithelial cells is less ambiguous, because this landmark is unaffected by traumatic changes in the sensory epithelium. The differences in the interpretation of labelling data in similar preparations could have resulted from the classification of the supporting cells in the damage zone at the edge as cuboidal cells that reside outside the epithelium. If this were true, there would be no need to postulate a second progenitor cell type in the same epithelium where both groups of investigators have found evidence for a role of supporting cells, or to postulate the occurrence of rapid sequential divisions and distant, directed migrations of progeny from outside the epithelium. In addition, the action of macrophages could account for the disappearance of some labelled cells, as will be discussed below (Jones & Corwin 1988, 1991).

Neuroepithelial cells

A third type of cell, termed a neuroepithelial cell, has been proposed as the progenitor of hair cells produced postembryonically in the sacculus of the oscar (*Astronotus ocellatus*), a teleost fish (Presson & Popper 1990). The evidence for this stems from quantitative analysis of autoradiographic labelling of cell nuclei in epithelia fixed at different times after a pulse injection of tritiated thymidine. The authors report that neuroepithelial cells are distinguishable from supporting cells on the basis of nuclear position. By their definition, neuroepithelial cells have nuclei at mid levels in the epithelium, and supporting cells have nuclei resting just above the basal lamina. The data show an initial high incidence of tritiated thymidine labelling in the neuroepithelial cell nuclei and lower levels of labelling in the basally located nuclei of 'basally located S-phase cells'. With increased survival time, the labelling of neuroepithelial cell nuclei and basally located S-phase cell nuclei decreases, while labelling increases in hair cells, supporting cells, Schwann cells, and cells classified as 'unknown'. The authors concluded that neuroepithelial cells are the progenitors that give rise to hair cells and supporting cells, that basally located S-phase cells are the progenitors

---

of the tail, where they eventually gave rise to cells that differentiate as hair cells and supporting cells of replacement neuromasts. *Panels on the right:* Sequential stages of recovery after individual ablation of all pre-existing hair cells in a neuromast by laser microbeam irradiation at time = 0 (Balak et al 1990). The positions of two internal supporting cells (1, 2) are shown one hour after hair cell ablation. Those cells divided at 48–56 h, then at 98 h one daughter cell (1a, 2a) from each of those divisions divided again. Over the course of the next few days one daughter cell (1b, 2b) from each of the secondary divisions differentiated as a hair cell (1c, 2c). The cells were brightly fluorescent after the salamander was immersed in a solution of DASPEI, a fluorescent dye that selectively stains hair cells. The orientations of mitotic spindles at the onset of anaphase are shown for each division.

of Schwann cells, and that mature supporting cells do not undergo mitosis at any time (Presson & Popper 1990).

These proposed progenitors were termed neuroepithelial cells apparently by analogy with cells in the embryonic neural tube (Presson & Popper 1990). Intermitotic nuclear migration occurs in the embryonic neural tube and has recently been demonstrated in the cells that form the sensory epithelium of the embryonic chicken cochlea (Fig. 3) (Katayama & Corwin 1991). In that epithelium, virtually all mitosis occurs at the lumen, but, after a brief pulse of tritiated thymidine, labelling is initially restricted to the basal and middle strata of nuclei. Label progresses into cells with visible mitotic figures only after significantly longer survival times. The documented occurrence of intermitotic nuclear migration in a developing hair cell epithelium suggests a more parsimonious alternative explanation for the data reported by Presson and Popper. Neuroepithelial cells may actually be supporting cells that differ in appearance from the basal supporting cells because they have re-entered the cell cycle and are exhibiting changes in nuclear position resulting from intermitotic nuclear migration similar to the nuclear migration demonstrated in the chick cochlea (Fig. 3) (Katayama & Corwin 1991).

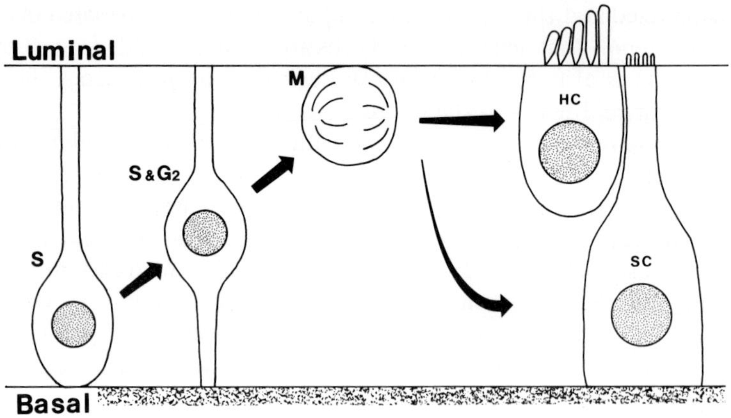

FIG. 3. Intermitotic migration of nuclei in cells of the developing sensory epithelium of the chick's cochlea. This diagram summarizes results from chartings of the positions of mitosing cells and results from pulse labelling with radioactive thymidine. Nuclei of cells that are in the synthetic phase (S) are in the basal and middle strata deep in the epithelium. These nuclei migrate toward the lumen as they progress through the cell cycle to the mitotic phase (M). The cells divide near the lumen, then their progeny can either re-enter the cycle or differentiate as hair cells and supporting cells. This form of intermitotic nuclear migration across the depth of the cochlear sensory epithelium resembles a comparable process that occurs in the developing neural tube. In each case it seems that nuclear migration may be necessary to allow columnar cells to round up in preparation for divisions, while providing the ionic isolation of the intraepithelial space by maintaining zonula occludens intercellular junctions near the lumen.

In the absence of convincingly rigorous evidence to the contrary, we currently hold that several of the cell types that other investigators have classed as distinct from supporting cells should still be considered somewhat questionable. More rigorous criteria for distinguishing between potential subtypes of supporting cells are required. As new evidence is presented, the identities of the postembryonic progenitors of hair cells should become firmly established.

Potential triggers of proliferation

The absence of tritiated thymidine labelling in control cochleas and in undamaged portions of overstimulated cochleas confirms that the hair cells and supporting cells of the avian cochlea do not normally divide during postembryonic life (Corwin & Cotanche 1988, Ryals & Rubel 1988). The cell production that underlies the avian cochlea's regenerative self-repair is not a postembryonic extension of the production of hair cells and supporting cells that occurs in embryos. Instead it is a response to trauma, in populations of cells that normally become mitotically quiescent even before hatching (Katayama & Corwin 1989). The initiation of this regenerative proliferation depends on a preceding loss of cells. What is the signal that triggers the regenerative proliferation? How do the cells that respond receive the signal that trauma has occurred?

We do not yet know what the signal is, but all cells receive information about their environment through the plasma membrane. In some cases, such as with steroid hormones, molecules convey signals by passing directly through the lipid bilayer. Other signals, in the form of changes in the concentrations of ions in the environment of a cell, result in changes in ionic currents that pass through protein channels in the membrane, thereby affecting both the ionic environment inside the cell and the potential gradient across the cell's membrane.

In other cases, signal molecules do not pass through the cell's membrane or ion channels; instead they bind to receptor sites on molecules that project from the cell surface into the extracellular environment. The binding between signal molecule and membrane receptor evokes a change in the receptor molecule, which is usually a protein or glycoprotein. The signal that triggers the regenerative response of cells in the cochlea of the chicken may be mediated by the latter form of information transmission, perhaps influencing the proliferative activity of cells through changes in glycoprotein receptors that are linked to the cytoskeleton (Corwin & Warchol 1991). Such a signalling mechanism appears to be involved in the normal proliferative response to soluble growth factors such as EGF (epidermal growth factor) and PDGF (platelet-derived growth factor), and in the uncontrolled proliferation that results from the oncogene-mediated transformation of tumour cells. It is to be expected that the mechanisms that regulate cell proliferation in hair cell progenitors will be similar to the overwhelmingly conserved mechanisms that control proliferation in other eukaryotic cells, most of which are little changed from yeast through to mammals (Lee & Nurse 1988, Lewin 1990).

The role of cell extrusion

At least five treatments can evoke the regenerative replacement of hair cells: compromising the metabolism of a preparation (Corwin 1985, 1986), acoustic overstimulation (Cotanche et al 1986, Cotanche 1987), aminoglycoside poisoning (Cruz et al 1987, Lippe et al 1989), phototoxic poisoning, and laser microbeam irradiation of existing hair cells (Balak et al 1990). All these treatments appear to result in the extrusion of dead or dying hair cells. Where hair cells have been damaged by overstimulation, but remain in the epithelium, regeneration does not appear to occur (Cotanche 1987). The destruction of dye-filled hair cells by laser irradiation has also shown that the hair cells do not have to be ruptured in order to trigger regeneration (Balak et al 1990). Hair cell extrusion itself may be capable of stimulating proliferation in the adjacent cells that remain in the sensory epithelium.

When a hair cell is extruded, the adhesive interaction of the membrane-spanning glycoproteins that had linked the hair cell to surrounding supporting cells at the adherens junctions will be uncoupled. The disruption of contact between hair cells and supporting cells could provide a signal that a cell has been lost. If the glycoproteins on the hair cell side of the junction are different from those on the supporting cell side, this signal should persist until new cells have been produced and one of those cells has progressed in differentiation to the point where it has expressed a hair cell-specific glycoprotein (see Fig. 5D). Both the initiation of proliferation and the control over which cells, and how many, differentiate as replacement hair cells, could be regulated through such inhibitory heterophilic binding between cell surfaces.

If supporting cell–hair cell contact acts as an inhibitor of proliferation, we can make predictions about the threshold of hair cell damage necessary to trigger a regenerative response. In the avian cochlea, hair cells and supporting cells appear to form a crystal lattice-like arrangement in which each supporting cell may contact a fixed number of hair cells. In the example illustrated in Fig. 5, two adjacent hair cells would have to be extruded before a single supporting cell would no longer be in contact with any hair cells.

The role of macrophages

Time-lapse microscopic observation of hair cell epithelia in the lateral line has revealed two forms of cell motility that occur during regeneration (Jones & Corwin 1988, 1991). The cells of the regenerative placode are individually motile, moving independently of each other via lamellipodia-mediated propulsion *in vivo*. Also, two classes of phagocytic cells move quickly around and within the lateral line epithelia (Fig. 4). These cells phagocytize new and pre-existing cells, both during regeneration and in undamaged lateral line neuromasts (Jones & Corwin 1988, 1991). Similar phagocytic cells have been observed by the same methods at sites of hair cell loss in explanted avian cochlear epithelia (M. E. Warchol, unpublished observations) and in explants of the mouse organ of Corti (M. W. Kelley, unpublished observations).

In the regeneration of lateral line neuromasts that follows amputation of the tip of the tail in the salamander, early phagocyte activity is significantly higher in the posterior half of the neuromast ($P<0.01$), and highest in the posteroventral quadrant that later becomes the site of formation of the regenerative placode. The elevated activity of phagocytes near the neuromast cells that will give rise to the regenerative placode could relate to the secretion of basic fibroblast growth factor (bFGF) that occurs in macrophages (Baird et al 1985). Transforming growth factor-α (TGF-α), transforming growth factor-β (TGF-β), platelet-derived growth factor (PDGF) and insulin-like growth

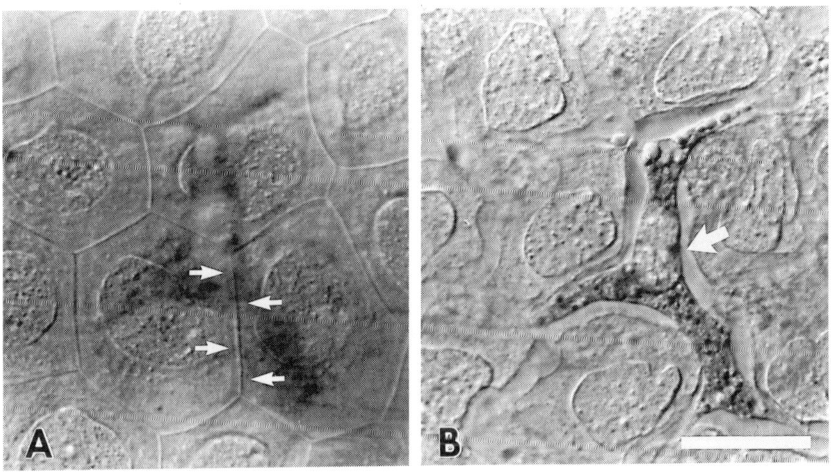

FIG. 4. Phagocytic cells appear to play a role in hair cell regeneration in the lateral line and are suspected to influence regeneration of hair cells in the avian cochlea. After amputation of the tip of the tail in axolotl salamanders, and prior to the onset of supporting cell mitotic activity that gives rise to the regenerative placode, phagocytic cells become significantly more active in contacting cells in the posterior half of the posteriormost neuromast. The phagocytes are especially noticeable in time-lapse recordings, because they are highly motile. Their identity has been verified histochemically. A. A differential interference-contrast microscope view of epidermal cells at the skin surface in an axolotl, with a phagocyte visible below the plane of focus. Arrows bracket the junctions between epidermal cells in the external layer overlying the stained phagocyte. B. The same field, with the histochemically stained phagocyte (arrow) at the plane of focus in the internal (second) layer of epidermal cells. Bar, 25 μm.

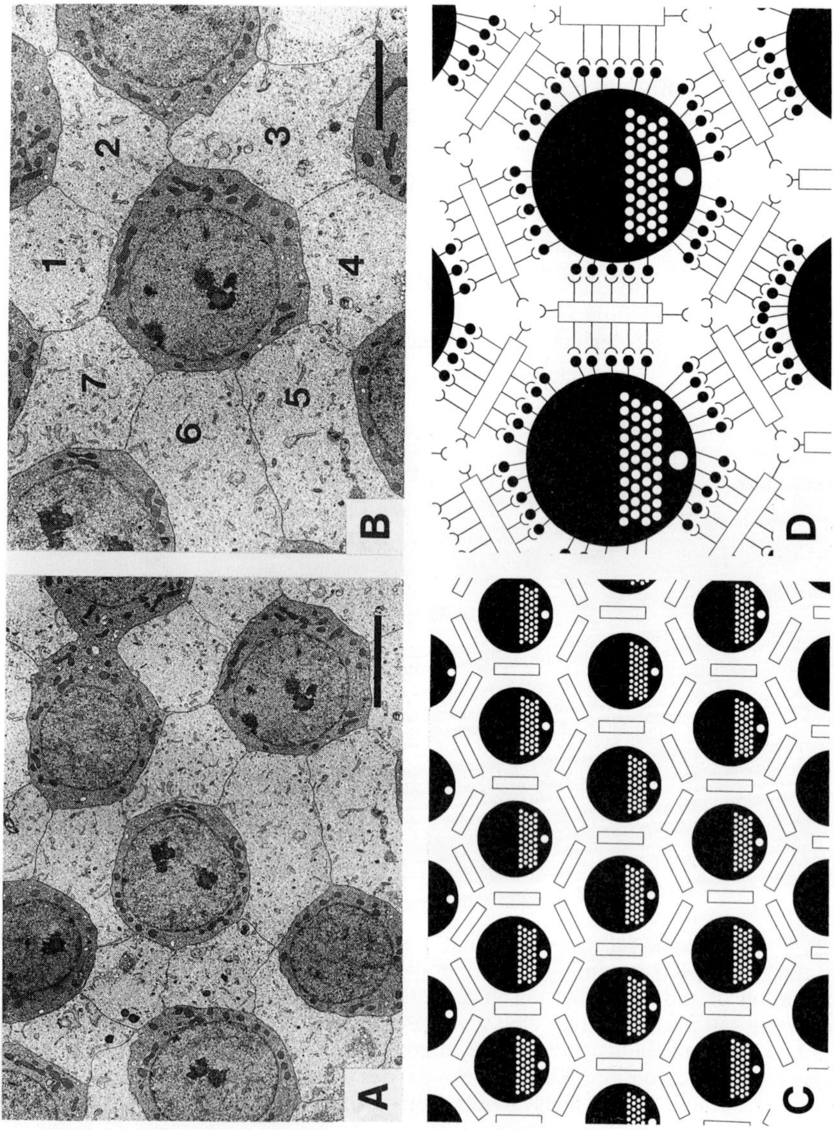
B
1
2
3
4
5
6
7
A
D
C

FIG. 5. The basilar papilla sensory epithelium in the avian cochlea has a regular mosaic or 'crystalline' pattern of hair cells and supporting cells. A. A transmission electron micrograph of a small segment of sensory epithelium that was sectioned in a plane parallel to its luminal surface. The circular outlines of electron-dense hair cells and the irregular outlines of the less densely stained supporting cells are visible. B. A higher magnification image of one hair cell, its seven neighbouring hair cells, and the seven supporting cell profiles that contact the cell (1–7). Each supporting cell profile contacts two or more hair cells. C. A Diagram illustrating the mosaic pattern of hair cells (dark circles with smaller unfilled circles representing stereocilia and kinocilia) and supporting cells (unfilled rectangles). The pattern develops during embryogenesis and is reconstituted during the regeneration of hair cells. After the pattern has developed, supporting cells each contact either two hair cells (as illustrated in this diagram) or more than two. Except at the epithelium's edge, they seldom if ever contact just one hair cell. D. During regeneration the proliferation and differentiation of the hair cells and supporting cells may be regulated by binding interactions between dissimilar surface glycoproteins, such as the cadherins, at the zonula adherens junction (Corwin & Warchol 1991). Heterophilic binding between a surface glycoprotein that is expressed early in the differentiation of new cells (represented as open half circles on thin shafts attached to the rectangles that represent supporting cells), and another surface glycoprotein (represented by filled circles on thin shafts) that is expressed by cells that have progressed further towards differentiation as hair cells, could result in the observed mosaic patterns of cell distribution. The proposed rule for determination is that a cell that is not in contact with a hair cell when it develops will become determined as a hair cell. If binding to a glycoprotein that is expressed by differentiating hair cells inhibits adjacent cells from differentiating as hair cells, then they will differentiate as supporting cells. Each supporting cell eventually would contact two or more hair cells after all the cells that could have differentiated as hair cells had done so. This hypothetical regulation of cell fate is analogous to the inhibitory surface glycoprotein interactions that are known to regulate the development of sensory cell fates in the ommatidia of the compound eye in *Drosophila* (cf Tomlinson 1988). Bar, 4 μm in A, 3 μm in B.

factor 1 (IGF-1) are also potential secretion products, as evidenced by mRNA transcripts in macrophages isolated from wounds (Rappolee et al 1988). The mantle-type supporting cells are distinguishable from internal supporting cells on the basis of differences in position, histochemical staining, and resting rates of cell division. These cells proliferate to give rise to the regenerative placode. Mantle-type supporting cells are rimmed by a glycocalyx that stains heavily after treatment with ruthenium red (Hama 1978), so phagocytic breakdown of that glycocalyx could have a stimulatory role. This either might release the mantle-type supporting cells from potential contact-mediated inhibition of proliferation, or it might allow greater exposure of receptors on the surface of these cells to mitogenic compounds that could initiate the placode-forming cell divisions. The validity of these proposed roles for phagocytic cells remain to be tested.

Other potential triggers

Hair cells in the ear are situated at an interface between two ionic environments: endolymph, which is high in potassium, and perilymph. These are separated by tight junctions at the apical surfaces of cells in the sensory epithelia. Disruption of the integrity of the tight junction separation is likely to occur during hair cell loss and should result in ionic currents flowing across the damaged epithelium. Injury-related ionic currents have been linked to regeneration in other tissues (Borgens et al 1989), so current flow might initiate or influence hair cell regeneration, through either electrical or ion-specific effects. To our knowledge, no experiments have addressed this issue.

Circulating growth factors could potentially trigger mitosis in otherwise quiescent supporting cells if open space were created by damage to other cells in the epithelium. Remaining supporting cells might then experience local increases in exposure to growth factors.

Potential guidance of hair cell differentiation

How is the differentiation of newly produced cells guided so that appropriate hair cells and supporting cells develop in damaged epithelia? The hair cell epithelium of the cochlea of the chicken has a highly regular appearance that is reminiscent of the mosaic patterns of cells in organs such as the dipteran compound eye (Fig. 5A). We hypothesize that a loosely constrained stereotyped geometry exists in this auditory epithelium (Fig. 5B). This could arise in development and be re-established during regeneration if the developing cells were to follow specific rules for determination, wherein the surface interactions between neighbouring cells might govern the final differentiated fates of the individual cells (Fig. 5C). This type of cellular interaction occurs during the development of different classes of geometrically positioned photoreceptors in the ommatidia of the *Drosophila* compound eye (Tomlinson 1988, Basler &

Hafen 1989). The regular geometric or 'mosaic' pattern of hair cells and supporting cells that develops both during normal embryonic formation and by reconstitution during the regenerative replacement of hair cells in the chicken's cochlea could result from a similar mechanism for the control of cell proliferation and differentiation. Immunocytochemical studies of the avian cochlea have reported differential spatiotemporal patterns of both the calcium-sensitive (A-CAM = N-cadherin, P-cadherin, and L-CAM = E-cadherin) and the calcium-insensitive (N-CAM and Ng-CAM) cell surface glycoproteins in developing hair cell epithelia (Nose & Takeichi 1986, Richardson et al 1987, Raphael et al 1987, 1988).

Asymmetrical cell divisions and the influence of position on cell fate

A monoclonal antibody, designated Zn-1, specifically labels hair cells in amphibians (Kornblum et al 1990). Observations of regenerating lateral line neuromasts that contain only a single immature hair cell that stains immunocytochemically with Zn-1 suggest that progenitor cell divisions can be asymmetrical. One daughter cell may differentiate as a hair cell, while the other continues in the cycle of proliferation or differentiates as a supporting cell. The occurrence of the first Zn-1-labelled cell in the centre of each regenerated neuromast (Kornblum et al 1990), as well as the first occurrence of regenerated hair cells in the centres of laser-treated neuromasts (Balak et al 1990), are consistent with the hypothesis that cell position relative to neighbouring cells and relative to the basal lamina may be important in determining which progeny are destined to differentiate as hair cells.

The influence of innervation on hair cell differentiation

Two recent reports provide experimental evidence that denervated chicken cochleas grown as transplants develop normal hair cells and normal gradients in hair cell ultrastructure, so we can conclude that innervation is not required to guide the differentiation of hair cells (Corwin & Cotanche 1989, Swanson et al 1990).

Summary

Direct evidence has demonstrated that at least some regenerated hair cells arise from the progeny of supporting cell divisions, but other cell types have been proposed as additional sources of postembryonically produced hair cells. The evidence for those other proposed progenitors is indirect and questionable in some regards. Additional investigations are needed but, at present, the evidence appears to be consistent with the proposal that divisions of supporting cells may give rise to all forms of postembryonically produced hair cells. There is a need to determine what subtypes of supporting cells may exist in hair cell epithelia. Specific markers for supporting cells should be sought. The triggers that initiate the regenerative replacement of hair cells have not been identified, but the extrusion of damaged hair cells from the epithelium consistently precedes the regenerative proliferation in the neighbouring cells. We have proposed that the disruption of intercellular junctional contacts that occurs with extrusion may play a role in triggering the proliferative response, but other potential triggers cannot be ruled out at present. Intercellular junctions or other forms of cell–cell interaction may also be required to guide the differentiation of hair cells and supporting cells. The sensory epithelium of the avian cochlea has a 'crystalline' arrangement of hair cells and supporting cells, that could result from lateral inhibition acting during the determination of cell fate. The predicted rule is that a cell developing in this epithelium will become a hair cell if it is not in contact with a hair cell. This could account for the alternating pattern of hair cells and supporting cells that develops in the avian cochlea during embryogenesis and becomes reconstituted during regeneration.

Acknowledgements

The original research reported in this article was supported by a Research Career Development Award and a grant from the National Institute on Deafness and Other Communication Disorders. M. E. Warchol is supported by the NBD training grant from the National Institutes of Health. We wish to acknowledge Jan Redick for wonderful assistance in electron microscopy. We also thank Arlene Trader and Christine Laverack for providing excellent technical assistance throughout these investigations.

References

Baird A, Mormède P, Böhlen P 1985 Immunoreactive fibroblast growth factor in cells of peritoneal exudate suggests its identity with macrophage-derived growth factor. Biochem Biophys Res Commun 126:358–364

Balak K, Corwin JT 1988 Hair cells originate from supporting cell progeny during regeneration in the lateral line system. Assoc Res Otolaryngol Abstr 11:107

Balak KJ, Corwin JT, Jones JE 1990 Regenerated hair cells can originate from supporting cell progeny: evidence from photoxicity and laser ablation experiments in the lateral line system. J Neurosci 10:2502–2512

Basler K, Hafen E 1989 Ubiquitous expression of *sevenless*: position-dependent specification of cell fate. Science (Wash DC) 243:931–934

Borgens RB, Robinson KR, Vanable JW, McGinnis ME 1989 Electric fields in vertebrate repair. Alan R Liss, New York

Bredberg G 1968 Cellular pattern and nerve supply of the human organ of Corti. Acta Oto-laryngol Suppl 236:1–135

Corwin JT 1981 Postembryonic production and aging in inner ear hair cells in sharks. J Comp Neurol 201:541–553

Corwin JT 1983 Postembryonic growth of the macula neglecta auditory detector in the ray, *Raja clavata*: continual increases in hair cell number, neural convergence, and physiological sensitivity. J Comp Neurol 217:345–356

Corwin JT 1985a Perpetual production of hair cells and maturational changes in hair cell ultrastructure accompany postembryonic growth in an amphibian ear. Proc Natl Acad Sci USA 82:3911–3915

Corwin JT 1985b Over two million hair cells in a single auditory epithelium. Assoc Res Otolaryngol Abstr 8:3–4

Corwin JT 1985c Early events in the regeneration of lateral line hair cell epithelia. Biol Bull (Woods Hole) 169:549–550

Corwin JT 1986 Regeneration and self-repair in hair cell epithelia: experimental evaluation of capacities and limitations. In: Ruben RJ, Van De Water TR, Rubel EW (eds) Elsevier, New York, p 291–304

Corwin JT, Cotanche DA 1988 Regeneration of sensory hair cells after acoustic trauma. Science (Wash DC) 240:1772–1774

Corwin JT, Cotanche DA 1989 The development of location-specific hair cell stereocilia in denervated embryonic ears. J Comp Neurol 288:529–537

Corwin JT, Warchol ME 1991 Auditory hair cells: structure, function, development, and regeneration. Annu Rev Neurosci 14:301–333

Corwin JT, Balak KJ, Borden PC 1989a Cellular events underlying the regenerative replacement of lateral line sensory epithelia in amphibians. In: Coombs S, Görner P, Munz H (eds) The mechanosensory lateral line: neurobiology and evolution. Springer-Verlag, New York, p 161–183

Corwin JT, Jones JE, Balak KJ, Cotanche DA 1989b Regenerated hair cells can originate from supporting cell progeny: evidence from laser ablations, acoustic trauma, and DNA labeling. Soc Neurosci Abstr 15:208

Cotanche DA 1987 Regeneration of hair cell stereociliary bundles in the chick cochlea following severe acoustic trauma. Hear Res 30:181–194

Cotanche DA, Saunders JC, Tilney LG 1986 Hair cell recovery after severe acoustic trauma to the chick basilar papilla. Assoc Res Otolaryngol Abstr 9:14

Cruz RM, Lambert PR, Rubel EW 1987 Light microscopic evidence of hair cell regeneration after gentamicin toxicity in chick cochlea. Arch Otolaryngol Head & Neck Surg 113:1058–1062

Duckert LG, Rubel EW 1990 Ultrastructural observations on regenerating hair cells in the chick basilar papilla. Hear Res 48:161–182

Girod DA, Duckert LG, Rubel EW 1989 Possible precursors of regenerated hair cells in the avian cochlea following acoustic trauma. Hear Res 42:175–194

Hama K 1978 A study of the fine structure of the pit organ of the common Japanese Sea Eel *Conger myriaster*. Cell Tissue Res 189:375–388

Jones JE, Corwin JT 1988 Cellular origin of the regenerative placode during lateral line regeneration in the axolotl. Soc Neurosci Abstr 14:1199

Jones JE, Corwin JT 1991 Regeneration of lateral line sensory organs: identification of progenitor cells and the action of leukocytes. In preparation

Jørgensen JM 1981 On a possible hair cell turn-over in the inner ear of the caecilian, *Ichthyophis glutinosus* (Amphibia: Gymnophiona). Acta Zool (Stockh) 62:171–186

Katayama A, Corwin JT 1989 Cell production in the chicken cochlea. J Comp Neurol 281:129–135

Katayama A, Corwin JT 1991 Cochlear cytogenesis visualized through pulse labeling of chick embryos in culture. Submitted

Kornblum HI, Corwin JT, Trevarrow B 1990 Selective labeling of sensory hair cells and neurons in auditory, vestibular, and lateral line systems by a monoclonal antibody. J Comp Neurol 301:162–170

Lee M, Nurse P 1988 Cell cycle control genes in fission yeast and mammalian cells. Trends Genet 4:287–290

Lewin B 1990 Driving the cell cycle: M phase kinase, its partners, and substrates. Cell 61:743–752

Lippe WR, Ryals BM, Westbrook EW 1989 Regeneration of hair cells in the chick basilar papilla following gentamicin toxicity. Assoc Res Otolaryngol Abstr 15:82–83

Nose A, Takeichi M 1986 A novel cadherin cell adhesion molecule: its expression patterns associated with implantation and organogenesis of mouse embryos. J Cell Biol 103:2649–2658

Popper AN, Hoxter B 1984 Growth of a fish ear. 1. Quantitative analysis of hair cell and ganglion cell proliferation. Hear Res 15:133–142

Presson JC, Popper AN 1990 Possible precursors to new hair cells, support cells, and Schwann cells in the ear of a post-embryonic fish. Hear Res 46:9–22

Raphael Y, Marshak GH, Barash A, Geiger B 1987 Modulation of intermediate filament expression in developing cochlear epithelium. Differentiation 35:151–162

Raphael Y, Volk T, Crossin KL, Edelman GM, Geiger B 1988 The modulation of cell adhesion molecule expression and intercellular junction formation in the developing avian inner ear. Dev Biol 128:222–235

Rappolee SA, Mark D, Banda MJ, Werb Z 1988 Wound macrophages express TGF-α and other growth factors *in vivo*: analysis by mRNA phenotyping. Science (Wash DC) 241:708–712

Raymond PA 1991 Retinal regeneration in teleost fish. In: Regeneration of vertebrate sensory receptor cells. Wiley, Chichester (Ciba Found Symp 160) p 171–191

Reh TA, Jones M, Pittack C 1991 Common mechanisms of retinal regeneration in the larval frog and embryonic chick. Wiley, Chichester (Ciba Found Symp 160) p 192–208

Richardson GP, Crossin KL, Chuong CM, Edelman GM 1987 Expression of cell adhesion molecules during embryonic induction. III. Development of the otic placode. Dev Biol 119:217–230

Ruben RJ 1967 Development of the inner ear of the mouse: a radioautographic study of terminal mitoses. Acta Oto-laryngol Suppl 220:1–44

Ryals BM, Rubel EW 1988 Hair cell regeneration after acoustic trauma in adult *Coturnix* quail. Science (Wash DC) 240:1774–1776

Stone LS 1937 Further experimental studies on the development of lateral-line sense organs in amphibians observed in living preparations. J Comp Neurol 68:83–115

Swanson GJ, Howard M, Lewis J 1990 Epithelial autonomy in the development of the inner ear of a bird embryo. Dev Biol 137:243–257

Tomlinson A 1988 Cellular interactions in the developing *Drosophila* eye. Development 104:183–193

DISCUSSION

Palay: You suggested a 'rule' of morphogenesis, that if a sustentacular cell is *not* next to a hair cell, then it will become one. But in your hexagonal arrangement (Fig. 5), the supporting cells are in contact with each other, on two of the four possible 'sides'. So the rule must be a little more complicated?

Corwin: I don't think so. The rule is that a cell that is not yet determined cannot become committed to differentiation as a hair cell if it is in contact with a hair cell. There can be another cell or cells in contact with the undetermined and undifferentiated cell, but if it is already in contact with even one hair cell, then the identities of the other cells do not really matter. The undetermined cell's fate is determined by the contact with a hair cell; further contacts are irrelevant to the determination of that fate. According to the proposed inhibitory 'rule', two uncommitted cells can meet and if they each contact a hair cell at some other points on their surfaces, neither will be 'allowed' to become a hair cell. Another cell, that may be next to those two non-hair cells, and surrounded by non-hair cells on all sides, can become a hair cell because it does not contact a hair cell at any point.

Rubel: This 'decision'—what we call 'determination' in embryology—is made during a short period of embryonic differentiation. Are you implying that when there are differentiated support cells and hair cells, and then the hair cell goes away as a result of trauma, the differentiated 'support' cell now transforms into a hair cell?

Corwin: No. Yet, you could envisage a mechanism whereby this type of contact-mediated interaction might be used as a trigger for regeneration as well, if the links between damaged hair cells and the surrounding supporting cells were broken during trauma. This might tell the supporting cell that it had lost its connection to a hair cell; breaking that mature connection could be an information-bearing event during regeneration (Corwin & Warchol 1991).

Dr Akiko Katayama conducted a cumulative DNA labelling study that showed that the first cells that are produced in the cochlea of the chick embryo form a band that runs longitudinally down the centre of the sensory epithelium. The next cells produced are just outside that first band of cells, and as more cells are produced they appear outside of the cells that were produced earlier (Katayama & Corwin 1989). If we consider what is happening across the transverse axis of the epithelium during that embryonic cell production, then it becomes clear that there is just the kind of spatiotemporal progression of cell production that would be desirable in a tissue where cell fate was determined in development by the simple inhibitory 'rule' that we have proposed here. That is, if the progression of determination followed the progression of cell births in the inferiorward and superiorward directions out from the central band where the first cells appeared, the cells would form the 'mosaic' or crystalline pattern that we've observed, if they just followed that simple rule. Obviously, we don't have experimental evidence to support this idea, but it's one of the potential rules for determination.

Pujol: Can you explain why you are excluding the nerve endings as a triggering possibility in your noise damage experiments?

Corwin: There are several reasons. From the lateral line system we know that hair cells can differentiate without any innervation. Also, Doug Cotanche and I denervated chick otocysts as early as embryonic day 3, which is three days before the final divisions normally give rise to hair cells and supporting cells in the cochlea. We transplanted the otocysts to the chorioallantoic membranes of hosts and allowed them to grow there for 12 days in the absence of innervation. Gavin Swanson and his coworkers did a similar study (Swanson et al 1990). In both studies there was normal development of sensory hair cells that developed in the absence of innervation (Corwin & Cotanche 1989).

Rubel: Remy, weren't you referring to regeneration?

Pujol: Either regeneration, or normal development.

Corwin: In development, the experimental evidence has shown that hair cells do not need innervation for differentiation to occur. That was also the conclusion reached in the 1976 paper in which Van De Water reported experiments with denervated mammalian otocysts.

Pujol: But in our laboratory, Gil-Loyzaga tried to grow a mammalian otocyst in culture after partial removal of Corti's ganglion. Only the part of the organ of Corti which is still in contact with the ganglion will develop hair cells; the other part will not (Gil-Loyzaga & Pujol 1987, Gil-Loyzaga 1991).

Corwin: It may well be different in the different classes of vertebrates.

Pujol: You haven't answered my question of why, in the avian papilla, you are excluding the nerve, when you still have nerves present within your epithelium?

Corwin: I am sure the nerve is still there in some cases of regeneration, so we cannot exclude its role on that basis; but I don't see how the nerve would be able to give rise to the crystalline pattern of the cells, with one supporting cell always bridging the gap between two hair cells.

Pujol: The nerve may not do anything directly, but may help the cell to differentiate.

Cotanche: We don't know for sure that the nerve ending is still there, after noise damage, and this is something we need to know. Nobody knows about the neural patterns after damage induced by noise.

Ryals: We have used serial transmission electron microscope analysis to examine the innervation pattern after noise exposure in mature quail. To date, we have only examined ears immediately after acoustic trauma, two weeks after trauma and 3–4 months after trauma.

Immediately after trauma the terminals on hair cells remaining in the epithelium are relatively normal in appearance. Two weeks later, both afferent and efferent terminals are swollen and/or abnormal in appearance. We have counted the number of synaptic specializations on hair cells 3–4 months after trauma (when afferent and efferent terminals appear normal) and have found a significant decrease in the number of afferent terminals on tall hair cells. The number of efferent terminals remained similar to that in non-noise-exposed animals. Although we did not positively identify regenerated hair cells, our acoustic trauma experiments show that more than 50% of the hair cells are lost after our levels of exposure (Ryals & Rubel 1988). So we are now pretty sure that regenerated hair cells eventually receive innervation. We think this probably reflects some active neural process, because ganglion cells of the VIIIth nerve degenerate over time after acoustic trauma in quail (Ryals et al 1989).

Jørgensen: I think Dr Van De Water (1986) has shown that you can grow the mouse otocyst in culture, with full cytodifferentiation, in the absence of the stato-acoustic ganglion.

Pujol: He hasn't done transmission electron microscopy, so his assumption is based only on light microscopy. Perhaps only very few nerve endings are needed to start the process of cell differentiation, and one should track them by transmission EM.

Rubel: This question clearly involves what one considers a hair cell to be. Orr showed during the 1960s that when you grow chick otocysts *in vitro*, hair cells will differentiate, but they won't be maintained without the nerve (Orr 1968).

Corwin: We find the same in the lateral line system of salamanders (*Ambystoma mexicanum*). If the nerve is not present over a long period (1–3 months), the lateral line neuromast actually disaggregates. But any nerve is capable of maintaining the neuromast; when they become reinnervated ectopically by spinal nerves they are maintained.

Presson: You discussed our proposal that the neuroepithelial cell is the progenitor of hair cells postembryonically, in the teleost that we study (*Astronotus ocellatus*, the oscar) (Presson & Popper 1990). In the oscar saccule, there is a continual addition of hair cells (in contrast to mammals). We did pulse labelling with [^{3}H]thymidine in oscars and killed them after different survival times, to determine the proliferating precursor cells and follow what happens to them.

After 30 minutes we see labelled nuclei that have a basal process that looks like an end-foot, as opposed to the broad basal process which support cells have. A process also reaches the lumen. The pattern of label with increasing survival time was exactly what you described for the chick. With increasing survival times, up to 5–6 hours, label appeared in cells at the lumen. At about nine hours I found what I believe to be labelled mitotic figures at the luminal surface. Later (12 or 15 h), I saw the first labelled hair cells and what I am calling supporting cells, which are down on the basement membrane. So my results suggest the presence of a cell type that labels and undergoes nuclear displacement.

There are two possibilities: either the cell which first labels is a separate cell type, or it is a supporting cell that is undergoing dedifferentiation. Those are difficult to discriminate. I called them a separate population, just to be conservative, and because they look like an embryonic neuroepithelial cell. I know that in teleost retina, brain or VIIIth nerve ganglion there are separate populations of undifferentiated precursor cells that give rise to new neuronal cells or photoreceptors. So the most conservative approach seems to be to call it a different cell type from the supporting cell.

In electron micrographs, the supporting cells have a cuboidal nucleus with very little cytoplasm and a broad base, whereas these neuroepithelial cells have a thin basal process and many arrays of microtubules under the nucleus. It's distinctly possible that a support cell rises up off the basement membrane and goes into the cell cycle and then becomes a precursor cell. I don't know which happens, but I do know that the supporting cell is never seen in S phase when it sits on the basement membrane.

Voyvodic: Dr Corwin, in your time-lapse cinematography studies you ablated hair cells and then watched the differentiation of new hair cells by playing the film backwards. You went back through two cell divisions and identified precursor cells arising from differentiated supporting cells. If they went through two cell divisions, there should be four cells from each supporting cell. If one was a hair cell, what happened to the other three cells? Do you know whether they survived?

Corwin: These are very new data, and that hasn't been traced out in many cases.

Fernald: In this study, how did you determine that the hair cell was derived from an ordinary support cell? Because there doesn't seem to me to be much difference between support cells; there may not be any difference between the story in teleosts and in birds. On the other hand, if you looked carefully, you might find some other characteristics which would identify these support cells as unique.

Corwin: That may be. I think the issue has arisen in part because of a semantic difference; I wouldn't have thought that naming a new cell class in these epithelia, the 'neuroepithelial cells', was a relatively 'conservative' approach, as Joelle Presson characterizes it. My feeling is that we should not split up cell types until we have done the hard work, such as serial sectioning at the level of electron microscopy, that is necessary to determine what subtypes of cells might exist in reality. I feel that convincing evidence needs to be provided before it's valid to use names such as 'neuroepithelial cells' as applied to hair cell epithelia. As participants in this symposium we may be using these terms with different amounts of caution, but others who are not present may not appreciate that significant uncertainties exist in relation to these possible subclasses of cells. I would like to consider all the proposed subclasses of non-sensory epithelial cells in hair cell epithelia as simply belonging to the category of 'supporting cells', until evidence has established that some may be different.

Jørgensen: Transmission EM serial sectioning of the regenerating axolotl neuromast by Dr Flock and myself showed the existence of basal cells that do not reach the surface of the neuromast. Basal cells were also found in a study of the utricular macula of the caecilian *Ichthyophis glutinosus* (Apoda, Amphibia) (see Jørgensen 1981).

Corwin: I don't reject this possibility; I would just like to see classification established on the basis of evidence such as that you have mentioned. We recognize mantle-type supporting cells, which you have already described as a separate category. Jay Jones has shown that they are distinguishable from the internal supporting cells in neuromasts on the basis of histochemistry. The mantle-type supporting cells could be equivalent in some fashion to the border cells that Ed Rubel was mentioning, and to the cells that make up the marginal growth zones that are present at the edges of the sensory epithelia that continue to add and accumulate hair cells in the ears of fish and amphibians during postembryonic life. The mantle-type supporting cells react to a nearby amputation in a manner different from the reaction of the internal supporting cells. The mantle-type supporting cells are almost quiescent, with less than one division in five days of time-lapse observations of the whole mantle-type supporting cell population of a control neuromast; whereas there are occasional divisions of the internal supporting cells in the same neuromast. Then, after the tip of the tail has been amputated, the mantle-type supporting cells in the neuromast that is nearest to the wound dramatically increase their frequency of proliferation, while the internal supporting cells decrease their frequency of proliferation at the same time.

Rubel: We need to decide what is a semantic problem and what is non-semantic, over the question of these 'support' versus 'non-support' cells. Our view is as follows. In the chick, basilar papilla, going from inferior to superior we have the following sequence: cuboidal cells; hyaline cells; then there is a junction that can be seen in both the basilar membrane proper and the cells underlying the basilar membrane. After this junction there are two or three border cells and then the first (most inferior) short hair cell (see Fig. 2, p 82). Under and surrounding the hair cells there are one *or more* types of 'supporting' cells. We seem to agree that the cuboidal cells may be spreading out, but are not playing a role as precursors to new hair cells; at least, none of their architectural features remains the same when proliferation is going on. We like to think that there are a number of cell types in the inferior region and there may be a *subset* of these cells that are kicked into proliferation; but, not all the cells become mitotically active.

The real difference in our views may be whether *any* cell in the epithelium that is not a hair cell can be induced into mitotic activity and can become a hair cell, or only a subclass of cells in this area have this capacity. Jeff Corwin, I think, would say that any cell in the region underlying the hair cells is capable of producing new hair cells, except for the existing hair cells. Does that represent your views correctly?

Corwin: Not quite! I would say that until we determine the different cell types there, I would prefer a 'lumping' rather than a 'splitting' approach. I feel that there may be subsets, but we don't know how to distinguish them as yet. I am *not* saying that any cell in that region is capable of dividing. But I would say that you haven't been able to distinguish between the different possible subsets; and I have not. I don't even know that the 'border' cell class, that you have proposed, is cytologically distinct. When we get to the edge of the sensory epithelium, somehow the production of hair cells has to be caused to stop during embryogenesis. In that regard, cochlear sensory epithelia are different from the postembryonically growing epithelia in the ears of fish and amphibians, which have a ragged edge; in the cochleas of birds and mammals the sensory epithelium, as defined by hair cells, ends abruptly at that point. You don't have any more hair cells, but maybe you have this extra 'supporting cell' that is outside the sensory epithelium proper and is only contacting one other supporting cell, and might therefore be more likely to be activated mitotically. I prefer to consider the cells that are at the edge, but otherwise look indistinguishable from supporting cells, as potential subsets of supporting cells, but I prefer not to give them names that imply a distinct difference until we have established whether there really are different subsets that can be distinguished on the basis of cytological markers.

Cotanche: From the histology, scanning electron microscopy and thymidine-labelling data on this proliferative zone, what you see in the frog and what Dr Rubel sees is structurally very similar. In your frog studies there is a lot of mitosis in the peripheral edge and also mitosis under the sensory epithelium; so I don't see why that is not exactly the same picture for Ed Rubel's regenerating epithelium.

Corwin: I agree, and I think Ed Rubel and coworkers have illustrated that point in a published figure (Girod et al 1989). Cytologically those cells look as if they belong to the major category that we might call 'supporting' cells. The 'edge' cells in the postembryonically growing fish and amphibian ears look the same as the supporting cells that are in the sensory epithelium, in terms of their cytoplasm, the number and the type of their organelles, and the density of their staining; and I suspect the same thing is true for the cells at the edge of the avian cochlea.

Fernald: But that is making the case that damage recapitulates ontogeny, and we haven't learnt much about ontogeny in this particular case. Maybe if you could explain to us what this candidate population of cells looks like during normal embryonic development, we could compare it with the response to damage.

Rubel: You have to be able to make the cell distinctions in order to explain that.

Fernald: Certainly; but if, say, during normal development, there were, amidst the support and easily identifiable hair cells, a third population of dividing cells, it might be easier to find these in a damage case.

Presson: We have heard comments from those working on the lateral line system that labelled hair cells are occasionally seen in a normal (non-traumatized) animal. If so, it raises the possibility that there is a small stem cell population in the adult that is stimulated to divide by trauma. If there were a couple of stem cells in the normal adult, they would kick in, and one would not have to argue for dedifferentiation of a mature cell to account for the addition of hair cells after trauma.

Corwin: But why postulate that it's an embryonic cell type, just because it is proliferatively competent? Fibroblasts are able to proliferate and are differentiated, functioning cells.

Presson: In the *Astronotus* saccule, only because the proliferating cell has a cell cycle, in terms of nuclear migration and mitosis, that resembles that of an embryonic cell. If you don't postulate an embryonic type, you have to argue that during development there is a shift in strategy from the embryo, where you have a cell which undergoes this nuclear migration, to a postembryonic animal which has a different mechanism for hair cell production.

Corwin: No; we also see the same type of nuclear migration in these lateral line neuromasts.

Presson: But are you saying that a different cell type is doing that in the embryo and in the adult?

Corwin: Not necessarily. I don't see that it has to be a distinction between *either* differentiation, *or* having competence to go on and divide.

Presson: An embryonic cell which is undergoing proliferation is not a support cell; we wouldn't call it that. Clearly, some support cells reach the stage where they don't normally undergo mitosis, so there is some point at which we would call them differentiated in the classical sense of postmitotic—at least in their normal (non-damage-inducing) environment—because we don't see them proliferating in the normal animal.

Reh: Where you use lasers to delete all the hair cells in the neuromast, Dr Corwin, do *all* the support cells undergo proliferation? Or do you see a small subset of support cells that can become mitotically competent?

Corwin: Not all the supporting cells divide, and there are two recognizably different populations of supporting cells in the neuromasts. One is a cell type that rims the epithelium, the mantle-type supporting cells. This type proliferates rapidly after a tail amputation, as I described earlier. That is, the population of mantle-type supporting cells up-regulates its frequency of division after amputation. Normally it is quite quiet. In contrast, the internal type of supporting cells have a low frequency of division during control situations, and that actually decreases significantly after nearby amputation.

Reh: Do all those support cells have an equal probability of up-regulating their rate of cell division?

Corwin: I don't know. You certainly see a lot of mitosis in there.

Lewis: Can you say whether it's more or less than 50% of that supporting population that divide?

Corwin: I would say that fewer than 50% are dividing, but it depends upon the period of time. It's less than 50% by the time we see two recognizable hair cells, but it might be 50% as a cumulative measure by the time the full number of pre-existing hair cells had been replaced.

Farbman: Is it possible that there is a low level of mitosis in physiological conditions in chick cochlea, and that the dividing cells never mature into hair cells, but die precociously? If this were so, then, when the organ is injured, the rate of mitosis could be up-regulated, and the newly formed cells could differentiate to fill the void caused by the damage. If there were a low level of mitosis under physiological conditions that hasn't been noticed because cells die soon after dividing, the numbers of hair cells or supporting cells would not be increased.

Rubel: I would doubt that very strongly!

Farbman: It's what happens in the olfactory system.

Rubel: With the continuous high level of labelling over a long period of time, we would have seen more examples of labelled cells, if that were so. In most of our experiments we kill the chicks at the end of the labelling. So I would expect to see more labelled cells if there was a low level of proliferation. It is only in one in 100 sections or one in 1000 sections that we see labelled cells.

Corwin: We have the same sort of result. Thousands of cells have been exposed to [^{3}H]thymidine for either seven or 10 days continuously, and yet we see almost no labelling in undamaged control sensory epithelia from chick cochleas.

Ryals: We occasionally see labelled cells in normal quail (Ryals & Westbrook 1990). You might expect to see just labelled support cells, if only support cells were being replaced. But we see [^{3}H]thymidine label over both hair cells and support cells.

Calof: In the corneal epithelium, in order to see the stem cell population in the corneal limbus, I believe you need continuous exposure to [^{3}H]thymidine (by mini-osmotic pump) for 14 days. One of the ways in which that stem cell population is defined is by its very long cell cycle, and by its retention of the [^{3}H]thymidine label; that is, if you 'chase' the label out by removing the [^{3}H]thymidine pump for many days after this long labelling period, the slow-cycling stem cells in the corneal limbus are the only cells in the corneal epithelium that retain label (cf. Cotsarelis et al 1989).

Rubel: We haven't done that experiment in the ear; maybe we should do it.

Calof: Where you have a precursor population with a very long cell cycle, your chances of hitting it with [^{3}H]thymidine while it is in S phase are very low for most of these labelling paradigms, unless those cells are induced to proliferate by something like one of the degeneration–regeneration paradigms.

Is it possible to tell, from any of these 'endpoint' ways in which we study these proliferating cells *in situ*, the difference between (1) one cell type that would divide and give rise to both a hair cell and a supporting cell, and (2) autonomously dividing supporting cell precursors and hair cell precursors? Suppose you gave a single pulse of [^{3}H]thymidine and looked at the tissue 10 days later. In both cases you would see labelled hair cells and labelled supporting cells. If you just looked at the end point, couldn't the two situations look the same? It seems to me there is a real problem in distinguishing between these two possibilities in this kind of *in vivo* experiment. Whereas, if you look at which cells are labelled immediately after a pulse of [^{3}H]thymidine, under conditions in which precursor proliferation is induced, you might pick up a rare precursor cell, that you wouldn't see under normal conditions.

Raymond: One would be deluding oneself to think that just because one sees only very rare incorporation of thymidine in normal conditions, there is not a population of cells that are capable of dividing. Remember how quickly you can go from one cell to many in an exponentially increasing population.

Lewis: That does not affect the issue of whether there is a separate population of stem cells, however. It could still be that there is just one sort of non-hair cell and that under normal conditions it very rarely divides.

Raymond: Exactly, and you do not need very many founder cells in an exponentially expanding population to produce the levels of mitotic division seen after damage.

Margolis: Isn't a major part of the problem the absence of molecular or immunological reagents for discriminating the different cell types?

Presson: For the supporting cell in my case (the *Astronotus ocellatus* saccule), which rises up off the basement membrane and changes into something else, even with reagents which distinguish those two types, I could not know if one were changing into the other unless I could label just one supporting cell and watch what happens to it over time. That is the only way to answer the question, because I could detect cytochemical differences, but it could still be the same cell in a different phase of its cycle.

Rubel: Let me challenge Jeff Corwin on the issue of why he typically sees proliferation only at the region between tall and short hair cells or only slightly inferior to this and not a lot of damage or proliferation at the inferior edge.

Corwin: In Doug Cotanche's earlier study on the damage caused by acoustic overstimulation, he reported that acoustic trauma from tonal stimuli produced damage at the inferior edge, and that a little further proximally from that point of damage there would also be a separate, additional region of damage more superiorly over the hinge point where the basilar membrane meets underlying cartilage (Cotanche et al 1987). We see most of our thymidine-labelled supporting cells there, and I think this explains why some of the time the labelling is in one place, and at other times the labelling is in the other place; but we see thymidine-labelled supporting cells all over both locations, at one time or another in these experiments.

References

Corwin JT, Cotanche DA 1989 The development of location-specific hair cell stereocilia in denervated embryonic ears. J Comp Neurol 288:529–537

Corwin JT, Warchol ME 1991 Auditory hair cells: structure, function, development and regeneration. Annu Rev Neurosci 14:301–333

Cotanche DA, Saunders JC, Tilney LG 1987 Hair cell damage produced by acoustic trauma in the chick cochlea. Hear Res 25:267–286

Cotsarelis G, Cheng S-Z, Dong G, Sun T-T, Lavker RM 1989 Existence of slow-cycling limbal epithelial basal cells that can be preferentially stimulated to proliferate: implications on epithelial stem cells. Cell 57:201–209

Gil-Loyzaga P 1991 Nerve fiber influence on cochlear hair cell differentiation. Cell Tissue Res, in press

Gil-Loyzaga P, Pujol R 1987 Trophic interaction between spiral ganglion neurons and the cochlear epithelium during *in vitro* development. Assoc Res Otolaryngol Abstr 10:223

Girod DA, Duckert LG, Rubel EW 1989 Possible precursors of regenerated hair cells in the avian cochlea following acoustic trauma. Hear Res 42:175–194

Jørgensen JM 1981 On a possible hair cell turn-over in the inner ear of the caecilian *Ichthyophis glutinosus* (Amphibia: Gymnonphiona). Acta Zool (Stockh) 62:171–186

Katayama A, Corwin JT 1989 Cell production in the chicken cochlea. J Comp Neurol 281:129–135

Orr MF 1968 Histogenesis of sensory epithelium in reaggregates of dissociated embryonic chick otocysts. Dev Biol 17:39–54

Presson JC, Popper AN 1990 Possible precursors to new hair cells, support cells, and Schwann cells in the ear of a post-embryonic fish. Hear Res 46:9–22

Ryals BM, Rubel EW 1988 Hair cell regeneration after acoustic trauma in adult *Coturnix* quail. Science (Wash DC) 240:1774–1776

Ryals BM, Westbrook EW 1990 Hair cell regeneration in senescent quail. Hear Res 50:87–96

Ryals BM, Ten Eyck B, Westbrook EW 1989 Continued ganglion cell loss after hair cell regeneration. Hear Res 43:81–90

Swanson GJ, Howard M, Lewis J 1990 Epithelial autonomy in the development of the inner ear of a bird embryo. Dev Biol 137:243–257

Van De Water TR 1976 Effects of removal of the statoacoustic ganglion complex upon the growing otocyst. Ann Otol Rhinol Laryngol 85 (suppl 33):1–32

Van De Water TR 1986 Determinants of neuron–sensory cell interaction during development of the inner ear. Hear Res 22:265–277

Structural reorganization of hair cells and supporting cells during noise damage, recovery and regeneration in the chick cochlea

Douglas A. Cotanche, Alicia Petrell and Daniel A. Picard

Department of Anatomy & Neurobiology, Boston University School of Medicine, 80 East Concord Street, Boston, MA 02118, USA

Abstract. Exposure of the chick cochlea to intense acoustic overstimulation leads to rapid changes in the structural organization of hair cells and supporting cells. During the noise exposure the surface areas of the hair cells are reduced and, concomitantly, the adjacent supporting cells expand. If the exposure continues long enough hair cells bleb out apically and are ejected from the sensory epithelium. Surviving hair cells begin to expand their surface areas again within 24 hours after the termination of noise exposure. By 48 hours of recovery most of the hair cells that have died are ejected from the epithelium and new, regenerating hair cells begin to sprout immature stereociliary bundles. The regenerating stereociliary bundles undergo steps in their maturation that are equivalent to those seen during the normal embryonic development of hair cells. The extent and timing of recovery in the chick cochlea correlates with the length of the noise exposure. Shorter exposures ($\approx$12 hours) result in only a few regenerating hair cells but the basilar papilla recovers almost completely within 14 days. Longer exposures ($\approx$24 hours) lead to the production of more new hair cells but the recovery of the basilar papilla is incomplete, even after 28 days of recovery.

1991 Regeneration of vertebrate sensory receptor cells. Wiley, Chichester (Ciba Foundation Symposium 160) p 131–150

Until recently it had been thought that hair cells in the auditory epithelium of birds and mammals were incapable of being replaced during the adult lifetime. This meant that once the initial population of sensory cells was established during embryonic development, any subsequent injury or trauma that killed the hair cells would result in a permanent structural and functional deficit in the cochlea. However, investigations from several laboratories over the last four years have shown that hair cells in the auditory sensory epithelium of the bird inner ear can be regenerated after acoustic trauma or ototoxic poisoning (Cotanche 1987b,c,

Cruz et al 1987, Henry et al 1988). This regeneration is brought about by the reinitiation of cell division in the supporting cells of the sensory epithelium and a subsequent differentiation and maturation of new hair cells to replace those that were lost (Corwin & Cotanche 1988, Ryals & Rubel 1988, Girod et al 1989). Studies of auditory function have shown that this structural regeneration is accompanied by a functional recovery, although it is not yet clear how much of the recovery is due to the regenerated hair cells (McFadden & Saunders 1989, Tucci & Rubel 1990).

In this paper we hope to address some of the structural mechanisms by which noise damage leads to injury and death of hair cells; how the epithelium recovers from the noise damage, both in the production of new hair cells and the recovery of the injured cells; and how the extent of noise damage correlates with the degree of recovery and regeneration.

Methods

Pairs of White Leghorn chick hatchlings at seven days of age were exposed to a 1500 Hz pure tone at 120 dB sound pressure level (SPL) for either 12 or 24 hours. At the end of the noise exposure period, one bird was killed immediately and processed for scanning electron microscopy (SEM). The other bird in each pair was allowed to recover, for periods ranging from one to 14 days, and was processed similarly for SEM. Control birds, not exposed to the pure tone, were processed at equivalent ages (seven to 21 days) in order to compensate for any age-related changes in the size of the hair cells and supporting cells. The basilar papillae were analysed qualitatively for the extent of hair cell injury and loss in the zero-recovery animals and for the amount of regeneration and repair to the epithelium in the animals allowed time for recovery. In addition, quantitative analyses were made of the changes in hair cell surface areas and stereociliary bundle orientations during regeneration. Surface areas were measured from hair cells along the inferior edge of the damaged region and compared to similar regions in control animals. In the recovered chick cochleas, the surface areas of surviving hair cells were measured separately from those of identifiable regenerating cells, so that the small surface areas of the regenerating cells did not artificially reduce the mean surface area for the surviving cells. The regenerating hair cells were identified by their immature stereociliary bundles. The mean surface area values for recovered hair cells include only the surviving cells.

Hair cell orientations were measured from cells in the middle of the damaged region, the area where most of the regenerated hair cells arise. Orientations were measured by drawing an arrow through the stereociliary bundle from the shortest to the tallest row and measuring the angle of the arrow as degrees on a circle. The ears were standardized so that 360° was directed toward the superior edge of the epithelium, 90° toward the proximal end, 180° toward the inferior edge, and 270° toward the distal end. The measurements of the individual hair cells were compiled into histograms with bin widths of 20°. These measurement techniques were established in a previous study of the development of hair cell orientation in the embryonic chick cochlea (see Cotanche & Corwin 1991).

Results and discussion

Response to acoustic overstimulation

Exposure of the chick cochlea to a pure-tone overstimulation results in a region of localized hair cell loss and injury when the basilar papilla is examined immediately after the exposure (Fig. 1a). In most of our experiments we have used a stimulus of 1500 Hz at an intensity of 120 dB SPL. We have previously established that the sound intensity required to produce noise damage detectable by scanning electron microscopy ranges from 115 to 125 dB SPL (Cotanche et al 1987). Below 115 dB SPL, damage is not invariably seen, whereas above 125 dB SPL the resulting damage is so severe that most of the cells in the sensory epithelium are destroyed. Initially, we used a 48 hour exposure period, but we have recently found that a 24 hour exposure results in a damaged region equivalent in size and extent of hair cell loss to that seen in the 48 hour exposure.

The 24 hour exposure produces a semicircular patch of damage in the sensory epithelium (Fig. 1a). The damage extends from the inferior edge of the basilar papilla up toward the superior edge, but normally does not reach it. Thus, most of the cells involved are short hair cells lying on the basilar membrane, and not the tall hair cells which are positioned over the superior fibrocartilaginous plate. Within the damaged region, the hair cells respond to the trauma in one of two ways. Most of the cells close to the inferior edge survive the trauma, but have very small surface areas and are separated from one another by expanded supporting cell surfaces. Many of the hair cells along the arched superior border of the damaged region are swollen and blebbed out, with the adjacent supporting cells expanding to cover the surface of the epithelium. It appears that the blebbing of the hair cells eventually results in the entire cell being squeezed out of the basilar papilla, for light and transmission electron microscopic studies show no remnants of the nuclei or basal parts of these cells in the epithelium (Cotanche 1987b). Occasionally, a few blebbed hair cells are seen among the surviving hair cells in the inferior part of the damaged region, but their numbers are not as extensive as those along the superior edge.

In order to understand how this pattern of hair cell damage evolves over the 24 hour exposure period, we have examined the extent of noise damage produced by exposures of four, eight and 12 hours (Cotanche & Dopyera 1990). The earliest detectable signs of damage occur after four hours of exposure as an expansion of some of the supporting cell surfaces along the inferior edge of the basilar papilla. At eight hours, the supporting cell surfaces expand even more and form lateral bands across the epithelium that run perpendicular to the length of the cochlea. After 12 hours, the expansion of the supporting cell surfaces spreads out to occupy a semicircular region equivalent in size to the 24 hour-damaged region. Moreover, there is a measurable decrease in the surface areas of the hair cells within this region and the first evidence of hair cells blebbing up and being ejected from the epithelium. Interestingly, the initial site of hair cell loss is located close to the inferior edge of the damaged region. When the exposure period is extended to 24 hours, the site of hair cell blebbing shifts to the superior edge of the damaged region, as described above.

The cellular mechanisms which lead to the contraction of the hair cell surface areas during noise exposure are not yet understood. Presumably, it involves the actin filaments in the junctional complexes at the apical ends of the hair cells (Hirokawa & Tilney 1982), but it must also encompass changes in the cuticular plate of the hair cells, since this normally fills most of the apical cytoplasm of the hair cell (Tilney et al 1983). It seems likely that this process also involves the junctional complexes of the supporting cells, which may help to actively expand the supporting cell surfaces. In fact, one cannot rule out the possibility that the hair cell contraction is a passive process and it is the active expansion of the supporting cells which constricts and, in some cases, ejects the hair cell from the epithelium. The reason why the hair cell surfaces contract is not clear, but one can hypothesize that it may be a protective response to the acoustic overstimulation. How this contraction may help to protect a hair cell, and what threshold is overridden to cause the death and expulsion of a hair cell from the epithelium, remain to be determined.

Recovery and regeneration

Structural changes begin to happen in the cochlea almost immediately after the noise exposure is terminated. Within the first 24 hours of recovery, the surviving hair cells in the damaged region begin to expand their surface areas once again and the supporting cell surfaces contract. In addition, the supporting cells begin to secrete new matrix for regenerating the tectorial membrane that was damaged by the acoustic trauma (Cotanche 1987c). The blebbed hair cells are removed from the apical surface of the basilar papilla by 48 hours of recovery and the supporting cells spread out to fill in the remaining surface area. The first evidence of regenerating hair cells can be seen as early as 48 hours of recovery among the supporting cells in the area where the blebbed hair cells had been. After five days of recovery (Fig. 1b) the surface areas of the hair cells that survive the noise damage expand again, sometimes even exceeding their pre-exposure size. In turn, the adjacent supporting cell surfaces contract once more to form thin rims around the hair cells. The small but prominent surfaces of the new hair cells are numerous at this stage of recovery, especially along the superior edge of the damaged region. From five to 14 days of recovery, the new hair

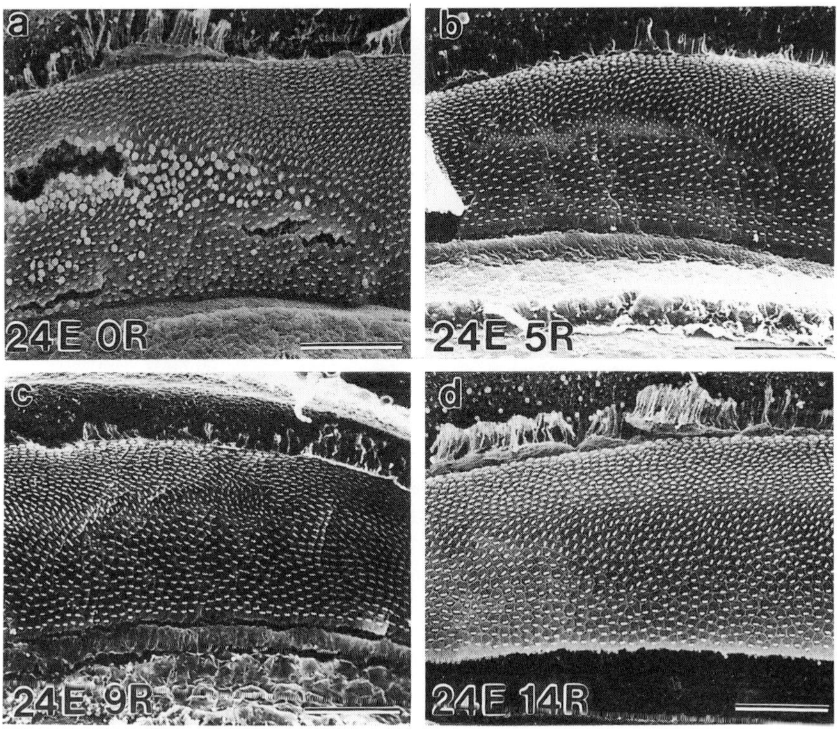

FIG. 1. Scanning electron micrographs illustrating recovery of the damaged region of the chick basilar papilla after a 24 hour exposure (24E) to a 1500 Hz tone at 120 dB SPL. (a) No recovery (0R); (b) five days of recovery (5R); (c) nine days of recovery (9R); (d) 14 days of recovery (14R). Bars, 100 μm.

cells expand their surfaces and the supporting cells around them contract, so that by 14 days the area of noise damage is barely identifiable on the basilar papilla (Fig. 1b,c,d). Even at 14 days of recovery, however, the new hair cell surfaces are not as large as normal cells in this region of the epithelium (data not shown).

As the new hair cells grow and expand in surface area, the stereociliary bundles on their surfaces undergo a process of maturation that is identical to that seen in the embryonic chick cochlea (Cotanche 1987a, Tilney et al 1986, 1988). When the new hair cells first appear, they have small stereocilia that are the same height as but slightly thicker in diameter than the microvilli on adjacent supporting cells. These new bundles also have a single, eccentrically placed kinocilium. After four days of recovery, the row of stereocilia closest to the kinocilium begins to grow in height. A second row begins to grow when the first row reaches the proper interval in height over the remainder of the bundle. These two rows then grow at equivalent rates until the second row reaches the proper height interval

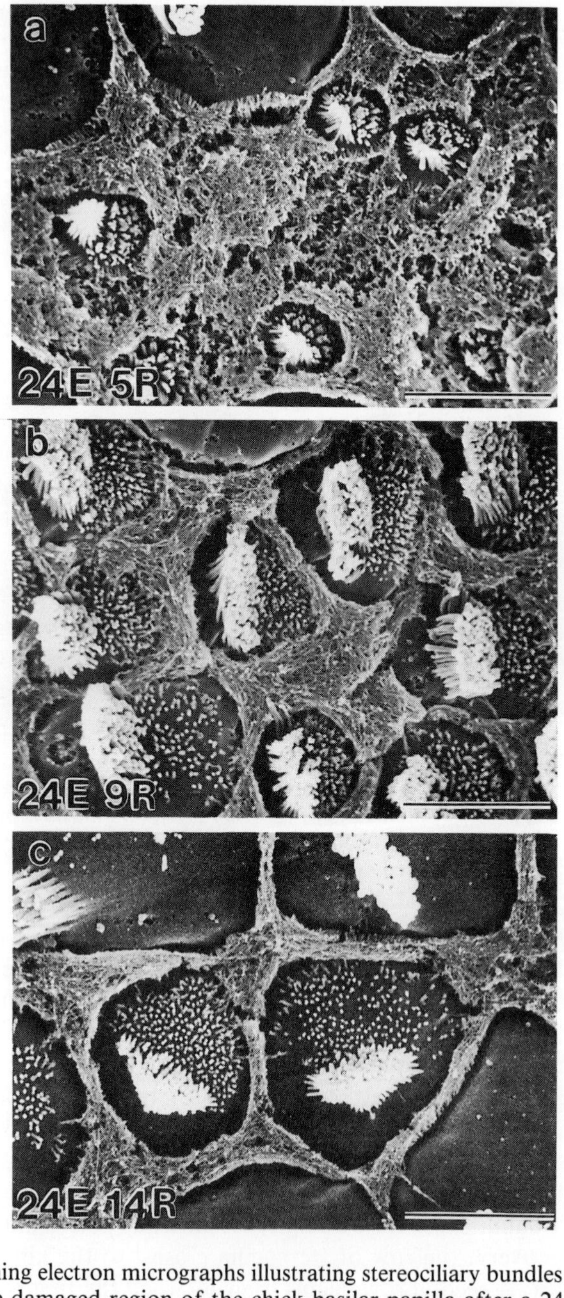

FIG. 2. Scanning electron micrographs illustrating stereociliary bundles of regenerating hair cells in the damaged region of the chick basilar papilla after a 24 hour exposure (24E) to a 1500 Hz tone at 120 dB SPL. (a) five days of recovery (5R); (b) nine days of recovery (9R); (c) 14 days of recovery (14R). Bars, 5 μm.

over the remaining bundle (Fig. 2a). This process continues until a bundle of 7 to 10 rows of stereocilia in a staircase-like gradient of heights is established (Fig. 2b). The rest of the short stereocilia on the cell surface that are not used to make the bundle are eventually resorbed into the hair cell (Fig. 2b,c). The remaining bundle will then grow as a unit until it reaches the proper height for its location on the basilar papilla (Fig. 2c). The recovering basilar papilla differs from the embryonic one in that the new hair cells have to establish the proper number and height of the stereocilia in their bundle relative to the existing hair cells around them. This suggests that information needed by the developing cells is derived from some positional cues continuously present in the cochlea, rather than from a transient gradient in a diffusible morphogen during a critical developmental period (Wolpert 1978).

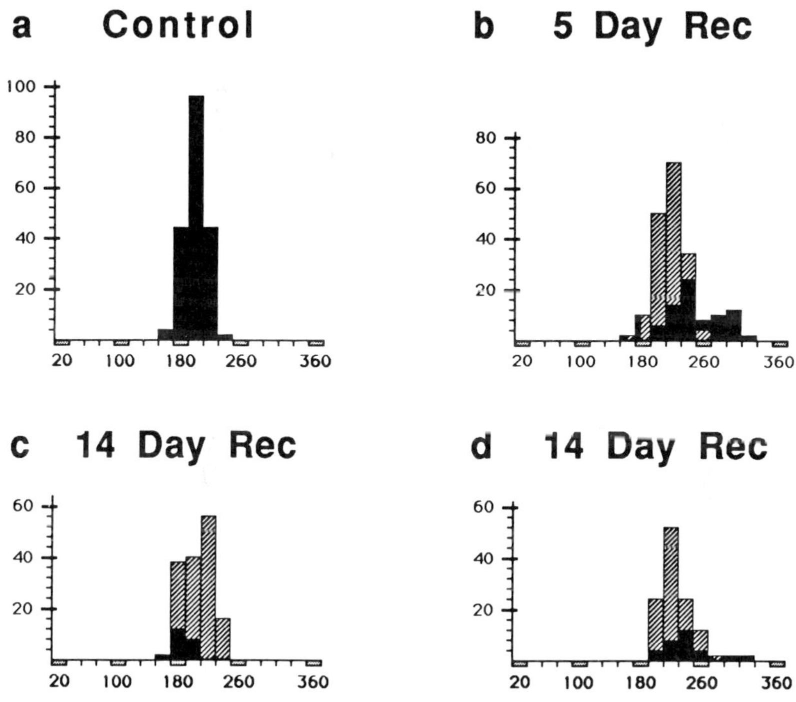

FIG. 3. Histograms of the orientations of the hair cells in the region of the regenerating cells. (a) Orientation in an equivalent region of a control chick; (b) orientation after five days of recovery; (c) orientation after 14 days of recovery in an ear which had a significant amount of recovery (i.e. Figs. 1d and 4b); (d) orientations after 14 days of recovery in an ear which still looked very abnormal (i.e. Fig. 4d). In b, c and d the black columns represent the orientations of the regenerating hair cells, while the cross-hatched columns represent the orientations of the surviving hair cells that surround the new ones. The ordinate indicates the number of cells at each orientation and the abscissa indicates the degrees in the circumference of a circle (see p 132).

The regenerating stereociliary bundles undergo a second aspect of maturation that is similar to the embryonic process. The bundles of newly differentiated hair cells in the embryo exhibit a broad range of orientations, as determined by the location of the kinocilium and tallest row of stereocilia. The initial orientations change as the hair cells mature, so that the mature basilar papilla has a very precise pattern of hair cell orientations. It has been proposed that these orientation changes are the result of interactions between the growing hair cells and the developing tectorial membrane (Cotanche & Corwin 1991). The regenerating hair cells in the recovering basilar papilla initially differentiate with a wide range of orientations (compare Fig. 3a and b). As they mature, their orientations shift to become more in line with those of the surrounding cells and with controls (Fig. 3c). Since tectorial membrane regeneration parallels that of hair cells (Cotanche 1987c), it is again possible that the shift in orientations is caused by interactions between the hair cells and the tectorial membrane.

Relationship between exposure and recovery

Previously, we studied the effect of duration of exposure on the extent of hair cell damage in chick cochleas fixed immediately after exposure (Cotanche & Dopyera 1990; see section on response to acoustic overstimulation, above). In a more recent series of experiments we have explored the relationship between the length of exposure and the extent of recovery and regeneration. Chicks were exposed for either 12 or 24 hours and the amount of damage was assessed either immediately after exposure or after 14 days of recovery. Our results on the animals examined immediately after exposure are consistent with our previous study (Figs. 1a and 4a). However, both the 12 hour and 24 hour exposure groups show some variability in the extent of damage. The 12 hour exposure groups range between very little damage with only limited hair cell loss to more extensive hair cell loss, especially along the superior edge of the damaged region. The 24 hour exposure groups range from moderate damage with hair cell loss only at the superior edge of the damaged region to severe damage with extensive hair cell loss throughout the damaged region. Comparisons between the two groups indicate that the most damaged 12 hour-exposed chicks resemble the least damaged 24 hour-exposed ones. More significantly, some animals in both groups exhibit differences in the extent of damage between the left and right ears.

When the animals exposed for 12 hours are allowed to recover for 14 days, the extent of recovery in all animals is quite extensive and consistent. In fact, it is often very hard to determine that they had been damaged at all (Fig. 4b). There are only a small number of identifiable new hair cells and the mosaic pattern of the hair cell surfaces appears normal. When the surface areas of hair cells within the damaged region are measured, however, the recovered hair cell surfaces are smaller than control values and only slightly larger than the surfaces immediately after exposure (Fig. 5a). Since we have not yet done a detailed study

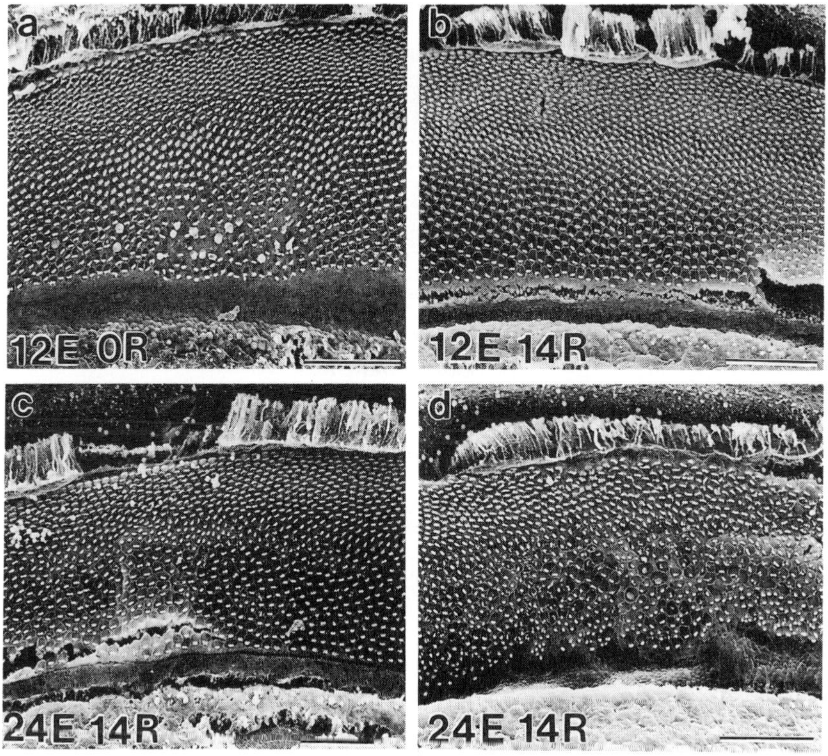

FIG. 4. Scanning electron micrographs illustrating recovery to the damaged region of the chick basilar papilla after 12 hours of exposure (12E) and 24 hours of exposure (24E). (a) 12 hours of exposure, no recovery period (12E, 0R); (b) 12 hours of exposure, 14 days of recovery (12E, 14R); (c) 24 hours of exposure, 14 day recovery (24E, 14R) showing a fairly high degree of recovery; (d) 24 hours of exposure, 14 day recovery (24E, 14R) showing an ear which is still abnormal in appearance. Bars, 100 µm.

of the intermediate stages of recovery in the 12 hour exposure group, the recovery at 14 days can be interpreted in two ways. It is possible that the regenerated hair cells have matured rapidly, so that we cannot identify them by their immature stereociliary bundles, yet they still have surface areas smaller than normal. Alternatively, the intitial damage may have stimulated only a limited amount of regeneration and the surviving cells have not regained their normal surface areas. Further experiements are under way to resolve this.

The animals exposed for 24 hours and allowed to recover for 14 days show a wide range in the extent of their recovery. In some, the mosaic pattern of the hair cells in the damaged area and the stereociliary bundles on surviving hair cells appear normal, although several regenerated hair cells are clearly identifiable (Figs. 1d and 4c). In other ears, the mosaic pattern remains severely

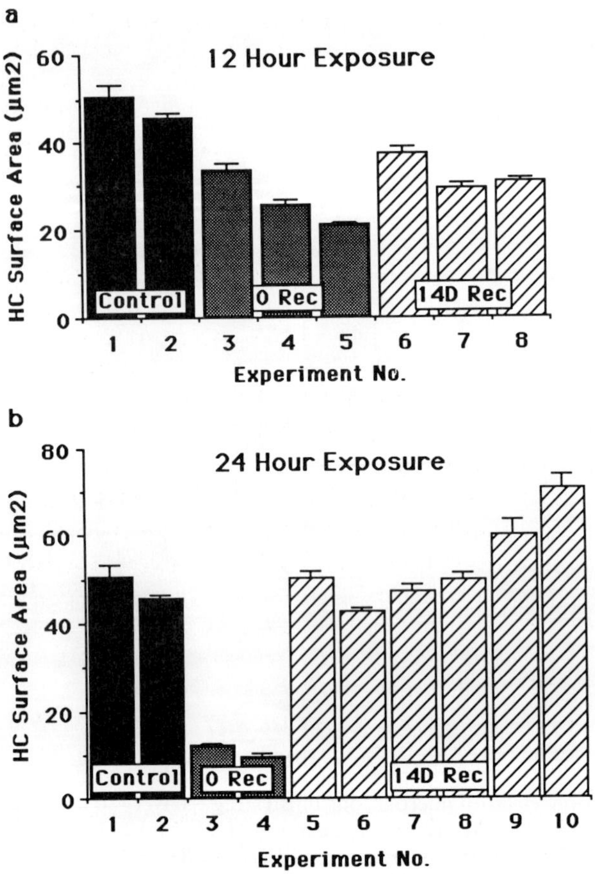

FIG. 5. Histograms showing the changes in the surface areas of surviving hair cells (HC) in the damaged region after a 12 hour exposure (a) and a 24 hour exposure (b). Each bar represents the mean value (±SEM) of 10 to 30 cells within the damaged area of an individual chick. Control values were obtained from equivalent areas of age-matched chicks not exposed to the pure tone.

disrupted and a large number of cells have damaged or abnormal stereociliary bundles (Fig. 4d). There are also a large number of regenerated hair cells in these cochleas. As in the zero-recovery group, some animals in this group show differences between the left and right ears. Surface area measurements of the hair cells in the 24 hour exposure group indicate that the surviving cells have very constricted surface areas immediately after the exposure, but by 14 days of recovery the surface areas have recovered to normal levels or beyond (Fig. 5b). It is important to note that the surface areas that resemble normal values are from ears that show more complete recovery, while the areas that are larger than normal are from cochleas which remain abnormal in appearance.

The most obvious, and not too surprising, conclusion from these experiments is that a 12 hour sound exposure produces less damage and leads to a more complete recovery than a 24 hour exposure. However, there are some other subtle but equally important conclusions. (1) There are significant variations in the extent of damage and recovery in all groups except the 12 hour exposure, 14 day recovery group. This group always exhibits fairly complete recovery, even though the initial extent of damage could vary. The variability in the other three groups between animals and even within the same animal suggests that studies which attempt to relate the extent of functional loss and recovery with levels of hair cell loss and regeneration should be interpreted with caution. (2) There appears to be a series of thresholds in the extent of recovery of the hair cell surfaces. A moderate level of damage (such as a 12 hour exposure) can lead to a fairly complete structural recovery of the basilar papilla. A more severe level of damage (such as the worst of the 24 hour exposures) can lead to recovery, but certainly an abnormal one, at least after only 14 days of recovery. It is not yet known whether this abnormal recovery will improve if given more time. The most severe level of damage (such as produced by some 48 hour exposures to 120 dB SPL or exposures to 125 dB SPL) destroys both the hair cells and the supporting cells in the damaged region of the basilar papilla, and the sensory epithelium is replaced by a thin layer of simple cuboidal epithelium which covers the basilar membrane. There is no recovery or regeneration in either the hair cells or the tectorial membrane in this region.

In future studies it will be important to try to relate these thresholds in the extent of structural recovery to the processes that stimulate supporting cells to re-enter the proliferative phase of their cell cycle and to the mechanisms which lead to the recovery of auditory function. Moreover, our studies have provided only a limited spectrum of observations on the regeneration of the apical surfaces of the hair cells. To clearly understand all the ramifications of regeneration, it will be necessary to study the entire hair cell and its connection to the auditory nerve and higher auditory pathways.

Acknowledgements

We would like to thank Jenny Stone for her comments on and critical reading of earlier versions of this manuscript. This work has been supported by a Deafness Research Foundation grant, a Boston University Biomedical Research Support grant, and National Institutes of Health/National Institute on Deafness and Other Communication Disorders grant # DC00412.

References

Corwin JT, Cotanche DA 1988 Regeneration of sensory hair cells after acoustic trauma. Science (Wash DC) 240:1772–1774

Cotanche DA 1987a Development of hair cell stereocilia in the avian cochlea. Hear Res 28:35–44
Cotanche DA 1987b Regeneration of hair cell stereociliary bundles in the chick cochlea following severe acoustic trauma. Hear Res 30:181–96
Cotanche DA 1987c Regeneration of the tectorial membrane in the chick cochlea following severe acoustic trauma. Hear Res 30:197–206
Cotanche DA, Corwin JT 1991 Stereociliary bundles reorient during hair cell development and regeneration in the chick cochlea. Hear Res 52:379–402
Cotanche DA, Dopyera CEJ 1990 Hair cell and supporting cell response to acoustic trauma in the chick cochlea. Hear Res 46:29–40
Cotanche DA, Saunders JC, Tilney LG 1987 Hair cell damage produced by acoustic trauma in the chick cochlea. Hear Res 25:267–286
Cruz RM, Lambert PM, Rubel EW 1987 Light microscopic evidence of hair cell regeneration after gentamicin toxicity in chick cochlea. Arch Otolaryngol Head & Neck Surg 113:1058–1062
Girod DA, Duckert LG, Rubel EW 1989 Possible precursors of regenerated hair cells in the avian cochlea following acoustic trauma. Hear Res 42:175–194
Henry WJ, Makaretz M, Saunders JC, Schneider ME, Vrettakos P 1988 Hair cell loss and regeneration after exposure to intense sound in neonatal chicks. Otolaryngol Head Neck Surg 98:607–611
Hirokawa N, Tilney LG 1982 Interactions between actin filaments and between actin filaments and membranes in quick-frozen and deeply etched hair cells of the chick ear. J Cell Biol 95:249–261
McFadden EA, Saunders JC 1989 Recovery of auditory function following intense sound exposure in the neonatal chick. Hear Res 41:205–216
Ryals BM, Rubel EW 1988 Hair cell regeneration after acoustic trauma in adult *Coturnix* quail. Science (Wash DC) 240:1774–1776
Tilney LG, Egelman EH, DeRosier DJ, Saunders JC 1983 Actin filaments, stereocilia and hair cells of the bird cochlea. II. The packing of actin filaments in the stereocilia and in the cuticular plate and what happens to the organization when stereocilia are bent. J Cell Biol 96:822–834
Tilney LG, Tilney MS, Saunders JC, DeRosier DJ 1986 Actin filaments, stereocilia, and hair cells of the bird cochlea. III. The development and differentiation of hair cells and stereocilia. Dev Biol 116:100–118
Tilney LG, Tilney MS, Cotanche DA 1988 Actin filaments, stereocilia and hair cells of the bird cochlea. V. How the staircase pattern of stereociliary lengths is generated. J Cell Biol 106:355–365
Tucci DL, Rubel EW 1990 Physiological status of regenerated hair cells in the avian inner ear following aminoglycoside ototoxicity. Otolaryngol Head Neck Surg 103:443–450
Wolpert L 1978 Pattern formation in biological development. Sci Am 239:154–164

DISCUSSION

Rubel: In your study on regeneration of the tectorial membrane (Cotanche 1987), what is the longest time you have looked after noise damage?

Cotanche: Twenty-eight days. It looks exactly the same then as at 10 days of recovery; you still do not have the lateral bars across the short hair cells. We see just a honeycomb mosaic pattern surrounding each hair cell.

Rubel: You haven't extended the period of recovery to several months?

Cotanche: No.

Lewis: Did the regenerated hair cells have just the right kinds of stereociliary bundles for their position in the cochlea?

Cotanche: Yes. A cochlear hair cell has a defined number of stereocilia and a precise height for its position in the cochlea. In the noise-damaged 1500 Hz region, hair cells grow back with their proper number and roughly proper height, for that position. They can therefore reproduce their morphology for their position in the cochlea; thus whatever gives them that information embryonically has still to be around many months later, because it then again provides this information to the cells when they regenerate.

Oakley: The hair cells reorient as well, as you described. Do you have any sense of how that process occurs?

Cotanche: The most logical idea is that as the tectorial membrane develops, growing across, and as the basilar papilla expands, the tectorial membrane pulls those hair cells into alignment. That is our hypothesis.

Pujol: But in your regenerating preparation, presumably you don't have the tectorial membrane; so what is acting there?

Cotanche: We do have the regenerating tectorial membrane growing back over the short hair cells in our regenerating cochleas.

Pujol: Do you mean that this regenerating tectorial membrane could play a role in aligning the new sensory cells?

Cotanche: Yes, it can. After the loss of hair cells in the damaged region, the supporting cells contract and the width of the basilar papilla is reduced. As the new hair cells grow in size and the tectorial membrane is regenerated, the basilar papilla expands in width once again. The expansion in width of the basilar papilla pulls on the tectorial membrane and this, in turn, pulls on the stereociliary bundles of the new hair cells to reorient them, just as it would in the embryo.

Farbman: We studied the reorientation of cilia in a non-sensory epithelium, the respiratory epithelium in the nasal cavity of the fetal mouse. In the differentiated condition, all cilia are structurally and functionally oriented so that they beat in the same direction, thus moving the surface mucus in that direction. Morphologically, this is reflected by directional orientation of central microtubules and basal feet. At the earliest stage of cilia formation in development, both of these morphological elements are totally disoriented. At some point, they become directionally aligned (Frisch & Farbman 1968). There is no structural equivalent of a tectorial membrane in the nasal epithelium, or anywhere else in the respiratory tract. Ciliated cells in the respiratory tract, and probably in the oviduct, seem to become aligned by some as yet unknown mechanism.

Pujol: This reminds me that in the mammalian cochlea, during normal development, there is an alignment of hair cells and of the stereocilia before these cells are covered by the tectorial membrane (Lenoir et al 1987, Lavigne-Rebillard & Pujol 1987).

Cotanche: This is not seen in the developing chick. I haven't studied the orientation of stereociliary bundles in a mammal.

Fernald: Have you considered what role the acoustic environment might play in the regrowth of hair cells? When the hair cells are allowed to regrow, have you looked at what kind of conditions you hold them in, and have you tried putting them in complete silence, for example?

Cotanche: I have not tried that. The chicks are in a fairly quiet room, but I can't control for the ambient noise level.

Rubel: It's very difficult to put animals in complete silence! In addition, they make quite a bit of noise themselves.

Fernald: I am wondering whether the sensory feedback might be facilitating this kind of reorientation process in particular.

Cotanche: This has to be considered. The other point is that in the embryo, this differentiation all occurs long before the chicks are able to hear. This suggests that the sound environment is not playing a role in establishing the gradients in hair cell morphology or orientation.

Steinberg: Has anyone tried to enhance the numbers of hair cells that survive the acoustic trauma, using survival factors, or by elevating body temperature to produce heat-shock proteins?

And secondly, is anything known about what is killing these cells, at the cellular level? Is it an increase in intracellular calcium, or what?

Cotanche: That has been a long-standing argument in relation to noise damage in particular—is it metabolic exhaustion, or some sort of ionic imbalance, or just the physical trauma of being bashed up and down for 48 hours at a very high intensity that kills the hair cells? I don't know.

Steinberg: What about ways of increasing cell survival?

Cotanche: I haven't thought of that; I have tried to stop cell survival, but not to enhance it.

Steinberg: One could try to protect the hair cells from dying, by using growth factors.

Cotanche: In fact, Canlon et al (1988) and Dr Y. Raphael (personal communication) have tried exposing the ears of guinea pigs and chickens, respectively, to a lower sound frequency and level, then re-exposing them, after a short recovery, to a higher frequency. They find much less hair cell loss after this pre-exposure of the cochlea to a different frequency, followed by a louder, higher-pitched sound, than with only the second sound.

Rubel: This is what is known as a 'toughening' phenomenon. The cell mechanisms whereby this happens are not understood.

With regard to mechanisms of noise damage, it's safe to assume that at least two processes are going on. First, cells are being forced out of the epithelium in what is clearly a mechanical problem. With high enough sound levels you almost always see some component of noise damage, right under the position of the columella, where a lot of mechanical force is being exerted.

Cotanche: The junction of the basilar membrane and superior fibrocartilaginous plate is where the basilar membrane moves up and down. The cells above that become traumatized and just one or two cells are lost, right above the joint. That may be purely mechanical motion crushing the cells.

Rubel: There is also something that goes on beyond this. After the noise trauma you can often observe a wave of hair cell loss that lasts a couple of days. That wave of hair cell loss may be at a very different frequency location from the first, mechanical damage. So, at least two phenomena are going on.

Cotanche: I should add that the hair cell loss appears to be over by 48 hours after the end of noise exposure. It's recovery from then on. All the hair cells that are going to die seem to have died by 48 hours after termination of the exposure. I don't see more 'blebbed out' hair cells after 48 hours, certainly.

Farbman: Are chick hair cells contractile? Do they contract on stimulation? In the mammal, the hair cells contract on stimulation.

Cotanche: Nobody has been able to show contraction of chick hair cells in *in vitro* conditions.

Farbman: Maybe there is a hypercontraction many times over, during the period of noise damage, and this might explain why the cells are dying, in these chicks.

Lewis: The calcium hypothesis would fit in quite neatly here, because you expect a calcium influx during transduction, and this influx might explain the contraction of the hair cells seen in mammals; if a lot of calcium was flowing into the noise-damaged hair cells, this would be likely to activate contraction of the cuticular plate.

Cotanche: It does appear that the cuticular plate must contract the surface area of the hair cell, and an influx of calcium could drive that. We need to quantify it and see if there is a change in the size of the cuticular plate. In normal chicks the cuticular plate covers the whole surface. After 24 hours of sound exposure, the cuticular plate is all at the middle of the hair cell and not out at the edge, so the cuticular plate must be depolymerizing, or contracting.

Corwin: There's an anti-parallel, bidirectional band of actin filaments just internal to the zonula adherens of these hair cells, so if there were a sliding filament mechanism operating, as proposed by Hirokawa & Tilney (1982), that might underlie the kind of contraction that you have reported.

Rubel: We are talking about different kinds of contraction. Al Farbman was referring to the kinds of slow and fast contractions in mammalian outer hair cells, first described by Brownell et al (1985), that may be related to the non-linearity of the inner ear. How these contractions might be related to the actual death of the hair cells is difficult to imagine. The other contraction being talked about by Dr Cotanche is some sort of slow contractility over minutes or hours, and we don't understand the mechanism there.

Corwin: Doug Cotanche also found that in some overstimulated regions of the chick cochlea, the hair cells have severe damage to their ciliary apparatus, but after some days the damaged-but-living hair cells have not been replaced by the type of regeneration that would have replaced dead hair cells. All the different mechanisms of inducing hair cell regeneration—including metabolic poisoning, phototoxicity, aminoglycoside poisoning, acoustic overstimulation and laser ablation—have the common feature of causing the *extrusion* of the damaged hair cells from the epithelium. In our laser ablation studies it appeared that relatively intact cells were being extruded, because the extruded hair cells outside the neuromast still contained DASPEI, in experiments where we prelabelled them with this cationic dye (in the cytoplasm or mitochondria) before the laser treatment. Therefore, extrusion of the hair cell may be a necessary event for triggering the regenerative response. If the cell is damaged, and possibly even non-functional, but remains in place, then that cell does not get replaced; there is no regeneration. There are indications that cell extrusion doesn't happen as frequently in mammalian ears after aminoglycoside poisoning. Dr Forge (1985) has shown that the mammalian reticular lamina seems to stay sealed up so there isn't an immediate extrusion of hair cells after aminoglycoside treatment, as there is in chicks.

Ryals: What provokes the extrusion?

Cotanche: We don't know if it's the hair cell contracting or the supporting cell pushing it out of the sensory epithelium. The hair cell doesn't just pop out; there is no hole left in the sensory epithelium. The supporting cell squeezes under the blebbing hair cell and closes up the surface of the sensory epithelium; whether that can actively push that hair cell out, I don't know.

Calof: If the supporting cells are the progenitors, and if you kill all the supporting cells, there is a whole chain of events that can occur, and you are left with this cuboidal epithelium and nothing that's regenerative, with your severe damage.

Steinberg: With epithelia, when you remove a cell, the cells around it will expand to fill the space.

Corwin: But after laser treatment, which is really the only treatment in current use where we can know exactly when we killed the cells, we do see a big hole in the neuromast for about four hours. What happens is that the hair cells are extruded within 15 minutes, leaving a large hole in the centre of the neuromast. Over the next four hours the remaining supporting cells jostle against each other and, as they move, the epithelium flattens down, so that the large cavity disappears over that period. When you overstimulate acoustically, for 18 to 48 hours, it's harder to say that there isn't actually a hole at some time after extrusion. We can't know how much time elapsed between the time when extrusion occurred and the time when the tissue was fixed.

Cotanche: There was a theory that as hair cells are lost, holes are left temporarily in the reticular lamina of the mammalian cochlea after noise damage. I believe that view is no longer held.

Rubel: Unfortunately, most of those interested in hair cell damage in the mammal have not looked at very short times after noise damage, or during noise damage.

Are we confident that in all cases of hair cell damage in chicks the whole hair cell is being extruded, as opposed to the apical surface of the cell being extruded?

Cotanche: I can be certain of this in my work in the chick, because, before we did autoradiography, I had to make sure that the new stereocilia bundles weren't forming on old cells. I spent a lot of time doing transmission electron microscopy, making sure that no hair cell left its nucleus and body below the epithelial surface, only to regrow a new bundle during recovery. I concluded that there are no 'half' hair cells under the surface of the epithelium during the early stages of recovery. But Cousillas & Rebillard (1988) have shown some strange cells in the damaged region of the chick basilar papilla after noise exposure; these could be the remains of hair cells, but I haven't seen many of these kind of cells in my noise trauma studies.

Pujol: You mean giant cells with two nuclei, as also shown by these authors?

Cotanche: No. Cousillas & Rebillard (1988) showed large cells with a flat surface and a single nucleus, but no stereocilia. They do not look like normal supporting cells. They are either atypical supporting cells, or nude hair cells.

Ryals: Might these be the giant hair cells you have seen after acoustic trauma?

Cotanche: Most of the noise-exposed cochleas that I have looked at in great detail by TEM have hair cells that aren't really traumatized, to the extent where one sees these unusual cells; that's probably why I haven't seen many of them. If I look at the papillae with a very distorted epithelium, I see all sorts of strange cells. But in the cochleas that appear to have just lost their hair cells and the supporting cells have expanded, we don't see many strange cells, in the upper part of the epithelium.

Ryals: But we have both seen whole, round cells with stereocilia, floating in the endolymph.

Cotanche: A strange thing I saw once was a large cell with two stereociliary bundles on it, in opposite orientations. I think Cousillas & Rebillard (1988) showed that this was actually two hair cells that had fused. So cell fusion is a possible response to the trauma.

Corwin: In the laser-treated lateral line system of the axolotl salamander, the hair cells are removed within 15 min and they always appear to be intact. And it's always within 15 min in the chick cochlea after laser treatment in Dr Mark Warchol's continuing experiments. Matt Kelley has used similar laser treatments in the mammalian organ of Corti, *in vitro*, and the whole cell also comes out, which surprised us in light of Dr Forge's (1985) findings relating to the response to aminoglycoside poisoning. From the first, the extrusion response that was brought on by laser treatment was surprising, because all we had done was to 'bake' the DNA inside the cell's nucleus, and yet the whole cell was extruded from the sensory epithelium within minutes.

Cotanche: Has extrusion of hair cells been seen after gentamicin treatment?

Rubel: Yes, I think so. We see a lot of debris above the hair cells, but the tissues are so distorted in the lumen that it's hard to say if they are whole cells.

Fernald: It seems to me that with the beautiful mapping of a specific lesion that then recovers it would be good if someone could do behavioural studies, or some kind of audiograms, to see whether this corresponds to real recovery of function.

Ryals: We have not done behavioural studies in birds, but I have examined VIIIth nerve evoked potentials after noise exposure. We find an initial 30–50 dB threshold shift which recovers to within 10 dB of normal within 2–3 days. McFadden has reported similar results (McFadden & Saunders 1989). So evoked potential thresholds are nearly normal, long before the regenerated hair cells are morphologically mature. This probably means that this technique is an inadequate measure of new hair cell function.

Rubel: Exactly. We've been trying to get at that with aminoglycoside ototoxicity and oto-acoustic emissions. Evoked potentials are a poor technique for determining frequency specificity. With noise trauma you are also confounding regrowth of hair cells with regrowth of tectorial membrane. That combination is inadequate for attacking the problem of when, and how selectively, the new hair cells are functioning.

However, with aminoglycosides, you get a wave of damage spreading from base to apex, which makes it very hard. Susan Norton and I have been studying oto-acoustic emissions; these are frequency-specific measures which are thought to reflect the sensitivity and tuning selectivity in the inner ear. This should be a more frequency-specific way of looking at functional properties after regeneration. In fact, the correspondence between the EM observations and the functional recovery using oto-acoustic emissions looks much better than what is observed using brainstem evoked potentials.

Powers: The problem with all these physiological techniques is not knowing what the animal perceives!

Rubel: Very true!

Pujol: Dr Cotanche, you talked in terms of 'levels' of recovery, but you were comparing the same time of recovery with a different time of exposure (e.g. 12 hours of exposure giving complete recovery in 14 days, but 24 hours of exposure leading to a range in the extent of recovery then). You referred to 'fairly complete' recovery and also 'abnormal' recovery, at 14 days. If you looked at a later group after a longer period, perhaps you would find less difference?

Cotanche: Those 'time windows' were very narrow. Probably, if we waited long enough, we would be able to reach a fairly normal recovery, except for the most severe damage, where there is total destruction of the sensory epithelium.

Ryals: You said that very loud stimuli (125 dB) destroyed all the cells on the basement membrane, including all the support cells. I have done that experiment too. What do you think the cells are that re-form the epithelial layer? They are not support cells, are they?

Cotanche: They could be support cells that are so traumatized that they have shrunk down and really regressed. We just don't know.

Ryals: They look like the cuboidal cells on the inferior edge.

Cotanche: They aren't, because the hyaline cells are still totally intact structurally, with their full cytoskeletal components; so unless the cuboidal cells pass through the epithelial layer without touching the hyaline cells, it's not the cuboidal cells re-forming the epithelial layer over the basilar membrane.

Burd: I've seen the same picture with the application of zinc sulphate to the olfactory epithelium in mice (Burd 1991). Zinc sulphate causes the olfactory epithelium to coagulate and slough off the underlying lamina propria. Two to four days later, the olfactory epithelium is replaced by a new, cuboidal epithelium that is quite different from the previous sensory epithelium. I did not determine the origin of these cells.

Calof: Are they the progeny of the remaining basal cells?

Burd: No, unless they have changed their structure dramatically. The size of these new cuboidal cells is 2–3 times that of the horizontal basal cells.

Cotanche: But the cell is also traumatized, so the nucleus may be responding to damage by a change in size.

Burd: It could be, but within one or two days after zinc sulphate administration, the olfactory epithelium separates completely from the underlying lamina propria. Not enough cells from the sensory epithelium remain to re-form this new, cuboidal epithelium.

Reasner: Perhaps these cuboidal cells which you see after zinc sulphate treatment originate from the lining of the Bowman's glands?

Burd: I think that's where they are coming from. Matulionis (1976) suggested that the new cells might be derived from the Bowman's glands.

Cotanche: There are some cells in the basilar membrane underneath the cochlea, but those are fibroblasts, and if they are re-forming the epithelial layer, it would mean fibroblasts jumping into the epithelium. I don't think that's it. I would imagine, from Doug Girod's work (Girod et al 1989), that some cells dedifferentiate, and it seems as if those cells are dividing to cover the basilar membrane but may not be differentiating.

Rubel: Clearly, these are tympanic border cells, but I don't think they are fibroblasts jumping into the epithelium.

References

Brownell WF, Bader CR, Bertrand D, de Ribaupierre Y 1985 Evoked mechanical responses of isolated cochlear outer hair cells. Science (Wash DC) 227:194–196

Burd GD 1991 Morphological effects of intranasal zinc sulfate irrigation on the mouse olfactory epithelium and olfactory bulb. J Electron Microsc Tech, in press

Canlon B, Borg E, Flock Å 1988 Protection against noise trauma by pre-exposure to a low level acoustic stimulus. Hear Res 34:197–200

Cotanche 1987 Regeneration of the tectorial membrane in the chick cochlea following severe acoustic trauma. Hear Res 30:197–206

Cousillas H, Rebillard G 1988 Morphological transformation of hair cells in the chick basilar papilla following an acoustic overstimulation. Hear Res 32:117–122

Forge A 1985 Outer hair cell loss and supporting cell expansion following chronic gentamicin treatment. Hear Res 19:171–182

Frisch D, Farbman AI 1968 Development of order during ciliogenesis. Anat Rec 162:221–232

Girod DA, Duckert LG, Rubel EW 1989 Possible precursors of regenerated hair cells in the avian cochlea following acoustic trauma. Hear Res 42:175–194

Hirokawa N, Tilney LG 1982 Interactions between actin filaments and membranes in quick-frozen and deeply etched hair cells of the chick ear. J Cell Biol 95:249–261

Lavigne-Rebillard M, Pujol R 1987 Surface aspects of the developing humn organ of Corti. Acta Oto-laryngol Suppl 436:43–50

Lenoir M, Puel J-L, Pujol R 1987 Stereocilia and tectorial membrane development in the rat cochlea. A SEM study. Anat Embryol 175:447–487

Matulionis DH 1976 Light and electron microscopic study of the degeneration and early regeneration of olfactory epithelium in the mouse. Am J Anat 145:79–100

McFadden EA, Saunders JC 1989 Recovery of auditory function following intense sound exposure in the neonatal chick. Hear Res 41:205–215

Regeneration of lateral line and inner ear vestibular cells

Jørgen Mørup Jørgensen

Department of Zoophysiology, Building 131, University of Aarhus, DK-8000 Aarhus C, Denmark

Abstract. Labelling experiments with [^{3}H]thymidine demonstrate a continuous production of cells in the mechanoreceptive lateral line organs of the eel (*Anguilla anguilla*) and butterfly fish (*Pantodon buchholzi*) as well as in the electroreceptive ampullary organ of the transparent catfish (*Kryptopterus bicirrhus*). Shortly after [^{3}H]thymidine injection many cells are labelled in the middle and basal parts of the sensory organ and after a few days' survival sensory cells are also labelled. The vestibular sensory organs of selected species of fishes, amphibians, reptiles and birds also show a continuous production of cells. In the budgerigar (*Melopsittacus undulatus*) labelled cells are found in the basal and middle layer of the sensory epithelium a few hours after injection with [^{3}H]thymidine. A few days after the injection labelled cells are found in non-calyceal hair cells. After one month the calyceal cells are also labelled. Similar experiments with the bat *Pipistrellus nathusii* and with normal and gentamicin-treated mice (*Mus musculus*) show no labelled cells in the inner ear sensory epithelia. The lateral line organs and vestibular epithelia of non-mammalian vertebrates all contain a small number of dark cells with the characteristics of apoptotic cells. Macrophages and inclusions in some cells, thought to be remnants of apoptotic cells, are occasionally seen. Fixation at different osmolarities has little effect on the number of dark cells. It is suggested that the continually produced cells replace apoptotic dying cells.

1991 Regeneration of vertebrate sensory receptor cells. Wiley, Chichester (Ciba Foundation Symposium 160) p 151–170

The octavolateralis organs of vertebrates consist of the lateral line organs and the inner ear. The lateral line organs may be divided into the mechanoreceptive neuromasts and the electroreceptive sensory organs. The neuromasts are situated superficially or in canals in lampreys, elasmobranchs and teleost fishes, and water-living amphibians. The electroreceptors are found as ampullary organs and tuberous organs in the skin of lampreys, other fishes, and water-living urodeles and apodans.

The inner ear may be divided into a *pars superior*, consisting of the semicircular canals, each with a crista, and the utricular macula; and a *pars inferior*, where the saccular and lagenar macula, and, in tetrapods, the basilar papilla, are situated.

Cell divisions are not seen frequently in the lateral line organs and are extremely rare in the inner ear of adult vertebrates. It was long believed that the octavolateralis organs in vertebrates with indeterminate growth may increase in size by the addition of new elements along the borders of the sensory epithelium. In vertebrates with determinate growth—birds and mammals—the formation of new elements was supposed to stop shortly after birth. In the following it will be shown that the picture is somewhat more complicated.

Formation of new cells in octavolateral epithelia

The production of new elements in animal tissue can be inferred from the presence of mitoses, and of a gradual series of differentiating cells leading to the mature cells. Experimentally, mitosis can be arrested by colchicine, vinblastine and related agents. Incorporation of tritiated thymidine in nuclei during synthesis of DNA prior to a mitosis indicates the possibility of mitosis, but is no proof of the formation of mature cells, because the cells with labelled nuclei may be eliminated before complete differentiation. Labelling of fully differentiated cells confirms the formation of new cells. All criteria for the formation of new sensory elements of octavolateralis organs are fulfilled for representatives of non-mammalian vertebrates, and will be discussed below.

During embryological development, hair cells of the inner ear first form a short apical cilium and microvilli. Later the cilium elongates and some of the microvilli take on the shape of slender stereovilli. After achieving their final height the stereovilli increase their diameter to the final size (Tilney & Tilney 1984). In postlarval fishes and amphibians, immature hair cells are seen along the borders of the sensory epithelia (Lewis & Li 1973, Corwin 1985).

Cell death

Cell elimination is found in most tissues. The appearance of dying cells varies, but at least two different sequences, leading to apoptosis and necrosis, can be defined morphologically.

Apoptosis, or programmed cell death, is supposed to occur during normal tissue turnover, embryogenesis and metamorphosis (Kerr & Searle 1973, Wyllie et al 1973). Apoptotic cells are characterized by an intensely staining cytoplasm with irregular outlines in the light microscope. Ultrastructurally the apoptotic cells are seen to lose all types of cell junctions and to form a dark, condensed cytoplasm; marginalization of nuclear chromatin occurs, and ultimately the cells break up into a number of apoptotic bodies. These fragments are removed by macrophages or nearby cells (Wyllie 1981). The occurrence of dark cells, presumed to be apoptotic cells, in octavolateralis organs is described below.

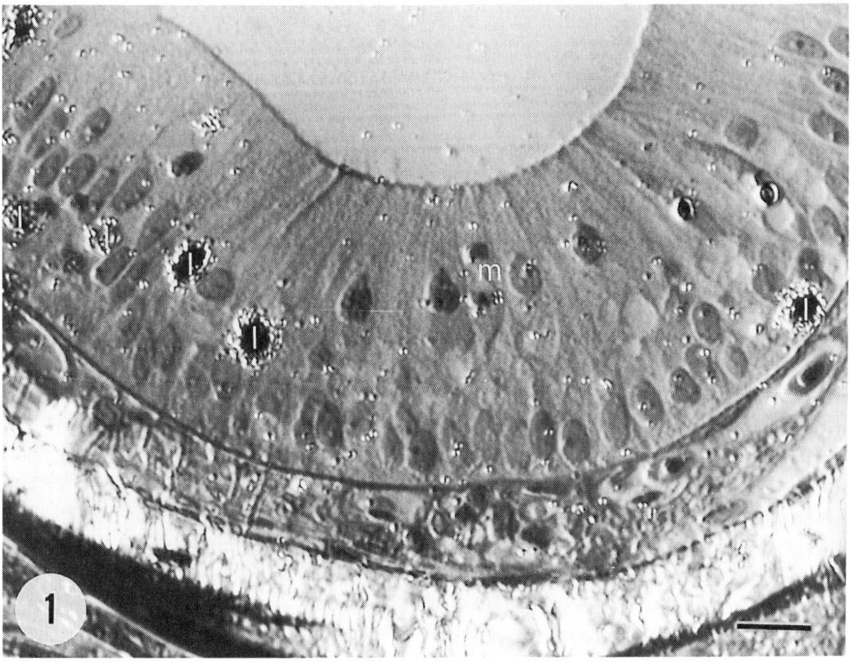

FIG. 1. Autoradiograph of a cross-section of the mechanosensory lateral line canal organ of a silver eel (*Anguilla anguilla*) injected for 14 days consecutively with [^{3}H]thymidine (1 mCi/kg body weight). Many labelled nuclei (l) as well as a mitosis (m) are seen. Interference-contrast optics. Scale bar, 10 µm.

Not all dark cells are necessarily apoptotic cells, however. Post mortem trauma on inadequately fixed tissue as well as the use of a low-osmolarity fixative vehicle may result in an increased number of dark cells (Cammermeyer 1972, Glasø et al 1989). Experiments involving the fixation of the budgerigar inner ear with glutaraldehyde in distilled water and in phosphate buffers of different strengths reveal no subsequent differences in the number of dark cells (author's unpublished experiments).

Necrosis occurs in circumstances very different from physiological conditions (Wyllie 1981). It has been observed after hypoxia, the inhibition of glycolysis and oxidative phosphorylation, hyperthermia, and exposure to a variety of toxins. It is morphologically characterized by cytoplasmic oedema and mitochondrial swelling, and, in the later stages, dissolution of cytoplasmic organelles and membrane rupture.

Mechanoreceptive lateral line organs

Butterfly fish (*Pantodon buchholzi*), injected with [^{3}H]thymidine (1 mCi/kg body weight) and killed 24 hours later, show a few labelled cells in the middle and basal part of the sensory epithelium. Similar injections each morning and evening for 14 days consecutively in silver eels (*Anguilla anguilla*) show many labelled cells in the canal organs (Fig. 1). Dark cells are occasionally seen (Jørgensen 1989). Mitoses and dark cells have previously been found in the canal organ of a shark (*Carcharhinus* sp.) (Tester & Kendal 1969), and dark cells in the neuromasts of the clawed toad *Xenopus laevis* (Shelton 1971).

Electroreceptive lateral line organs

Ampullary organs of the transparent catfish *Kryptopterus bicirrhus* generally have many labelled cells in the middle and basal part of the sensory hillock when the fish is killed 16 and 24 hours after the injection of [^{3}H]thymidine (Fig. 2). If the fish is killed 18 days after injection, some of the sensory cells also have labelled nuclei. Light and dark cells as well as cells of intermediate staining intensity have been observed in this fish (author's unpublished observations) and also in the ampullary organs of the bichir *Polypterus* and the African lungfish *Protopterus* (see Jørgensen 1989). Degenerating cells in the ampullary organs of the mormyrid *Gnathonemus* were found by Szamier & Bennett (1974).

Inner ear vestibular organs

Fishes

In elasmobranchs, injections of [^{3}H]thymidine resulted in the labelling of hair cell nuclei primarily in the marginal zone of the sensory epithelium, while labelled supporting cell nuclei were seen throughout the sensory epithelium of the macula neglecta. As elasmobranchs continue to grow throughout life, and constantly increase the number of sensory cells, most new hair cells are considered to be added marginally (Corwin 1981, this volume: Corwin et al 1991).

Bony fishes also have indeterminate growth. In an investigation of the saccular macula in the ear of the teleost *Astronotus ocellatus*, Popper & Hoxter (1990) found ^{3}H-labelled hair cell nuclei in all parts of the sensory epithelium within 24 hours. This rapid development is not surprising, because this fish increases its hair cell population by more than 150 cells per day (Popper & Hoxter 1984). Dark hair cells, thought to be degenerating, were occasionally found in the lagenar macula of the fish *Colisa labiosa* (Wegner 1982).

Amphibians

In a study of the saccular macula of the toad *Bufo marinus*, Corwin (1985) found an increase in the hair cell population from about 400 cells at metamorphosis to more than 1600 in the adult toad. The addition of hair cells was greatest during the first year, when about 15 new hair cells were added per

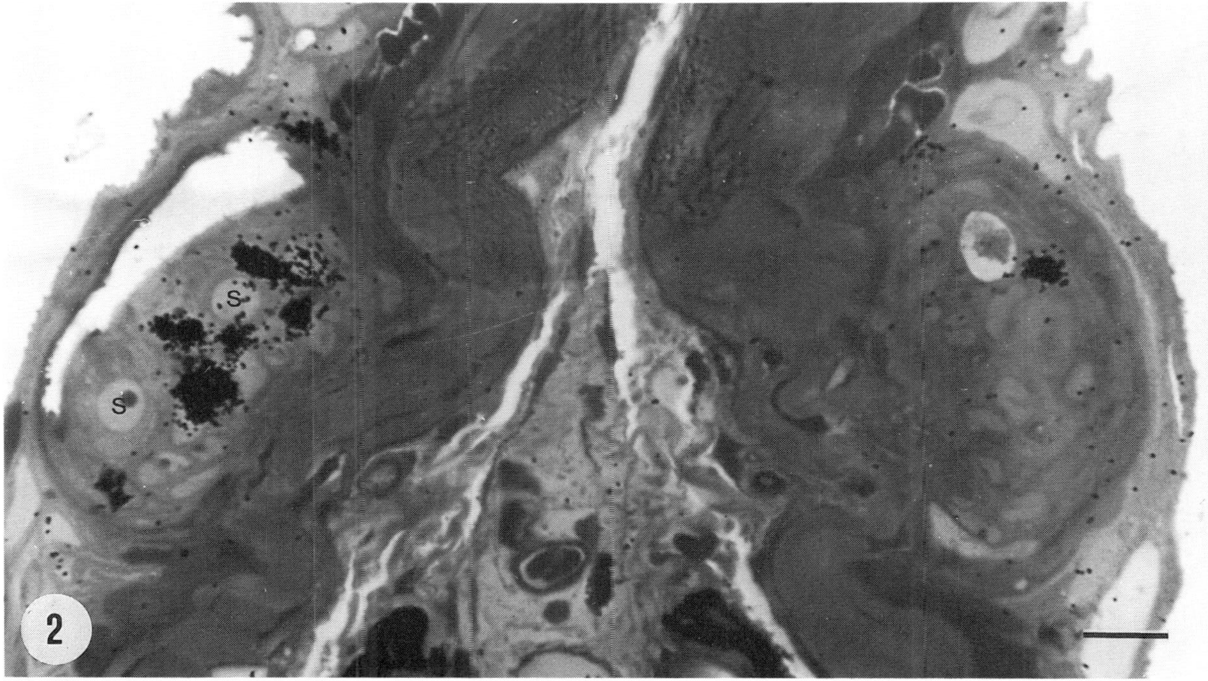

FIG. 2. Autoradiograph of two electrosensory ampullary organs of the transparent catfish (*Kryptopterus bicirrhus*), killed 16 hours after injection with [^{3}H]thymidine. The left-hand organ has many labelled nuclei, while only a single labelled nucleus is apparent in the right-hand organ. The sensory cells (s) are not labelled. Scale bar, 10 μm.

10 days, while only one new hair cell was formed in the same timespan in the fourth year of life. The vast majority of new hair cells was found at the margins of the sensory organ. On the basis of labelling with [^{3}H]thymidine, the supporting cells were thought to turn over every 24 days. However, what Corwin (1985) considered to be labelled supporting cell nuclei may in fact be nuclei of undifferentiated cells; moreover, degenerating supporting cells were not described in the toads. Dark basal/supporting cells are extremely rare in any examined animal, to the best of my knowledge.

In an ultrastructural study of the utricular macula of the caecilian (apodan amphibian) *Ichthyophis glutinosus*, Jørgensen (1981) found a pleomorphic population of hair cells. Newly formed hair cells, as well as grey to dark hair cells, with increasing signs of degeneration, led to a suggestion of hair cell turnover. Other hair cells had voluminous mitochondria and large vacuoles and other signs of necrosis, so these hair cells might be the result of some sort of inner ear disease, or of exposure to low temperatures. The presence of granular cells with lobular nuclei, probably macrophages, as well as inclusion bodies in some cells, considered to be apoptotic bodies (Fig. 3), strengthens the idea that the dark cells are apoptotic cells.

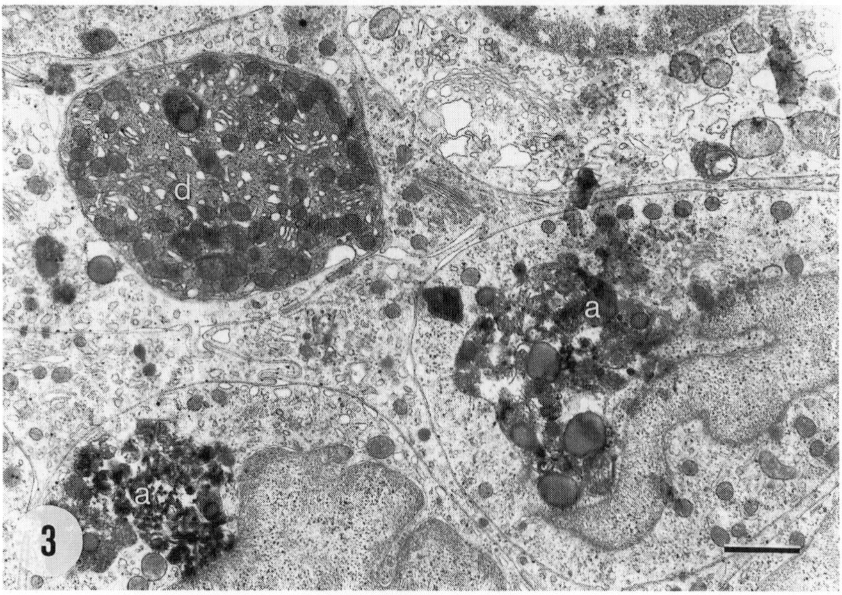

FIG. 3. Electron micrograph of a section parallel with the surface of the utricular macula of the caecilian *Ichthyophis glutinosus*. The basal part of a dark hair cell (d) and two other cells with presumed engulfed apoptotic bodies (a) are seen. Scale bar, 1 μm.

Reptiles

A series of experiments with [^{3}H]thymidine injections (1 mCi/kg body weight) in adders (*Vipera berus*), with survival times ranging from hours to 19 days, did not demonstrate any labelled nuclei in the sensory epithelium. The adders were injected in the posterior part of the body, and the failure to demonstrate any labelled nuclei may be due to an unfortunate injection site, or may mean that DNA synthesis occurs irregularly. Labelled blood cells were noted. However, similar injections in the neck of two specimens of the turtle *Pseudemys elegans* resulted in labelling of both basal/supporting and sensory cell nuclei (author's unpublished experiments).

Birds

Incorporation of [^{3}H]thymidine into basal and sensory cell nuclei in the budgerigar (*Melopsittacus undulatus*) was demonstrated by Jørgensen & Mathiesen (1988). These, as well as my later experiments, involve survival times of 16 and 24 hours after a single dose (four specimens); of 19 days, with consecutive injections each morning and afternoon (two specimens); and of 300 days after two injections (two specimens) (Table 1).

Autoradiography of 2 μm serial sections of plastic-embedded inner ear sensory organs of these budgerigars revealed labelled nuclei in all vestibular organs, but not in the hearing organ, the basilar papilla. Generally, the short survival experiments showed labelling of nuclei situated basally and in the middle part of the epithelium. The 19 days' survival experiment showed labelled nuclei of the bouton-innervated (Type II) hair cells as well as basal and medially located nuclei. Only the birds that were allowed to survive for 300 days had labelled calyceal (Type I) hair cells (enclosed in a nerve calyx) (Fig. 4).

TABLE 1 Numbers of [^{3}H]thymidine-labelled nuclei in the utricular macula of adult budgerigars, counted from 2 μm serial sections

Survival time	*Undifferentiated/ supporting cells*	*Bouton-innervated hair cells*	*Calyceal hair cells*	*Total*
16 h after single dose	21	0	0	21
24 h after single dose	18	0	0	18
19 days after injection each morning and evening	136 542	46 87	0 0	182[a] 629
300 days after two doses	8 11	2 5	1 3	11 19

[a]Series not complete.

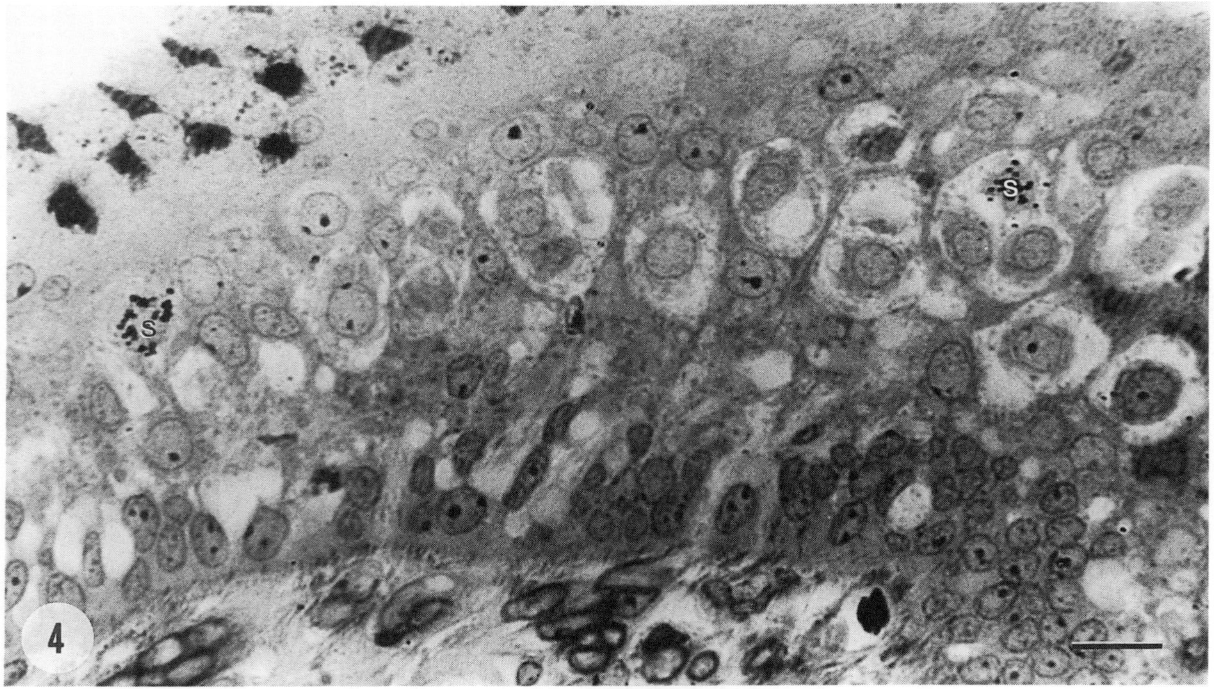

FIG. 4. Autoradiograph of a section of the utricular macula of a budgerigar that received an injection of [^{3}H]thymidine 300 days before sacrifice. To the left is seen a labelled nucleus of a bouton-innervated (Type II) hair cell (black s), to the right a labelled calyceal (Type I) hair cell nucleus (white s). Scale bar, 10 μm.

Mitoses in the inner ear sensory epithelia of adult birds (and mammals as well) have, to the best of my knowledge, never been described. But they can be demonstrated with the metaphase blocking agents colchicine and vinblastine (Fig. 5), administered in a concentration of 1–2 mg/kg body weight, as recommended by Tannock (1965). The majority of mitoses are seen rather luminally, on a level with the highest positioned hair cell nuclei. Examination of complete series of plastic sections usually reveals only 1–3 mitotic figures (author's unpublished observations). However, more may exist, since the presence of many dark staining inclusions makes it easy to overlook mitoses. So far, we have no information on the time it takes for a mitosis to occur in inner ear sensory epithelia.

Dark cells are occasionally seen in non-mammalian inner ear epithelia. In order to examine whether the number of dark cells can be affected by the osmolarity of the fixative, as suggested for the epidermis by Glasø et al (1989), or by delayed fixation after sacrifice, or rough treatment of the inner ear epithelia, as suggested for nervous tissue by Cammermeyer (1972), we did a

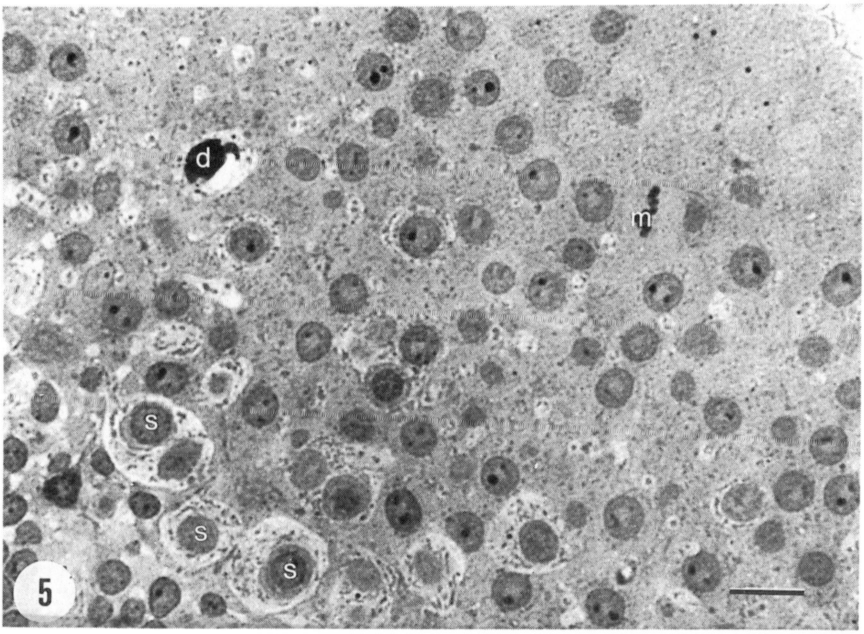

FIG. 5. Autoradiograph of an oblique section of the utricular macula of a budgerigar that had been injected with colchicine. An arrested mitotic figure (m) is seen in the right part of the photograph. To the left is seen a dark cell (d). Calyceal hair cells (s) are seen in the lower part of the picture, just above the supporting cell nuclei. Scale bar, 10 μm.

series of experiments with budgerigars. Examination in the electron microscope showed that delayed fixation, or maltreatment, resulted in swelling of the mitochondria and other cytoplasmic constituents as well as numerous membrane ruptures, but not in an increased number of dark cells. The total number of dark hair cells after fixation with glutaraldehyde in distilled water, and in phosphate buffers of different strengths, counted from 2 μm serial plastic sections, was found to vary from four to 62 in the entire utricular macula, while the number of dark undifferentiated/supporting cells was 0–3 (author's unpublished observations). It appears that only a very small fraction of the total hair cell population consists of dark cells, irrespective of the strength of the fixative vehicle.

The dark cells fulfil all morphological criteria for designation as apoptotic cells. Apoptotic bodies, believed to represent fragments of these dying cells, are seen occasionally. We have no indication of the time it takes for hair cells to pass through an apoptotic sequence.

The number of [^{3}H]thymidine-labelled hair cells and apoptotic cells should be related to the total number of hair cells. Table 2 shows the results of cell counts from the utricular macula of the budgerigar. Unfortunately, no very old birds were examined. The counts do not demonstrate any significant differences in the total sensory area and number of hair cells in this sensory organ between juvenile ($n=5$) and adult birds ($n=4$).

Although the number of birds examined is low, and counts from a larger series of experiments should correct or confirm the results, the numbers given in Table 1 above may indicate that within the time for which the [^{3}H]thymidine circulated in the blood, about 11–21 nuclei are in the phase of DNA synthesis, leading to the formation of about 3–8 hair cells. The number of apoptotic hair cells from tissue fixed in 0.1 and 0.2 M phosphate buffer is 3–33. The numbers may be taken as an indication of a slow turnover, although we have no idea whether all hair cells are replaced after a given time, or whether only some, perhaps damaged, hair cells are replaced.

Mammals

In an autoradiographic study of the developing inner ear of the mouse, Ruben (1967) found no formation of new sensory elements to occur after the 17th gestational day. A study of vestibular differentiation in the rat showed terminal mitoses at about the 14th–19th gestational day, with a slight difference in timing between the maculae and cristae (Sans & Chat 1982).

Injections of [^{3}H]thymidine (1 mCi/kg body weight) in a series of mice did not reveal any labelled nuclei in the sensory areas of the inner ear. Some of the mice had previously been injected with gentamicin (Garamycin, 100 or 200 mg/kg body weight) in order to destroy some of the sensory cells (author's unpublished observations).

TABLE 2 Morphometric measures of the utricular macula of the budgerigar

	Total area (mm²)	*Number of hair cells per 0.01 mm²*	
		Striola	*Extrastriola*
Juvenile birds	0.5234	177	258
(3–4 weeks old)		177	245
	0.5260	180	252
		199	216
	0.5244	197	234
	0.5724	199	199
Mean ($n = 5$)	0.5366	188	234
Percentage of total area		32	68
		29	71
		27	73
Mean percentage		29.5	70.5
Number of hair cells		2956	8877
Total number of hair cells		11833	
Adult birds	0.5540	198	216
(more than 6 months old)		188	233
	0.4947	194	233
		203	206
	0.5691	157	208
		145	204
Mean ($n = 4$)	0.5393	181	216
Percentage of total area		24	76
		25	75
Mean		24.5	75.5
Number of hair cells		2392	8795
Total number of hair cells		11187	

If any mammal could be expected to show post partum formation of sensory cells in the inner ear, it ought to be a bat. Bats are flying species, they have a small number of hair cells (1300 hair cells in the utricular macula of the little brown bat *Myotis lucifugus*; Ramprashad et al 1980), and they are considered to be primitive mammals. Injections with [^{3}H]thymidine (1 or 2 mCi/kg body weight) in *Pipistrellus nathusii* (three specimens) did not, however, reveal any labelling of inner ear cells.

The total number of hair cells in the utricular macula of very old mice and humans is about 14–21% lower than that for young individuals. The reduction of sensory elements in the cristae ampullares was found to be about 24% in mice (Park et al 1987) and 40% in humans (Rosenhall 1973).

Summary

In non-mammalian vertebrates, the mechanosensory and electrosensory lateral line as well as the inner ear vestibular sensory organs have pleomorphic sensory cells. A few hair cells are immature, while others show signs of apoptosis. Studies of the incorporation of tritiated thymidine show a continuous production of sensory cells in these octavolateralis organs of non-mammalian species, whereas no incorporation has been found in the inner ear sensory organs of a species of mouse and a bat. The number of newly produced cells in the utricular macula of a bird, the budgerigar, seems not to be very different from the number of dying cells.

Acknowledgements

The skilful technical assistance of S. Bisgaard Christensen, B. Dahl, P. Holm Hansen and G. Friis is acknowledged. Thanks are due to Dr J. Tønnes Nielsen and Dr H. Walhowd for the supply of mice and bats. The many valuable suggestions and corrections to the manuscript by Mrs J. Whelan are acknowledged. Parts of the project were supported by the Danish Natural Science Research Council.

References

Cammermeyer J 1972 Nonspecific changes of the central nervous system in normal and experimental material. In: Bourne GH (ed) The structure and function of nervous tissue. Academic Press, New York & London, vol 6:131–251

Corwin JT 1981 Postembryonic production and aging of inner ear hair cells in sharks. J Comp Neurol 201:541–553

Corwin JT 1985 Perpetual production of hair cells and maturational changes in hair cell ultrastructure accompany postembryonic growth in an amphibian ear. Proc Natl Acad Sci USA 82:3911–3915

Corwin JT, Jones JE, Katayama A, Kelley MW, Warchol ME 1991 Hair cell regeneration: the identities of progenitor cells, potential triggers and instructive clues. In: Regeneration of vertebrate sensory receptor cells. Wiley, Chichester (Ciba Found Symp 160) p 103–130

Glasø M, Iversen OH, Hovig T 1989 The influence of fixation on the relative amount of cytoplasmic ribosomes in mouse epidermal basal keratinocytes. Virchows Arch B Cell Pathol 56:221–235

Jørgensen JM 1981 On a possible hair cell turn-over in the inner ear of the caecilian *Ichthyophis glutinosus* (Amphibia: Gymnophiona). Acta Zool (Stockh) 62:171–186

Jørgensen JM 1989 Evolution of octavolateralis sensory cells. In: Coombs S, Görner P, Münz H (eds) The mechanosensory lateral line: neurobiology and evolution. Springer-Verlag, New York & Berlin, p 115–145

Jørgensen JM, Mathiesen C 1988 The avian inner ear. Continuous production of hair cells in vestibular organs, but not in the auditory papilla. Naturwissenschaften 75:319–320

Kerr JFR, Searle J 1973 Deletion of cells by apoptosis during castration-induced involution of rat prostate. Virchows Arch Abt B Zellpathol 13:87–102

Lewis ER, Li CW 1973 Evidence concerning the morphogenesis of saccular receptors in the bullfrog (*Rana catesbeiana*). J Morphol 139:351–361

Park JC, Hubel SC, Woods AD 1987. Morphometric analysis and fine structure of the vestibular epithelium of aged C57BL/6NNia mice. Hear Res 28:87–96
Popper AN, Hoxter B 1984 Growth of a fish ear. I. Quantitative analysis of sensory hair cell and ganglion cell proliferation. Hear Res 15:133–142
Popper AN, Hoxter B 1990 Growth of a fish ear. II. Location of newly proliferated sensory hair cells in the saccular epithelium of *Astronotus ocellatus*. Hear Res 45:3340
Ramprashad F, Landolt JP, Money KE, Laufer J 1980 Neuromorphometric features and dimensional analysis of the vestibular end organ in the little brown bat (*Myotis lucifugus*). J Comp Neurol 192:883–902
Rosenhall U 1973 Degenerative patterns in the aging human vestibular neuro-epithelia. Acta Oto-laryngol 76:208–220
Ruben RJ 1967 Development of the inner ear of the mouse: a radiographic study of terminal mitoses. Acta Oto-laryngol Suppl 222:1–44
Sans A, Chat M 1982 Analysis of temporal and spatial patterns of rat vestibular hair cell differentiation by tritiated thymidine radioautography. J Comp Neurol 206:1–8
Shelton PMJ 1971 The structure and function of the lateral line system in larval *Xenopus laevis*. J Exp Zool 178:211–232
Szamier RB, Bennett MVL 1974 Special cutaneous receptor organs of fish. VII. Ampullary organs of mormyrids. J Morphol 143:365–384
Tannock IF 1965 A comparison of the relative efficiencies of various metaphase arrest agents. Exp Cell Res 47:345–356
Tester AL, Kendall JI 1969 Morphology of the lateralis canal system in the shark genus *Carcharhinus*. Pac Sci 23:1–16
Tilney LG, Tilney MS 1984 Observations on how actin filaments become organized in cells. J Cell Biol 99:76s–82s
Wegner N 1982 A qualitative and quantitative study of a sensory epithelium in the inner ear of a fish (*Colisa labiosa*; Anabantidae). Acta Zool (Stockh) 63:133–146
Wyllie AH 1981 Cell death: a new classification separating apoptosis from necrosis. In: Bowen ID, Lockskin RA (eds) Cell death in biology and pathology. Chapman & Hall, London & New York, p 9–34
Wyllie AH, Kerr JFR, Currie AR 1973 Cell death in the normal neonatal rat adrenal cortex. J Pathol 111:255–261

DISCUSSION

Burd: Does water from the external environment contact the lateral line hair cell, Dr Jørgensen?

Jørgensen: The hair cells are protected by a cupula, but numerous studies have shown that ions of different sorts penetrate the basal cupula very easily.

Presson: Among the mammals, you have looked at mice and bats. There are some hints from other kinds of studies, for example dorsal root ganglia counts in rats (Devor & Gorvin-Lippmann 1985), where an argument was made that the occurrence of normal growth in the mammalian CNS might be related to the indeterminate growth of the animal. Perhaps kangaroos should be studied, for example, because they continue to grow throughout life. It might be interesting to look at the rat, as a mammal that continues to grow in body size throughout its lifetime. Rats grow slowly as adults, but do they add hair cells?

Rubel: We compared normal and aminoglycoside-damaged gerbils while they were still growing, at a few weeks old. We gave them two daily injections of either 5-bromo-2′-deoxyuridine or [^{3}H]thymidine, for seven days. We waited a month and saw no labelled hair cells in cochlea or vestibular labyrinth (unpublished information).

Calof: In the aminoglycoside-treated gerbils, was it clear that the hair cells are destroyed, just as in the chick?

Rubel: Yes. We did some noise damage experiments too and looked at the cochlea. We found no labelled hair cells.

Corwin: Dr Jørgensen, is it possible that your dark cells are less turgid than the lighter cells? It looked as if the dark ones were conforming to the outline of their neighbours, whereas the lighter cells were rounder.

Jørgensen: Yes, in the last stages of what I believe is an apoptotic process, the dark cells shrink considerably.

Presson: Relative to Jeff Corwin's earlier comments, do you have any evidence that these dying cells are ejected out of the epithelium?

Jørgensen: No. But when I count the dark cells, I only very rarely see a dark non-sensory cell, or supporting cell. They are almost exclusively hair cells.

Ryals: In our serial sections of the chick and quail basilar papilla, we saw such dark cells and assumed that they were some artifact of staining. As I recall, they were rather thin, and were not being extruded from the epithelium.

Rubel: We see dark cells, particularly after aminoglycoside treatment.

Jørgensen: Most inner ear morphologists, I think, have seen dark calls here and there, but have seldom paid much attention to them.

Watt: I would like to make a general comment about using [^{3}H]thymidine, which relates to the earlier papers as well. It is obviously a useful technique for identifying cells that are synthesizing DNA, but there are problems with it which we see in cell culture and which people who work on epidermis have noticed *in vivo*. For example, in cell culture, it's clear that giving cells high specific activity [^{3}H]thymidine can cause them to arrest in G2 phase, apparently while they carry out DNA repair (Watt et al 1988). Are there controls in your systems, to check that after you remove the tritiated thymidine the labelled cells continue to progress through the cell cycle?

Two other points from the *in vivo* studies are, first, that different types of cells can have different-sized intracellular pools of thymidine, so two different cell types may incorporate different amounts of [^{3}H]thymidine, even though their S phase parameters are the same (Chwalinski & Potten 1986). Then there is always the problem of diffusion; if you are injecting at the posterior end of an adder, do you know that all of the cells that you are interested in are receiving the same amount of tritiated thymidine?

Jørgensen: No, I don't know that.

Calof: If you do a thymidine block of mitosis in culture, don't you have to go to high levels (millimolar) of thymidine? Surely none of these injections result in levels as high as that.

Watt: I have observed growth arrest of cultured keratinocytes in 2 μCi/ml [^{3}H]thymidine (Watt et al 1988).

Potten: In the small intestinal system, doses of [^{3}H]thymidine such as Dr Jørgensen used (e.g. 1 mCi per kg body weight) will kill some cells. We have to go down to very small doses, 2.5 μCi per mouse (0.1 mCi per kg) to reach levels where we feel we are not killing any cells (K. Ijiri & C. S. Potten, unpublished data).

Reh: What specific activity are you using?

Potten: About 5 Ci per millimole.

Corwin: I think there are two issues here. One is the question of how much *thymidine* is present, and hence the potential thymidine block of cell division. That has been studied, and we refer to relevant technical studies before we decide on the doses used.

The other issue is radioactivity. When we are using autoradiographic exposures that require two weeks or a month to yield detectable evidence of β-particles hitting the silver grains in the liquid emulsions employed, we can state with confidence that there is only a small amount of radioactivity there. If we are using the highest specific activity of [^{3}H]thymidine that's available, I think we are doing the right thing. We are using the least amount of thymidine with the highest likelihood of radioactive decay; yet we are seeing only a few β-particles emitted over a month. So I don't think that the radioactivity is likely to be much of a danger.

Watt: Have you looked at your animals? Some of these thymidine injection regimes seem very severe—injecting every day for 20 days. Have you let the animals live for another couple of months and seen whether there is any sign of damage?

Rubel: Yes; in most of these studies we let animals survive for anywhere from one month to six months. In every way, in terms of cross-sections of plastic-embedded tissue, or looking at stereocilia in light or electron microscopy, the hair cells look normal. That's not to say that some cells weren't arrested in S phase and then died; we don't know that.

Reh: That is probably the most worrying aspect. If you don't see a result—for example, if you don't see a labelled hair cell—perhaps you are just seeing the hair cells that incorporated the least amount of thymidine and were close to completing S phase when they divided; those that incorporated more thymidine are gone.

Corwin: In the shortest post-injection survivals, then we should see *very* hot cells, and we don't.

Raymond: There must be enormous variation from one tissue to the next in sensitivity to the toxic effects of thymidine. To determine the proper dose empirically, we increased the amount of thymidine injected until we demonstrated cellular damage, and then reduced the amount until we were confident that we were in a range where we had long-term survival of differentiated retinal cells with labelled nuclei.

Potten: There must be variation between different species, too. For example, in the adder and the bat, you didn't see incorporation, Dr Jørgensen, but are you sure that these species have effective thymidine salvage pathways? Some species do not incorporate tritiated thymidine at all, because they have either a deficient thymidine salvage pathway, or very efficient thymidine catabolic pathways. They make all their thymidine through the *de novo* pathway and you cannot get tritiated thymidine into them. You need to look at other tissues to be sure that the adder is capable of incorporating tritiated thymidine.

Jørgensen: We did that. Occasionally blood cells were found labelled in the adder, indicating that the tritiated thymidine had reached at least some organs.

Corwin: Can you give an example of animals that lack the salvage pathway, Dr Potten?

Potten: Some American ground squirrels (*Marmota* and *Citellus* species) do not incorporate thymidine (Adelstein et al 1964, Adelstein & Lyman 1968). There may be other non-mammalian species that incorporate poorly or not at all. The efficiency with which the *de novo* and salvage pathways function in different tissues may also vary, and the catabolic efficiency in the liver may also vary; and finally the endogenous thymidine pools may vary, and so you get different thymidine uptakes. We even see that in the mouse, an intravenous or intraperitoneal injection of tritiated thymidine is distributed round the body very quickly but is taken up rather differently in different tissues.

Rubel: We almost always run a piece of chick gut along with the inner ear epithelium in any experiment, as a control for false negatives.

Potten: That is a good control. But I would reiterate the point that Fiona Watt was making; some of the tritiated thymidine doses and regimes that have been described are worrying to people who work with rapidly turning over mammalian systems, because they certainly will cause some cell death and other cell kinetic disturbances.

Jørgensen: Yes, but but do you think that we are overdosing the animals, when we find good-looking labelled cells after a survival time of 300 days?

Presson: The question may be of how many dying cells you see in an animal shortly after giving it one of these large doses of thymidine.

Jørgensen: The total number of dark (apoptotic) cells varies between four and 60 in the total utricular macula of the budgerigar, but it is difficult to compare that with the number of labelled cells in the same sensory organ, since we don't know the timespan of a possible apoptotic sequence.

Corwin: Have you looked with scanning electron microscopy? Using SEM I observed cells that had very small numbers of cilia which were short and looked disarrayed, in the macula neglecta sensory epithelia of adult sharks (Corwin 1981).

I thought that those might have been degenerating cells. I wonder whether you may have seen something like that, and whether the numbers of abnormal ciliary bundles correlate with the numbers of dark cells.

Jørgensen: I have not seen that, but I believe that the hair cells die out in the central part of the shark utricular sensory epithelium. Dr Vilstrup (1951) showed this in the shark *Squalus acanthias*.

Rubel: You have done some serial sectioning from the surface of the epithelium downwards in the utriculus of birds and have tried to reconstruct some cells. This problem of whether there is a homogeneous or non-homogeneous population of non-hair cells is best studied in the vestibular epithelium.

Jørgensen: The non-hair cell population of cells is non-homogeneous in at least some octavolateral sensory epithelia. Evidence come from serial transmission electron microscopical sections of the axolotl neuromast and the caecilian inner ear. Here we find, besides supporting cells which reach from the basal lamina to the sensory surface, some cells that do not reach the surface. However, we can't just look at a section and say which cell type is which, because we have found no clear cytological markers for basal cells so far.

Rubel: On another issue, do you see any difference in the cell body, or in cytological markers, between Type I and Type II hair cells, or do you distinguish them only on the basis of the afferent nerve ending?

Jørgensen: They also have a different shape; most of the calyceal hair cells (Type I) have a very narrow neck region, filled with microtubules, as shown by the Montpellier group. The bouton-innervated cells are much more cylindrical.

Pujol: Dr A. Sans and colleagues have shown in the mammalian vestibular epithelium that the density of the cytoplasmic inclusions differs between the two types of cells, with a greater density of microtubules in Type I (Favre & Sans 1983). So even if you can't see the synaptic pole, you can distinguish Type I and Type II, at transmission EM level.

Presson: In teleost fish, everyone has thought that there's only one type of hair cell in the vestibular end organs. In our lab we have evidence for two types. An interesting point, that might be useful if people are looking for cell markers, although I don't know how far it would apply to other vertebrates, is that we find differences with the S-100 (calcium-binding protein) antibodies, which label striolar hair cells (which would be the calyceal types in other vertebrates) and don't mark the extrastriolar cells (Saidel et al 1990). There certainly is a possibility for finding cytochemical markers that differentiate between cell types.

Jørgensen: My unpublished observations show that in the utricular macula of the dab (a flat fish, *Limanda limanda*) and a frog (*Rana temporaria*), there are hair cells with a different innervation, but never a calyceal one. The striolar hair cells are innervated in a way that reminds us very much of what Dr Lowenstein and his coworkers described as candelabra-shaped nerve endings

(1968). They are formed from thick-diameter nerve endings which send up processes along the sides of the hair cells.

Pujol: In the mammalian vestibular system this is just a stage in ontogeny, before the complete calyx is formed (Favre & Sans 1978).

Jørgensen: Yes; in the fish and in the frog you don't see further development.

Pujol: The evolution stops at one stage in the fish and goes further in mammals.

Corwin: In sharks the same thing occurs; the typical innervation of a hair cell looks like a 'hand' in the centre of the macula neglecta epithelium, and at the edge often only one bouton will come to a new hair cell.

Can you speculate on why you see the Type I hair cells being labelled so late? Is it a progression from Type II to Type I, which would also imply a progression from peripheral to central location? You mentioned that the Type II hair cells (those with the bouton type of innervation) were in the periphery, and the Type I cells were in the centre.

Jørgensen: We find labelled nuclei after injection of tritiated thymidine in the avian striolar middle zone of bouton-innervated hair cells (Type II), so there is no progression from peripheral to central sensory cells with respect to innervation.

Corwin: So it's not exclusively one cell type or the other in those two areas?

Jørgensen: No.

Rubel: In the ampulla of birds, most of the Type I cells are in the centre of the epithelium and most of the Type II cells are along the edges; but you also see Type II cells in the centre, and after labelling you first see only Type II cells. Then, 30 days later, you see Type I cells.

Corwin: So the hypothesis is that the the new Type II cells that label in the centre of the epithelium may be incorporated into a calyx after a time and therefore become Type I hair cells by definition?

Jørgensen: Well, some sort of bouton-innervated hair cell seems to be incorporated into an afferent nerve calyx and then, by definition, it becomes a calyceal hair cell.

Pujol: Dr Sans doesn't believe in a transformation of Type II into Type I cells! He thinks there may be a common primitive type of vestibular cell, which early on in evolution differentiates into Type I and Type II (Sans & Chat 1982, Pujol & Sans 1986).

Presson: Maybe at the light microscope level one can't tell the difference?

Zakon: Couldn't there be different rates of turnover? The Type II may be turning over faster than the Type I cells.

Rubel: I don't see how that explains the data. We give a few days (three) of [^{3}H]thymidine to newly hatched chicks. One week later, we see only Type II hair cells. We have stopped giving the thymidine. Two months later, we see Type II *and* Type I hair cells.

Zakon: But presumably the support cells are still labelled, and are quiescent; then the labelled precursor gives rise to Type I cells at a later time.

Rubel: You are suggesting that the mature support cells are turning into hair cells?

Zakon: Yes. If the support cell divides and gives rise to *any* hair cell type, and if the type of progeny depends on demand, then if the ear 'needs' the Type II hair cells, it will produce Type II cells more frequently; if the Type I cells are 'needed' less frequently, the precursor cell will make a new Type I cell less frequently. Or there could be separate populations of precursor cells dividing at different rates.

Presson: There is another explanation. You are talking about base rates here—that if you have 10 Type I cells labelled per month and only one Type II cell labelled in six months, you have different base rates. If so, over a short time period, you are more likely to see Type I cells. This is a different hypothesis from the idea that the precursors are sitting around longer, interrupted in some stage of their cell cycle, before they turn into Type I.

Rubel: I think I can reject the hypothesis based on 'need' from our damage studies. When we damage the vestibular epithelium of the chick by giving aminoglycosides we get rid of both hair cell types. So you 'need' them both! We still see a long delay before we see Type I hair cells coming back. So there could be 'support' cells that are generated and sit around for a while and then differentiate into Type I hair cells, as Julian Lewis suggested. I don't know how we could show this.

Cotanche: Is there evidence that a cell would double its DNA and then wait for 30 days before dividing?

Presson: There is evidence for this from other systems.

Corwin: Yes, they stop before the S phase in G0; this is the normal arrest point. There are some cells that are polyploid, in the central nervous system, but it is not thought that they are pausing in a polyploid state before dividing.

Reh: Is it possible that it takes longer for one hair cell type to differentiate than another?

Rubel: If so, it looks like a Type II cell while it's differentiating!

Reh: Are there any intermediate phenotypes?

Corwin: It would be like the development that occurs postembryonically in the continuously growing sensory epithelia in sharks, where you have a young cell analogous to what you described as the bouton type. Those young cells are in the periphery of the epithelium and are innervated by a few small terminals. When they have matured they have a substantial innervation that looks like a hand enclosing the basal part of the cell. This has been described as a progression in time, so it could be analogous to the progression in time from Type II to Type I cells suggested by Dr Rubel.

Pujol: Again referring to work on mammalian vestibular development by Sans' group, the two types of vestibular hair cell differ quite clearly from the very beginning. They also differ in some immuno-labelling, such as for calbindin (Sans et al 1987). So the Type I hair cell, in its ontogeny, seems to go through a phase in which it looks like a Type II cell, but it is not.

Corwin: We are discussing two very different systems: the mammalian system that develops embryonically and is complete at birth with all the cells present, and fish like the shark, where in the macula neglecta you start off in the newborn free-swimming animal with 20 000 hair cells, and in the adult there are a quarter of a million hair cells in the same organ. The organ is now 12 times larger than it was at birth, yet there is only a small band of new, young hair cells at the edge. There is clearly a difference between what is occurring there and what might have been run through very quickly in a mammal, where the development of these epithelia is complete by birth.

Presson: I think that might be the key!

References

Adelstein SJ, Lyman CP 1968 Pyrimidine nucleoside metabolism in mammalian cells: an *in vitro* comparison of two rodent species. Exp Cell Res 50:104–116

Adelstein SJ, Lyman CP, O'Brien RC 1964 Variations in the incorporation of thymidine into the DNA of some rodent species. Comp Biochem Physiol 12:223–231

Chwalinski S, Potten CS 1986 Cell position dependence of labelling thymidine nucleotides using the *de novo* and salvage pathways in the crypt of small intestine. Cell Tissue Kinet 19:647–659

Corwin JT 1981 Postembryonic production and aging of inner ear hair cells in sharks. J Comp Neurol 201:541–553

Devor M, Gorvin-Lippmann R 1985 Neurogenesis in adult rat dorsal root ganglia. Neurosci Lett 61:189–194

Favre D, Sans A 1978 The development of vestibular efferent nerve endings during cat maturation: ultrastructural study. Brain Res 142:333–337

Favre D, Sans A 1983 Organization and density of microtubules in the vestibular sensory cells in the cat. Acta Oto-laryngol 96:15–20

Lowenstein OE, Osborne MP, Thornhill RA 1968 The anatomy and ultrastructure of the labyrinth of the lamprey (*Lampetra fluviatilis* L.). Proc R Soc Lond B Biol Sci 172:113-134

Pujol R, Sans A 1986 Synaptogenesis in the mammalian inner ear. In: Aslin R (ed) Advances in neural and behavioural development. Ablex Press, Norwood, NJ, p 1–18

Saidel WM, Presson JC, Chang JS 1990 S-100 immunoreactivity identifies a subset of hair cells in the utricle and saccule of a fish. Hear Res 47:139–146

Sans A, Chat M 1982 Analysis of temporal and spatial patterns of rat vestibular hair cell differentiation by tritiated thymidine radioautography. J Comp Neurol 206:1–8

Sans A, Brehier A, Moniot B, Thomasset M 1987 Immuno-electron microscopic localization of 'vitamin D-dependent' calcium binding protein (CaBP-28K) in the vestibular hair cells of the cat. Brain Res 435:293–304

Vilstrup T 1951 Structure and function of the membranous sacs of the labyrinth in *Acanthias vulgaris*. Munksgaard, Copenhagen (dissertation)

Watt FM, Jordan PW, O'Neill CH 1988 Cell shape controls terminal differentiation of human epidermal keratinocytes. Proc Natl Acad Sci USA 85:5576–5580

Retinal regeneration in teleost fish

Pamela A. Raymond*

Department of Anatomy & Cell Biology and the Neuroscience Program, The University of Michigan Medical School, 4610 Medical Sciences Building II, Ann Arbor, Michigan 48109-0616, USA

Abstract. The neural retina in teleost fish can regenerate after surgical or neurotoxic destruction. Whereas in amphibians the retina regenerates by transdifferentiation of pigmented retinal epithelial cells, in goldfish (*Carassius auratus*) the source of regenerated retinal cells is a population of scattered proliferating cells located in the outer nuclear layer within the differentiated retina. These proliferating cells are modified neuroepithelial cells termed 'rod precursors' because in the intact retina they produce only rod photoreceptor cells and do so continuously, inserting new rods into the growing adult retina. Although rod precursors normally exhibit a restricted developmental fate they appear not to be committed to the rod lineage. When retinal neurons are destroyed, rod precursors cease producing rods and give rise to clusters of primitive neuroepithelial cells which divide vigorously and reconstitute the retina in an orderly temporal pattern that mimics the process of normal development. Only after production of cones and other retinal neurons has ceased do rod precursors again appear and resume the generation of rods. We conclude that rod precursors respond to local cues in their environment that regulate the differentiation and choice of cell fate by their progeny.

1991 Regeneration of vertebrate sensory receptor cells. Wiley, Chichester (Ciba Foundation Symposium 160) p 171–191

Regeneration of the neural retina has been investigated in amphibians and to a more limited extent in teleost fish. There is an extensive international literature on the regeneration of lens and neural retina in adult urodele and larval anuran amphibians (reviewed by Reyer 1977, Stroeva & Mitashov 1981). In amphibians the source of cells for retinal regeneration in central regions is the pigmented retinal epithelium (Keefe 1973a,b, Reh & Nagy 1987, Reh et al 1987) which, like the neural retina, is derived embryonically from primitive neuroepithelial cells of the neural tube (Walls 1967). The process whereby pigmented epithelial cells give rise to neural retina will be described by Dr Reh (this volume: Reh et al 1991). This process has been termed 'transdifferentiation' and involves recruitment of differentiated pigmented cells into the mitotic cycle, with

*Pamela A. Raymond has published previously as P. R. Johns.

subsequent proliferation and eventual differentiation into neuronal phenotypes.

In contrast to amphibians, the source of proliferating cells from which the regenerated retinal neurons are derived in teleost fish lies within the neural retina itself. The difference between teleosts and amphibians is that an internal source of undifferentiated, proliferating cells in the neuroepithelial lineage exists in the intact adult teleost retina: these cells are called rod precursors because they divide continuously in the growing adult retina and give rise to new rod photoreceptors that are inserted into the retinal mosaic (Sandy & Blaxter 1980, Johns & Fernald 1981, Johns 1982, Raymond & Rivlin 1987). Rod precursors are derived from residual neuroepithelial cells that remain behind after the retinal neurons differentiate (Raymond & Rivlin 1987). When retinal neurons in adult goldfish are destroyed with a cytotoxin, for example ouabain (Maier & Wolburg 1979) or 6-hydroxydopamine (Negishi et al 1987), in some cases they regenerate. A careful examination of the conditions under which regeneration does and does not occur supports the proposal that the rod precursors are responsible for the regenerative response, and that their fate is altered—that is, they revert to a multipotent developmental pathway—only when the local microenvironment around them is disturbed (Raymond et al 1988, Braisted & Raymond 1990a). The results of experiments designed to evaluate this hypothesis are described below.

In both teleost fish and amphibians, a circumferential germinal zone of primitive neuroepithelial cells surrounds the peripheral edge of the retina, where annuli of new retina are generated continuously by the growing eye (Johns 1977, Keefe 1973a,b, Reh & Nagy 1987). As long as the germinal zone is not damaged, production of new retinal annuli continues at the margin (Raymond 1988). 'Regeneration', as discussed above, refers to cells lost from central retinal regions that are subsequently replaced by mechanisms not activated except under abnormal circumstances. This distinction between normal growth at the periphery and true regeneration in the centre has sometimes been blurred in the retinal regeneration literature, and in some cases the conclusion that the peripheral retina regenerated but the centre did not (e.g. Kurz-Isler & Wolburg 1982) could probably be more accurately thought of as continued 'normal' growth at the edges.

There are several types of proliferating cells in the adult teleost retina

The histological organization of the teleost retina is like that of all vertebrate retinas: there is an outer tunic consisting of a single layer of pigmented epithelial cells, the retinal pigmented epithelium (RPE), surrounding a multilayered neural retina consisting of an outer layer of photoreceptors, a middle layer of interneurons and an inner layer of ganglion cells whose axons leave the eye as the optic nerve (Walls 1967). The RPE and neural retina are separated by the subretinal space, created as a result of morphogenetic movements that shaped the eye during embryonic development. The pattern of histogenesis of the neural retina is likewise similar across vertebrate species: pluripotent (primitive) neuroepithelial cells divide to give rise to ganglion cells, interneurons, photoreceptors and Müller glia in a sequential but overlapping order (Raymond 1985a, Turner & Cepko 1987, Wetts & Fraser 1988, Holt et al 1988). In teleost fish, cone photoreceptors are produced along with other retinal neurons, but the production of rod photoreceptors is delayed and, once initiated, continues into adulthood (Müller 1952, Lyall 1957, Blaxter & Staines 1970, Johns 1982, Raymond 1985b). Proliferating cells synthesize DNA and therefore can be detected either autoradiographically after the injection of [^{3}H]thymidine or immunocytochemically after injecting the thymidine analogue, 5-bromo-2′-deoxyuridine (BrdU). The proliferation of residual populations of neuroepithelial cells clustered in the circumferential germinal zone and scattered across the outer nuclear layer is associated with retinal growth and the insertion of new rods, respectively (Figs. 1 and 2A).

In addition to neuroepithelial cells, there are other proliferating cells in the retina, including vascularly derived cells, Müller glial cells and RPE cells. We felt it was important to characterize all the potential sources of mitotic cells that exist in the intact retina in order to identify those that might be responsible for the regenerative response. We identified the various types of proliferating cells by their unique profile of staining with cell-specific antibodies (Table 1). Vascularly derived cells, including microglia and endothelial cells, and RPE cells are specifically stained with the monoclonal antibodies NN1 and NN2 generated in our laboratory from mice immunized with goldfish retinal homogenate (Fig. 2B; Raymond & Barthel 1988, Wagner & Raymond 1991). The NN-labelled (NN$^+$) cells are phagocytic: they ingest derivatized (fluorescent) latex beads injected into the eye (Wagner & Raymond 1991). Müller glia cells and RPE cells in the goldfish retina are stained with commercially obtained polyclonal antibodies to glial fibrillary acidic protein, GFAP (Fig. 2C; Wagner & Raymond 1991). Müller glia, but not RPE cells, are recognized by another of our goldfish monoclonal antibodies, RET1 (Fig. 2D). The antigen recognized by RET1 is primarily localized to nuclei and is found in a subset of retinal neurons and Müller glia (Raymond & Barthel 1988). Finally, neuroepithelial cells in the germinal zone, and rod precursors, do not stain with any of these markers (Table 1). All of these cell types can proliferate in the normal retina, but except for the neuroepithelial cells, most do so at a low rate.

Retinal degeneration is accompanied by increased proliferation of phagocytic microglia/macrophages but not RPE cells

When ouabain, which binds to and blocks Na^+, K^+-ATPase, is injected intraocularly in adult goldfish at 5–20 μM (estimated intraocular concentration),

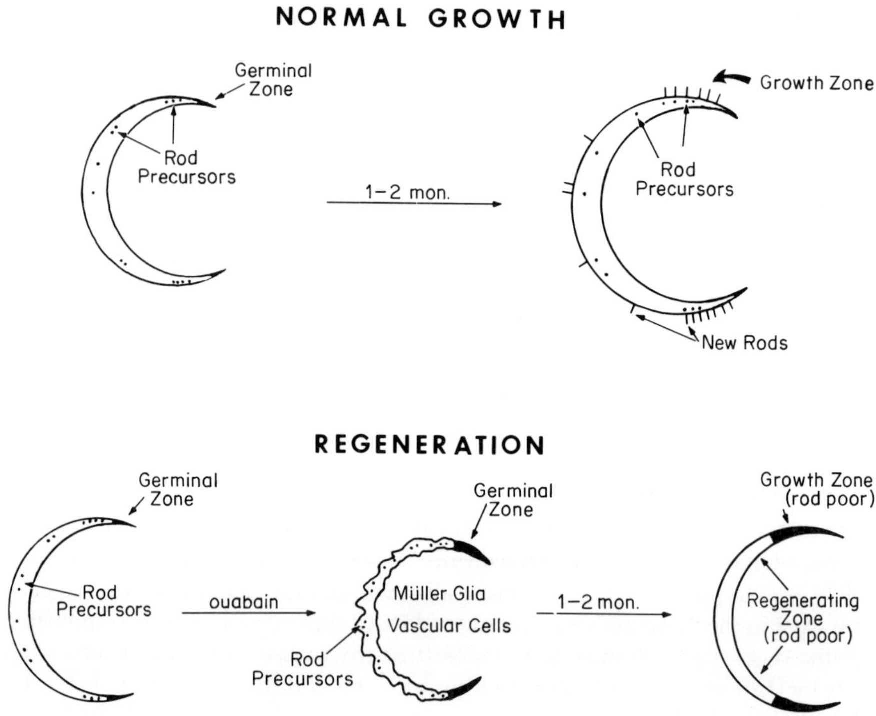

FIG. 1. (*Upper*) There are two sites of neurogenesis in the goldfish (*Carassius auratus*) retina. A 'germinal zone' of primitive neuroepithelial cells at the circumference produces annular increments of new retina. Rod precursors are modified neuroepithelial cells scattered across the retina in the outer nuclear layer, but are especially abundant in the newly added peripheral 'growth zone' just central to the germinal zone; some of the daughter cells differentiate as rod photoreceptors and are inserted into the retinal mosaic while other daughters continue to divide and thus maintain the population of rod precursors. (*Lower*) When ouabain is injected into the eye, retinal neurons are destroyed. Mitotic cells survive including neuroepithelial cells in the germinal zone, rod precursors, Müller glia, microglia and vascular cells. The central regions regenerate from foci of primitive neuroepithelial cells which are derived by mitotic division from rod precursors. At the peripheral margin, annular increments of new retina are added by the germinal zone as part of the ongoing normal growth process.

retinal neurons are killed (Raymond et al 1988). During the first 1–2 weeks after injection the retina degenerates. Some cells survive, however, including all the various types of proliferating cells: neuroepithelial cells in the germinal zone, rod precursors, Müller glia, RPE and vascular cells (Fig. 1). To identify proliferating cells we injected BrdU into the eye at various intervals after ouabain, and prepared the retina for immunocytochemistry the following day (Barthel & Raymond 1990). Most of the proliferating cells ($BrdU^+$) in the retina during this initial period are also NN^+ and are therefore

TABLE 1 Identification of proliferating cells in goldfish retina with cell-specific monoclonal antibodies

	Antibodies used		
Proliferating cells	*MAb NN*	*GFAP*	*MAb RET1*
Germinal zone	−	−	−
Rod precursors	−	−	−
Vascularly derived cells	+	−	−
Müller glial cells	−	+	+
Retinal pigment epithelium (RPE)	+	+	−

probably phagocytes, either microglia or macrophages (Wagner & Raymond 1991).

Only a few RPE cells ($NN^+/GFAP^+$) are labelled with BrdU, and they are rarely found within the degenerating retina. This is an important observation because, in amphibians, the RPE cells detach from the basal lamina (Bruch's membrane), invade the degenerating neural retina, and begin to proliferate, depigment and transform into neuroepithelial cells (Reh & Nagy 1987). In goldfish, the RPE cells become engorged with the debris of the dead photoreceptor cells that they have phagocytosed (Fig. 3A); they then move in the opposite direction, through Bruch's membrane and into the vascular choroid layer at the back of the eye.

Neural regeneration follows proliferation of neuroepithelial cells

By two weeks after ouabain injection, clusters of $RET1^-/NN^-/GFAP^-/BrdU^+$ neuroepithelial cells are found scattered across the retina (Figs. 2E, 3B). Retinal neurons then begin to differentiate, following the same pattern as during normal development with respect to the order of production of cells: cones and other neurons are generated first (Fig. 3C), and subsequently rods begin to be produced (Raymond et al 1988). Cone density is restored to normal levels by about 8–10 weeks, at which time rod density is still increasing (Raymond et al 1988). (The time required for complete replacement of rods and other neurons has not yet been determined.) The laminar organization of the regenerated retina is not without errors, however, and fusions between the nuclear layers remain as permanent manifestations of the scattered foci of neurogenic cells from which the regenerate was derived (Fig. 3D). The distribution of these neurogenic foci is similar to that of the rod precursors, which is consistent with the hypothesis that the neurogenic foci derive from rod precursors. Further evidence for the putative link between rod precursors in the intact retina and neurogenic clusters in the regenerating retina is presented in the next section.

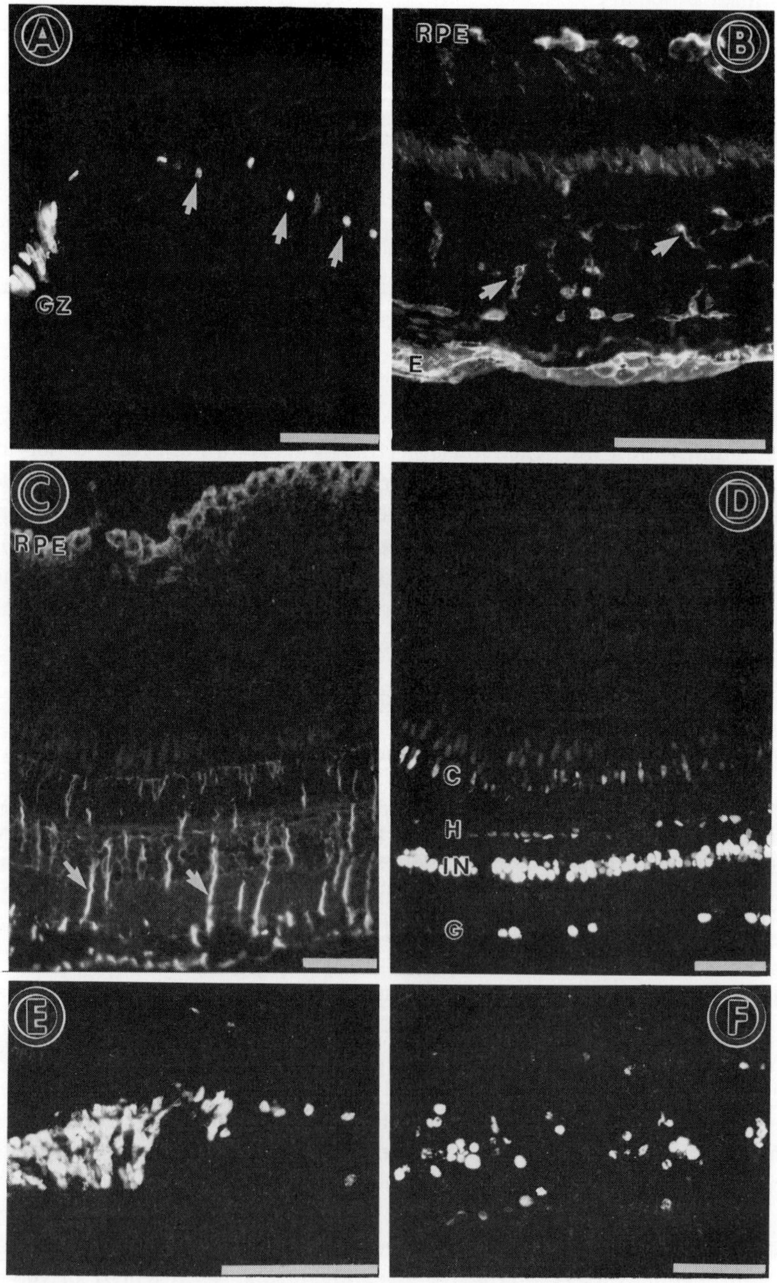
A
GZ
B
RPE
E
C
RPE
D
C
H
IN
G
E
F

Meanwhile, at the retinal margin the germinal zone remains intact, and proliferation there is greatly enhanced (Raymond et al 1988). We considered the possibility that neuroepithelial cells in the germinal zone might migrate toward the centre to colonize the degenerated regions, thus giving rise to the neurogenic foci. To examine this, we did the following experiment. Neuroepithelial cells in the germinal zone of both eyes were marked with a pulse of [^{3}H]thymidine, and simultaneously ouabain was administered to one eye. We found no evidence of centralward migration of labelled neuroepithelial cells in the ouabain-treated retina. Instead the labelled progeny of the germinal cells marked at the time of injection remained in a wedge spanning all three retinal layers, just as they did in the intact eye (Fig. 4). Furthermore, we could show that after one month, the increment of new retina added to the ouabain-treated eye was on average 3.6 times as wide as in the control eye (300 μm versus 82 μm, $n = 3$), although the overall retina was slightly smaller in the experimental eye, as a result of partial collapse of the degenerated central region (Raymond 1988).

FIG. 2. A. Immunofluorescent labelling of mitotic cells with 5-bromo-2′-deoxyuridine (BrdU). Labelled cells include neuroepithelial cells in the germinal zone (GZ) and scattered rod precursors in the outer nuclear layer (arrows). Indirect immunofluorescence (FITC conjugated secondary antibody) of a 3 μm cryosection (Barthel & Raymond 1990) of peripheral retina from a juvenile goldfish injected intraocularly with 50 μM BrdU and fixed 24 h later.

B. Immunofluorescent labelling of non-neural cells in goldfish retina with monoclonal antibody NN1. This and the remaining panels illustrate indirect immunofluorescence on 3 μm cryosections. Labelled cells include endothelial cells (E) on the vitreal surface, microglia (arrows) within the retina, and retinal pigmented epithelial cells (RPE). The processes of cone photoreceptors are endogenously fluorescent and form a faint horizontal band separating neural retina from pigmented retinal epithelium.

C. Immunofluorescent labelling of goldfish retina with a commercial (DAKO, Carpinteria, CA) polyclonal antibody against glial fibrillary acidic protein (GFAP). Müller glial cells are stained, including their prominent radial fibres (arrows). RPE cells are also labelled.

D. Immunofluorescent labelling of goldfish retina with monoclonal antibody RET1. Nuclei of cones (C), horizontal cells (H), selected cells in the inner nuclear layer (IN) and ganglion cells (G) are labelled.

E. Immunofluorescent labelling of mitotic cells in a regenerating goldfish retina three weeks after intraocular injection of ouabain. The eye was injected with BrdU one day before fixation. Some sections were processed for both BrdU (visualized with a secondary antibody conjugated to Texas Red) and a second (cell specific) antibody (visualized with the green fluorochrome, FITC; see panels B, C, D). The BrdU-labelled cells were not double-labelled and are therefore neuroepithelial cells (see Table 1).

F. Immunofluorescent labelling of regenerated retinal neurons with RET1. Ouabain was injected intraocularly and, beginning at three weeks, BrdU was injected three times at weekly intervals; the retina was fixed three weeks after the last injection. When sections were processed for both BrdU and RET1, some of the RET1$^+$ cells were also BrdU$^+$, indicating that these were regenerated neurons.

Calibration bars for all panels, 50 μm.

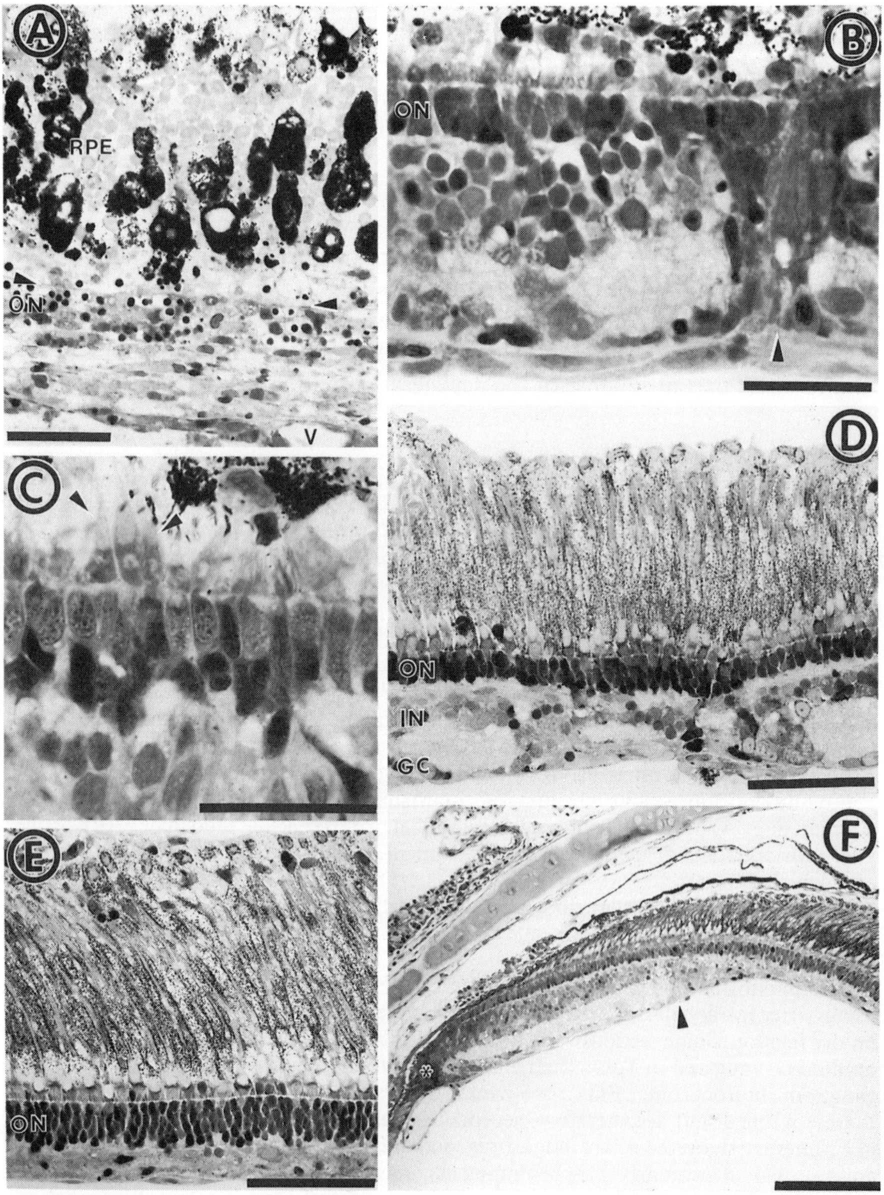
A
RPE
ON
V
B
ON
C
D
ON
IN
GC
E
ON
F

Thus the enhancement of cell proliferation in the germinal zone of the ouabain-treated eye resulted in an increase in the amount of new retina added. The annulus of new growth at the margin was completely normal in histological organization and lacked the laminar disruptions characteristic of regenerated central regions. This further supports the conclusion that the retina added peripherally by the germinal zone after ouabain-induced damage represents a normal growth process, albeit somewhat enhanced.

Regeneration of neural retina does not occur unless the outer nuclear layer is damaged

At lower doses of ouabain (1–5 μM, estimated intraocular concentration) the retina is often only partially destroyed; inner retinal layers are affected first because they are closest to the site of injection of the toxin (Raymond et al 1988). The appearance of these retinas is striking. When the outer nuclear layer of photoreceptors is undamaged, the inner retinal layers do not regenerate (Fig. 3E), although growth at the peripheral margin continues normally (Fig. 3F). Furthermore, there is a strong correlation between loss of photoreceptor cells and increased [^{3}H]thymidine labelling at 3–4 weeks, the period of most intense regenerative activity (Fig. 5). In addition, regeneration of cells in the inner retinal layers (including the inner nuclear and ganglion cell layers)

FIG. 3. A. Goldfish retina seven days after intraocular injection of ouabain. The laminated neural retina is destroyed; blood vessels (V) at the vitreal surface are intact; some pyknotic rod nuclei remain in the outer nuclear layer (ON); the retinal pigmented epithelial cells (RPE) have detached from the basal lamina and are engorged with photoreceptor debris, but have not penetrated into the neural retina; the outer limiting membrane is intact (arrowheads). Calibration bar, 50 μm.

B. A focal site of regeneration (arrowhead) four weeks after ouabain injection. The outer nuclear layer (ON) adjacent to the column of neuroepithelial cells consists of a single row of immature cones. Calibration bar, 25 μm.

C. Immature cones (arrowheads) in another retina four weeks after ouabain treatment. Calibration bar, 25 μm.

D. Regenerated retina at 10 weeks after ouabain injection. The outer nuclear layer (ON) is reconstituted across the entire retina; the density of cones is restored to normal levels, but the density of rods remains low. The inner retinal layers are not as well organized, and there are periodic fusions between the inner nuclear layer (IN) and the ganglion cell layer (GC). Calibration bar, 50 μm.

E. Another retina also at 10 weeks after ouabain, but in this case the inner retinal layers were destroyed and have not regenerated. The outer nuclear layer (ON) is intact and the photoreceptors are normal. Calibration bar, 50 μm.

F. The margin of the retina shown in E. The arrowhead indicates a zone of sharp transition between unregenerated central retina to the right and normal retina to the left which has been added as part of the normal growth process resulting from proliferation of neuroepithelial cells in the germinal zone (*). Calibration bar, 100 μm.

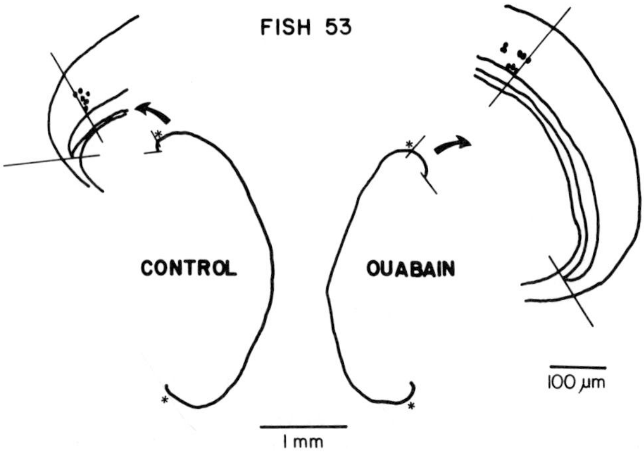

Total retina	4.50 mm	3.90 mm
Original retina	4.38	3.17
Annulus	0.09	0.44

FIG. 4. Comparison of the amount of new growth added at the periphery as a result of mitotic activity in the germinal zone in the left (intact) retina and the right (experimental) retina of a goldfish, four weeks after injection of ouabain into the right eye. The fish was injected intraocularly in both eyes with [^{3}H]thymidine 24 h before ouabain was injected into the right eye only. Methacrylate sections were processed for autoradiography. The semicircular lines in the centre of the figure represent camera lucida tracings of the linear extent of the retina in meridional sections through the centre of each globe. Asterisks mark the position of a wedge of heavily labelled cells—immediate progeny of the neuroepithelial cells in the germinal zone which incorporated [^{3}H]thymidine.

The labelled cells at one margin of each eye are shown at higher magnification to the right and left. The increment of new retina added in the four weeks between injection and death was determined by measuring the linear extent of retina from the peripheral boundary of the labelled wedge of cells to the origin of the inner plexiform layer adjacent to the germinal zone at the retinal margin. (Lines orthogonal to the retinal layers and intersecting the point of origin of the inner plexiform layer and the edge of the labelled wedge are shown, and delimit the annulus of new retina added.) The linear extents (in mm) of the total retina, original retina (central to the labelled wedge) and annulus of new retina (peripheral to the labelled edge) are given for each retina.

is strongly correlated with the presence of regenerating photoreceptors (Raymond et al 1988). A plausible interpretation of these observations is that rod precursors located in the outer nuclear layer are the source of the neuroepithelial foci that are responsible for replenishing retinal neurons. The implication of this suggestion is that the developmental programme of rod precursors is altered only when the local microenvironment around them is disturbed.

This hypothesis, that the outer nuclear layer must be damaged in order to provoke regeneration of any retinal neurons, independent of their phenotype or the layer in which they are destined to reside, was examined further by killing retinal neurons with more selective neurotoxins. For example, 6-hydroxydopamine (6-OHDA) at low doses is selectively toxic to catecholaminergic neurons, especially dopaminergic cells, which in the goldfish retina constitute a subset of the interplexiform cells (IPCs) (Dowling & Ehinger 1978). Intraocular injections of 6-OHDA have been used to destroy dopaminergic IPCs in goldfish (Negishi et al 1990). Negishi and colleagues (1987) had reported that when a minimal dose of 6-OHDA was administered, the dopaminergic IPCs were permanently ablated, but when high doses of the drug were used, the dopaminergic neurons regenerated within several months. We and others (Heikkila & Cohen 1975) have reasoned that at high concentrations non-specific cellular damage might occur, since 6-OHDA is chemically unstable, and in solution it is oxidized to its semiquinone with the release of hydrogen peroxide and free radicals (Cohen & Heikkila 1974), highly reactive compounds known to be damaging to neural tissue (Hall 1989). Results presented by Negishi et al (1987) already provided some support for this proposal, since at the high doses of 6-OHDA, indoleaminergic neurons were lost in addition to the dopaminergic cells.

To test this idea we quantified the number of cells in the inner (INL) and outer (ONL) nuclear layers after injecting high (0.58 μg/μl, estimated intraocular concentration) and low (0.14–0.29 μg/μl) doses of 6-OHDA and looked for a relationship between cell loss and replacement of dopaminergic IPCs (Braisted & Raymond 1990a). In order to unambiguously identify dopaminergic IPCs that had regenerated, we gave two injections of BrdU, at one and two weeks after the 6-OHDA injection. Dopaminergic cells were recognized immunocytochemically with antibodies against tyrosine hydroxylase (TH), the synthetic enzyme for dopamine (Negishi et al 1990). By three weeks, TH^+ cells had reappeared in the retinas exposed to the high dose of 6-OHDA (but not the low dose) and some of these were also $BrdU^+$, proving that they had regenerated. The density of cells in both the INL and the ONL in the area surrounding the regenerated dopaminergic IPCs was about 35% lower than in control (intact) retinas, whereas the density of cells in retinas exposed to the low dose was not different from controls. (Dopaminergic IPCs had gone in both experimental conditions, of course, but they represent such a small fraction of the cells in the INL, less than 1% [Negishi et al 1990], that their absence alone has no measurable effect on total cell number.)

Finally, in a study by Hitchcock (1989), ganglion cells in the goldfish retina were selectively destroyed by retrograde ('suicide') transport of propidium iodide inserted into the optic nerve. They were not replaced, even after six months. The results of all these experiments are consistent with our earlier ouabain studies, and further support the hypothesis that damage to the ONL is necessary to provoke the regenerative response that leads to replacement of neurons in inner layers.

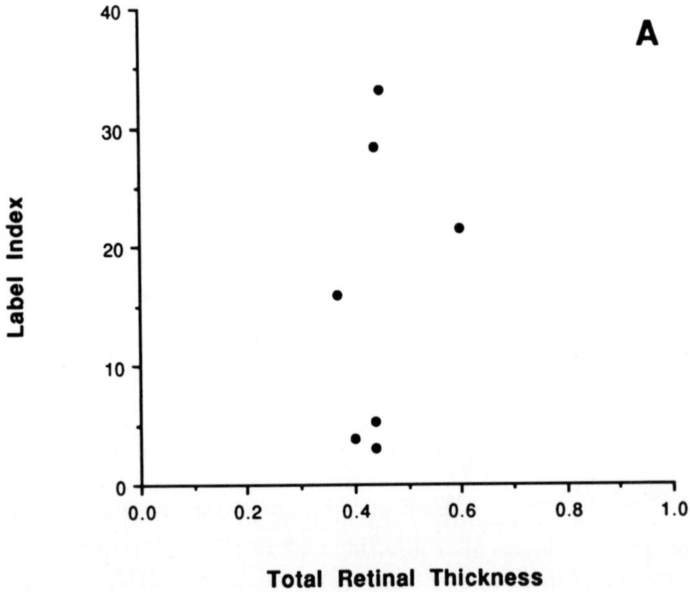
A
Label Index
0
10
20
30
40
0.0
0.2
0.4
0.6
0.8
1.0
Total Retinal Thickness

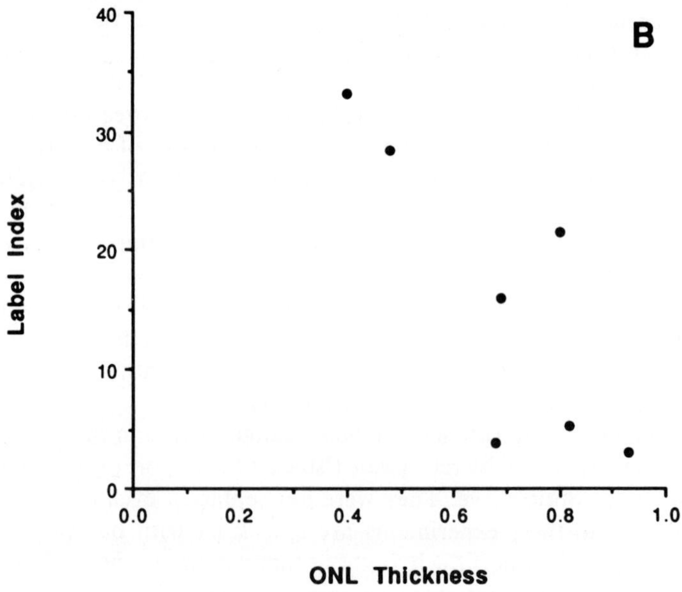
B
Label Index
0
10
20
30
40
0.0
0.2
0.4
0.6
0.8
1.0
ONL Thickness

Summary and conclusions

Loss of cells in the outer nuclear layer appears to be a necessary condition for retinal regeneration in goldfish. If only a selected population of retinal neurons in the inner layers is destroyed (e.g. ganglion cells or interplexiform cells), they are not replaced. However, these (and other) neurons can be regenerated, but only when more extensive damage is done, and the damage must include loss of photoreceptor cells in the outer nuclear layer.

Why must photoreceptors be destroyed in order for other neurons to be regenerated?

We suggest that the explanation for this at first perplexing observation is that the source of proliferating neuroepithelial progenitor cells, from which the regenerated retinal neurons must be derived, is the scattered population of proliferating rod precursor cells located in the outer nuclear layer. Although in the intact retina the rod precursors are morphologically distinct from the demonstrably pluripotent, primitive neuroepithelial cells in the germinal zone at the margin of the retina (the former are not elongated and are not attached to the ventricular surface), they appear *not* to be developmentally restricted to the rod pathway of development. The apparent lineage restriction of rod precursors in the intact retina is probably a consequence of positional cues and cell–cell interactions in the local microenvironment and does not represent an autonomous or intrinsic commitment to the rod fate (Raymond 1990). Further evidence for their lack of commitment to the photoreceptor fate is that they do not express the visual pigment, opsin (Knight & Raymond 1990). Their

---

FIG. 5. Stimulation of cell proliferation as a function of retinal damage. The right eyes of seven goldfish were injected with ouabain and three weeks later the fish were injected intraperitoneally with [^{3}H]thymidine. After another week, methacrylate sections of right (experimental) and left (control) retinas were processed for autoradiography. All labelled cells were counted in six sections from each retina. The thickness of the entire neural retina (from outer to inner limiting membrane) and of the outer nuclear layer (ONL) was measured. The labelling index, defined as the ratio of the number of labelled cells per section on the experimental versus the control side, was calculated for each fish. This index (ordinate) is plotted as a function of the ratio (experimental/control) of total retinal thickness in A and as a function of the ratio of ONL thickness in B. In A, note that in six out of seven fish the ouabain-treated retina was slightly less than half as thick as the intact retina; in the seventh fish the retina was not so badly damaged (ratio of 0.6). The labelling index, however, varies over a ten-fold range with 3–30 times as many labelled cells on the experimental side as on the control side. Panel B shows that the labelling index is inversely proportional to the ratio of ONL thickness: the greatest stimulation of proliferation was seen in the retinas with the most severe loss of cells (ratios of 0.5 or less), whereas retinas with little damage to the ONL (ratios close to 1.0) tended to have a lower labelling index. (Data replotted from Raymond et al 1988.)

somewhat altered morphology is likewise a consequence of their residence in the differentiated retina; when the cellular structure around them is destroyed, they revert to the more primitive, neuroepithelial cell phenotype.

Regeneration requires a complete developmental programme to be turned on

Whenever regeneration of retinal neurons is observed in the goldfish retina, a sequence of events takes place that in some respects mimics embryonic development. Of course, the overall spatiotemporal pattern differs, in that regeneration occurs asynchronously from scattered foci of neuroepithelial cell clusters, whereas during development the presumptive retina is a continuous, homogeneous sheet of neuroepithelial cells from which neurons differentiate progressively in a circular wave that sweeps from centre to periphery. However, in spite of differences in the broad, overall pattern of retinal cell production and maturation, at each individual locus on the regenerating sheet of retinal tissue, the developmental programme appears to be replayed in a manner that is faithful to the original scheme. In particular, the production of cone photoreceptors precedes that of rods, and rods are not generated until other retinal neurons have become postmitotic. At a very local level, then, it is quite accurate to say that regeneration recapitulates development.

These observations together suggest that rod precursors produce rods only when the opportunity to produce other types of retinal neurons, including cones, is lost. An alternative interpretation is that selection of the rod developmental fate requires inductive influences produced by or in response to cohorts of retinal cells generated earlier. The molecular signals that regulate selection of cell fate and that control the location and the numbers of different cell types produced in the vertebrate retina during development and regeneration are at present completely unknown (Braisted & Raymond 1990b). In *Drosophila* the mechanisms underlying the selection of neuronal (Knust & Campos-Ortega 1989) and photoreceptor (Zipursky 1989) cell fate are beginning to be unravelled with the aid of the powerful combined tools of genetics and molecular biology. In vertebrates the search has only just begun in earnest for developmentally regulated cell surface receptors and molecular signals that alter gene expression and thereby determine cell fate.

Acknowledgements

Supported by Public Health Service National Institutes of Health grant EY04318. I thank Ms Linda Barthel for help with the illustrations.

References

Barthel LK, Raymond PR 1990 Improved method for obtaining 3-μm cryosections for immunocytochemistry. J Histochem Cytochem 38:1383–1388

Blaxter JHS, Staines M 1970 The development of the retina and retinomotor responses in the herring (*Clupea harengus*). J Mar Biol Assoc UK 50:677–697

Braisted JE, Raymond PA 1990a 6-Hydroxydopamine and proliferation in goldfish retina. Soc Neurosci Abstr 16:173

Braisted JE, Raymond PA 1990b Lineage versus environment as a determinant of neuronal phenotype. In: Raymond PA, Easter SS, Innocenti GM (eds) Systems approaches to developmental neurobiology. Plenum Press, New York, p 29–39

Cohen G, Heikkila RE 1974 The generation of hydrogen peroxide, superoxide radical and hydroxyl radical by 6-hydroxydopamine, dialuric acid, and related cytotoxic agents. J Biol Chem 249:2447–2452

Dowling JE, Ehinger B 1978 The interplexiform cell system. I. Synapses of the dopaminergic neurons of the goldfish retina. Proc R Soc Lond Biol Sci 201:7–26

Hall ED 1989 Free radicals and CNS injury. Crit Care Clin 5:793–805

Heikkila RE, Cohen G 1975 Cytotoxic aspects of the interaction of ascorbic acid with alloxan and 6-hydroxydopamine. Ann N Y Acad Sci 258:221–230

Hitchcock PF 1989 Exclusionary dendritic interactions in the retina of the goldfish. Development 106:589–598

Holt CE, Bertsch TW, Ellis HM, Harris WA 1988 Cellular determination in the *Xenopus* retina is independent of lineage and birth date. Neuron 1:15–26

Johns PR 1977 Growth of the adult goldfish eye. III. Increase in retinal cell number. J Comp Neurol 176:331–342

Johns PR 1982 Formation of photoreceptors in larval and adult goldfish. J Neurosci 2:178–198

Johns PR, Fernald RD 1981 Genesis of rods in teleost fish retina. Nature (Lond) 293:141–142

Keefe JR 1973a An analysis of urodelian retinal regeneration. I. Studies of the cellular source of retinal regeneration in *Notophthalmus viridescens* utilizing ^{3}H-thymidine and colchicine. J Exp Zool 184:185–206

Keefe JR 1973b An analysis of urodelian retinal regeneration. IV. Studies of the cellular source of retinal regeneration in *Triturus cristatus carnifex* using ^{3}H-thymidine. J Exp Zool 184:239–258

Knight JK, Raymond PA 1991 Time course of opsin expression in developing rod photoreceptors. Development 110:1115–1120

Knust E, Campos-Ortega JA 1989 The molecular genetics of early neurogenesis in *Drosophila melanogaster*. BioEssays 11:95–100

Kurz-Isler G, Wolburg H 1982 Morphological study on the regeneration of the retina in the rainbow trout after ouabain-induced damage: evidence of dedifferentiation of photoreceptors. Cell Tissue Res 225:165–178

Lyall AH 1957 The growth of the trout retina. Q J Microsc Sci 98:101–110

Maier W, Wolburg H 1979 Regeneration of the goldfish retina after exposure to different doses of ouabain. Cell Tissue Res 202:99–118

Müller H 1952 Bau und Wachstum der Netzhaut des Guppy (*Lebistes reticulatus*). Zool Jahr Abt Allg Zool Physiol Tiere 63:275–324

Negishi K, Teranishi T, Kato S, Nakamura Y 1987 Paradoxical induction of dopaminergic cells following intravitreal injection of high doses of 6-hydroxydopamine in juvenile carp retina. Dev Brain Res 33:67–69

Negishi K, Teranishi T, Kato S 1990 The dopaminergic system of the teleost fish retina. In: Osborne N, Chader J (eds) Progress in retinal research. Pergamon Press, Oxford, vol 9:1–48

Raymond PA 1985a The unique origin of rod photoreceptors in the teleost retina. Trends Neurosci 8:12–17

Raymond PA 1985b Cytodifferentiation of photoreceptors in larval goldfish: delayed maturation of rods. J Comp Neurol 236:90–105

Raymond PA 1988 Evidence that the cellular source for retinal regeneration in goldfish is within the differentiated retina. Soc Neurosci Abstr 14:1198

Raymond PA 1990 Cell determination and positional cues in the teleost retina: development of photoreceptors and horizontal cells. In: Lam DMK, Shatz C (eds) Development of the visual system. MIT Press, Cambridge, MA (Retina Res Found Symp 3) p 59–78

Raymond PA, Barthel LK 1988 Retinal regeneration in goldfish: use of cell specific monoclonal antibodies to assess the reestablishment of normal cytoarchitecture. Soc Neurosci Abstr 15:808

Raymond PA, Rivlin PK 1987 Germinal cells in the goldfish retina that produce rod photoreceptors. Dev Biol 122:120–138

Raymond PA, Reifler MJ, Rivlin PK 1988 Regeneration of goldfish retina: rod precursors are a likely source of regenerated cells. J Neurobiol 19:431–463

Reh TA, Nagy T 1987 A possible role for the vascular membrane in retinal regeneration in *Rana catesbienna* tadpoles. Dev Biol 114:463–469

Reh TA, Nagy T, Gretton H 1987 Retinal pigmented epithelial cells induced to transdifferentiate to neurons by laminin. Nature (Lond) 330:68–71

Reh TA, Jones M, Pittack C 1991 Common mechanisms of retinal regeneration in the larval frog and embryonic chick. In: Regeneration of vertebrate sensory receptor cells. Wiley, Chichester (Ciba Found Symp 160) p 192–208

Reyer RW 1977 The amphibian eye: development and regeneration. In: Crescitelli F (ed) The visual system of vertebrates. Springer-Verlag, New York (Handbook of sensory physiology) vol 7/5:309–390

Sandy JM, Blaxter JHS 1980 A study of retinal development in larval herring and sole. J Mar Biol Assoc UK 60:59–71

Stroeva OG, Mitashov VI 1981 Developmental potential of vertebrate eye tissues: regeneration of retina and lens. In: Kruschcov N (ed) Problems of developmental biology. MIR Publishers, Moscow, p 168–207

Turner DL, Cepko CL 1987 A common progenitor for neurons and glia persists in rat retina late in development. Nature (Lond) 328:131–136

Walls GL 1967 The vertebrate eye and its adaptive radiation. Hafner, New York

Wagner EC, Raymond PA 1991 Müller glial cells of the goldfish retina are phagocytic *in vitro* but not *in vivo*. Exp Eye Res, in press

Wetts R, Fraser SE 1988 Multipotent precursors can give rise to all major cell types of the frog retina. Science (Wash DC) 239:1142–1145

Zipursky SL 1989 Molecular and genetic analysis of *Drosophila* eye development: *sevenless*, *bride of sevenless* and *rough*. Trends Neurosci 12:183–189

DISCUSSION

Steinberg: The pigment epithelial cell is not the cell that is directly involved here in the regeneration of the neural retina, so what is the need for the pigment epithelium to be present? It is possible that it could be supplying growth factors; it might be responding to the damaged cells, themselves; and it could be an essential cell in the regeneration sequence.

Raymond: Those are certainly possibilities. We have not destroyed the pigment epithelium in our experiments, so we do not know whether it plays a necessary, albeit indirect, role in the regeneration of the neural retina.

Reh: Along the same lines, perhaps that is one of the other explanations why, if you don't damage the outer nuclear layer and get degeneration of rods, you don't get concomitant activation of the pigment epithelium, and you wouldn't get secretion of one of these putative factors. The pigment epithelial cells are really associated very closely with the retina and are almost reactive-looking in these regenerating stages.

Steinberg: It's also possible that oubain and 6-hydroxydopamine might activate the pigment epithelium in some way.

Raymond: That's right. They are both depolarizing agents and potentially have effects on ionic conductances, and may thereby alter levels of intracellular messengers.

With regard to your information about fibroblast growth factor, Dr Steinberg (Faktorovich et al 1990), it has been demonstrated that RPE cells can synthesize basic fibroblast growth factor (bFGF) (Schweigerer et al 1987). It has also been shown that bFGF may be implicated in retinal regeneration in the chick embryo (Park & Hollenberg 1989). It seems entirely possible that there could be some effect of bFGF in the regenerating fish retina.

We have looked for bFGF; with anti-FGF antibodies we find localization not in the RPE but rather in the horizontal cells in goldfish retina (Raymond 1991). We do not know what that means yet, but we suspect that growth factors are involved in regeneration. As to which growth factors are involved, what cells are producing them, and how the process of regeneration is controlled, we are only now learning the right questions to ask in trying to understand the mechanisms.

Steinberg: Do ganglion cell axons regrow into the optic nerve? How far have you followed that?

Raymond: We have not looked at axonal regrowth. Stuermer et al (1985) suggest that axons from regenerated ganglion cells in goldfish regrow into the optic tectum, but because there is continual new growth at the retinal margin, it will require more careful studies to demonstrate that the axons entering the tectum are from regenerated ganglion cells, as opposed to new ones added at the margin.

Steinberg: You could 'back label' the retinal ganglion cells from the tectum and determine where the cell bodies are located in the retina.

Raymond: Yes. We haven't done those experiments. Clearly, it is important to demonstrate that the optic tectum becomes reinnervated from these regenerating retinas, in order to assess the functional recovery of vision.

Lewis: Can you selectively destroy the photoreceptors? And if so, what happens?

Raymond: That's an excellent idea, and of course that is what we are trying to do! A number of agents have been shown to kill photoreceptors, while sparing most other retinal cells. We would especially like to be able to just kill cones, and see whether the rod precursors replace them. Because cones are generated first, during the early stages of retinal development, whereas rods are added much later, we think that cones may be the preferred pathway of photoreceptor differentiation, and that selection of the rod pathway of differentiation requires inductive influences and interactions with other cell types (Raymond 1991). If cones are the 'preferred' pathway, and one could destroy the cones selectively, would the rod precursors revert to producing the cone phenotype? We are trying to do those experiments but, so far, we have not been able to specifically target cones. We can destroy both rods and cones and new photoreceptors are generated in the correct sequence: first cones, then rods (Raymond et al 1988a).

Rubel: So you have a lesion that kills *all* photoreceptors?

Raymond: It is not as good as that. A number of agents, including tunicamycin and iodoacetic acid, have been used to kill photoreceptors in several species, but the loss of cells is never complete. Whereas constant light kills photoreceptors in the albino rat, for example, it does not do so in the goldfish (Raymond et al 1988b).

Costanzo: Could you use laser irradiation?

Raymond: That has been tried by Steve Easter (unpublished work). It is hard to focus the laser beam into the goldfish eye because of its small size and, furthermore, the melanin of the RPE absorbs the light, and the result is a heat lesion. You can kill photoreceptors in that way, but it is difficult to confine the damage to the layer of photoreceptors; cells in other layers are also damaged.

Rubel: So if you kill photoreceptors and also the RPE with laser irradiation, what happens in terms of the regeneration?

Raymond: I don't know.

Margolis: Are there albino goldfish? Could you do the light damage experiment in them, if their PE is also albino?

Raymond: There are albino strains of goldfish which are pink-eyed, but I don't know whether this would work. There are also albino zebrafish.

Pujol: Certain reptiles (such as crocodiles and geckos, for example), which live in darkness, are said to have only rods in their retinas. Can you speculate on developmental events in such retinas? You suggest that the cone developmental pathway is the first pathway, so what happens to make the cones disappear in these species? What is the regulation of that?

Raymond: Many retinas that are supposedly all rods have a few cones in the adult. I do not know about the course of events in embryonic development in the species that you mentioned.

Fernald: A good precedent is the ultraviolet-sensitive cone, which is almost certainly present in many species of vertebrates during embryogenesis, but is then eliminated (Hawryshyn & Beauchamp 1985). I think this phenomenon is more widespread than people realize; in development, cones always appear first, and then, much later, rods.

Raymond: Yes; wherever you look, whether in mammals, or in many species of fish, there are first cones in the embryonic retina, and rods appear later.

Cotanche: Why do you not find any kind of cone-generating precursors in the adult retina, comparable to the rod precursors?

Raymond: There is a teleological explanation for that! The cone is typically thought of as a 'photopic photoreceptor', active in daylight, interested in distinguishing shape and form, and responsible for the acuity of the eye. Acuity is a measure of resolution and is dependent upon the cone spacing in the photoreceptor mosaic. As the eye grows, the retina is stretched, and cone cells decrease in density as they are pulled apart. One might therefore expect acuity to fall, with a resulting decrease in resolution, but the lens grows as well, so the image that is projected onto the retina is magnified in the bigger eye. The theoretical acuity—the limit of the resolution of the eye—is not compromised, because of the growth of the lens.

In contrast, rods are 'scotopic photoreceptors'. In dim light, the strategy of the eye is to trap all available photons coming in; in other words, to maximize sensitivity. If the rods were to spread apart as the eye grows, 'holes' would be created through which photons could escape, and the sensitivity of the eye would be compromised. To counteract the effect of retinal stretching a constant density of rods is maintained by continued addition. Powers et al (1988) have measured the visual sensitivity of small and large goldfish, and found that it is approximately constant. We think this explains why new rods are added but not new cones.

Voyvodic: You made the point that the rod precursor cells are morphologically different from the dividing cells of the developing neuroepithelium. In regeneration, when you believe that the rod precursors reiterate the full developmental programme and give rise to all of the retinal cell types, do they re-extend radial processes, to resemble immature neuroepithelial cells?

Raymond: Yes. We have not done serial reconstruction, so I cannot say this with the same degree of confidence as I can for the developing system, but from looking at random sections in the electron microscope and in the light microscope, it appears that the proliferating cells from which regenerated neurons derive are elongated and spindle-shaped. I do not know if these cells extend the full width of the retina, as they do in the germinal zone, but they are clearly no longer the round or tear drop-shaped precursors that we see in the intact retina. However, the morphology of the rod precursors is probably just a reflection of mechanical constraints of the environment that they find themselves in. In the intact retina they are surrounded by differentiated neurons and neuronal processes, and they simply cannot span the whole width of the retina as neuroepithelial cells do during early development, when there is more space available. We nevertheless believe that rod precursors are neuroepithelial cells with a full range of developmental potentials.

Corwin: How frequent are the rod precursors? What is their distribution in the intact retina?

Raymond: It is enormously variable from one individual to the next, and there is also a circadian rhythm to their proliferation (Stell et al 1988). We suspect that the frequency of occurrence of rod precursors is also related to the rate of growth of the individual fish. These goldfish are growing at different rates, and the rod precursors are proliferating at a rate that keeps the density of rods constant in the adult. We showed this by counting rods in different-sized fish (Johns & Easter 1977). Also, if goldfish are starved they of course do not grow, and in a population of starved fish there are fewer mitotically active rod precursors than in a population of fish that are growing rapidly (P. A. Raymond, unpublished work).

Corwin: Have you compared regeneration in starved and non-starved goldfish, and looked at the number of foci?

Raymond: No, but it is a good idea.

Corwin: A technical question: you use 5-bromo-2′-deoxyuridine (BrdU) to label cells passing through S phase. One worries about it being eliminated through the cell's DNA repair mechanisms, whereas [^{3}H]thymidine should not be eliminated, because the chemistry of the tritiated molecule shouldn't be different; there is no difference in the number of electrons.

Raymond: That is so; BrdU is bigger than thymidine, so there are steric hindrances to putting a lot of BrdU molecules into a DNA molecule. We have not studied extremely long survival times with BrdU as we have using [^{3}H]thymidine as a marker. After 2–3 year survival times with the latter, there are still labelled cells. Up to now, we have looked only up to a couple of months with BrdU.

Burd: The other problem with BrdU is that cell fate can be changed, because of effects on differentiation.

Raymond: We worried about that as well. We believe that the toxic effects of BrdU, in particular its ability to block differentiation, are a function of concentration. The concentrations of BrdU at which effects on differentiation have been seen are 100–1000 μM for rat embryos in organ culture (Biggers et al 1987). Using intraocular injection, we get good labelling at 5 μM (estimated intraocular concentration), and we typically work at around 10 μM; so the dose we use is one to two orders of magnitude below the toxic dosage.

Watt: You use a monoclonal antibody, RET1, to identify neurons and proliferating Müller glial cells. Are there any other monoclonal antibodies that see cell type-specific nuclear antigens?

Raymond: Not that we have generated.

Watt: Do you have any evidence that this reactivity changes during the cell cycle?

Raymond: No. In general, RET1 only labels postmitotic cells. Immunoreactivity for RET1 appears during the development of the retina only after differentiation begins. There are immature cells that are postmitotic but still not yet expressing RET1, so this marker is certainly not the earliest evidence of differentiation.

Watt: Outside the eye, do any other tissues stain with this antibody?

Raymond: Not that we have seen. We looked in muscle, liver, and brain in the goldfish. An occasional cell in brain is stained, but none stains in liver or muscle.

References

Biggers WJ, Barnea ER, Sanyal MK 1987 Anomalous neural differentiation induced by 5-bromo-2′-deoxyuridine during organogenesis in the rat. Teratology 35:63–75

Faktorovich EG, Steinberg RH, Yasumura D, Matthes MT, LaVail MM 1990 Photoreceptor degeneration in inherited retinal dystrophy delayed by basic fibroblast growth factor. Nature (Lond) 347:83–86

Hawryshyn C, Beauchamp B 1985 Ultraviolet photosensitivity in the goldfish: an independent UV retinal mechanism. Vision Res 25:11–20

Johns PR (Raymond), Easter SS 1977 Growth of the adult goldfish eye. II. Increase in retinal cell number. J Comp Neurol 176:331–342

Park CM, Hollenberg MJ 1989 Basic fibroblast growth factor induces retinal regeneration *in vivo*. Dev Biol 134:201–205

Powers MK, Bassi CJ, Rone LA, Raymond PA 1988 Visual detection by the rod system in goldfish of different sizes. Vision Res 28:211–221

Raymond PA 1991 Cell determination and positional cues in the teleost retina: development of photoreceptors and horizontal cells. In: Shatz C, Lam DMK (eds) Development of the visual system. MIT Press, Boston (Retina Research Foundation Symposium) vol 3:59–78

Raymond PA, Reifler MJ, Rivlin PK 1988a Regeneration of goldfish retina: rod precursors are a likely source of regenerated cells. J Neurobiol 19:431–463

Raymond PA, Bassi CJ, Powers MK 1988b Lighting conditions and retinal development in goldfish: photoreceptor number and structure. Invest Ophthalmol & Visual Sci 29:27–36

Schweigerer L, Malerstein B, Neufeld G, Gospodarowicz D 1987 Basic fibroblast growth factor is synthesized in cultured retinal pigment epithelial cells. Biochem Biophys Res Commun 143:934–940

Stell WK, Karkhanis A, Bremner L 1988 Rod neuroblast proliferation in goldfish: circadian rhythm and detection. Invest Ophthalmol & Visual Sci (suppl) 29:348 (abstr)

Stuermer CAO, Niepenberg A, Wolburg H 1985 Aberrant axonal paths in regenerated goldfish retina and tectum following intraocular injection of ouabain. Neurosci Lett 58:333–338

Common mechanisms of retinal regeneration in the larval frog and embryonic chick

Thomas A. Reh, Michael Jones and Catrin Pittack

Department of Biological Structure, SM-20, University of Washington, Seattle, WA 98195, USA

Abstract. Amphibians and embryonic chicks possess the ability to regenerate retinal neurons by the transdifferentiation of pigment epithelium into neuronal progenitors. Recent studies have begun to identify the molecular factors involved in this process. Laminin (a component of the extracellular matrix) has been shown to be important in the process of retinal regeneration in the larval frog both *in vitro* and *in vivo* and basic fibroblast growth factor (bFGF) stimulates the same process in chicks *in vivo*. To determine the mechanisms by which these factors induce retinal regeneration we studied their effects on cultured chick pigment epithelium cells. bFGF was added to enzymically dissociated chick embryo pigment epithelial cells plated at several different densities on various substrates including laminin. We found that bFGF stimulated proliferation but although the cells lost pigmentation and demonstrated distinct morphological changes, no definitive transdifferentiation could be demonstrated using several neuron-specific antibodies as markers. When the pigment epithelium was cultured as aggregates on a shaker device which prevented flattening and spreading on the substrate a large number of retinal progenitor cells were generated from the pigment epithelium treated with bFGF. The ability of chick pigment epithelium to undergo transdifferentiation thus appears to be dependent on the physical configuration of the cells.

1991 Regeneration of vertebrate sensory receptor cells. Wiley, Chichester (Ciba Foundation Symposium 160) p 192–208

The regeneration of sensory receptor organs is nowhere more dramatic than in the sensory retina of amphibians. When the retina is surgically removed or destroyed by severing its blood supply, complete structural and, frequently, functional restoration of this structure occurs within a few months (see Stone 1960 and Reyer 1977 for review). This remarkable process has attracted the attention of several generations of biologists, yet we are still a long way from explaining why amphibians possess this capability while most other vertebrates do not. Some insight into this question has been obtained by comparing the response to retinal damage in adult amphibians with the response that occurs in embryonic or larval stages of other species. In this review we shall describe tissue culture experiments in both the chick embryo and the larval frog that suggest that common cellular and molecular mechanisms are responsible for the regenerative response observed in these two species. In addition, the understanding of this regenerative process has given rise to important insights into the role of cellular interactions in the normal developmental decisions in ocular histogenesis.

The sensory retina and the pigmented epithelium develop from a region of the anterior neural plate that has been determined at gastrulation to give rise to ocular tissue (Clarke 1936; see Lopashov & Stroeva 1964 for review). Experiments in both frog and chick embryos have indicated that the state of determination of the eye primordia is not immediately fixed; instead, there is a certain amount of lability in whether the cells will adopt a pigmented epithelial phenotype or a sensory retinal phenotype. Even after the cells of the pigmented epithelial layer have become clearly different morphologically and biochemically from the cells of the sensory retina, they retain the capacity to replace the sensory retinal cells after retinal damage by a process of phenotypic switching known as transdifferentiation (Okada 1980).

Although the molecular basis of this switch in phenotype is not yet known, it presumably involves the types of transcription factors that have been found to be involved in the transdifferentiation of fibroblasts to myoblasts that occurs after 5-azacytidine treatment (Lassar et al 1986). It is also not known what factors in the local microenvironment regulate this shift in pigment cell phenotype after retinal damage; however, evidence from both the frog and the chick indicates that two of the molecules important in this process are laminin and basic fibroblast growth factor (bFGF).

Laminin and solubilized basement membrane promote transdifferentiation of frog retinal pigment epithelium into neural retinal progenitors

Several lines of evidence suggest that the extracellular matrix is important in the transdifferentiation of pigment epithelial (PE) cells into neuronal progenitors during retinal regeneration in various species. In regenerating *Rana catesbeiana* retina, after the retina had been destroyed by devascularization, we observed that the first new neuronal progenitors always arose in association with the vitreal vascular membrane (Reh & Nagy 1987). This association was investigated using a monoclonal antibody against frog N-CAM (neural cell adhesion molecule), 2D3, which we generated using *R. catesbeiana* retinal membranes (Reh & Nagy 1987). When the retina was labelled with this antibody during the early stages of regeneration the first N-CAM-positive neuronal progenitor cells were almost always associated with the vitreal vascular membrane, which was visualized by immunolabelling the same sections with either monoclonal antibody 1C2 or anti-fibronectin antiserum.

These results suggested that some component(s) of the vitreal vascular membrane were important in determining the phenotype of retinal cells. We proposed that the composition of the vitreal vascular extracellular matrix might somehow differ from the matrix that the PE cells were normally in contact with on Bruch's membrane, and that when they were exposed to these factors during regeneration, the pigmented epithelial phenotype shifted to the neuronal progenitor phenotype. Initial biochemical characterization of this membrane revealed that it contained a considerable amount of laminin (Reh et al 1987). This correlation was then extended by *in vitro* experiments in which dissociated larval *R. catesbeiana* pigment epithelial cells were cultured on a variety of different substrates, including solubilized basement membrane prepared from the EHS sarcoma (Matrigel) and laminin that had been purified from the same source. The results of these culture experiments are summarized in Table 1. Whereas laminin and the EHS basement membrane were effective promoters of neuronal transdifferentiation, the other substrates that we tried were

TABLE 1 Growth and transdifferentiation of dissociated retinal pigment epithelial cells from *Rana catesbeiana* on various substrates

Substrate	*% of wells containing 2D3-positive neuronal clusters (no. of measurements)*	*No. of γ-crystallin-positive cells per well*	*[^{3}H] thymidine incorporation as d.p.m. per well*
Plastic	10.5 (19)	146 ± 79 (6)	432.4 + 224 (3)
Polylysine	5.2 (19)	–	326.5 + 22.3 (3)
Collagen I	8.0 (27)	172 ± 80 (6)	501.1 + 227 (3)
Fibronectin	0 (15)	–	413 + 108.8 (3)
Laminin	86 (26)	1 (6)	268 + 60.5 (6)
EHS membrane (Matrigel)	94 (27)	7 ± 5 (6)	217.5 + 79.9 (6)

Values are means ± SEM where appropriate. Neuronal clusters were identified as aggregates of small phase-bright cells, bearing long processes (≈ 10 × the cell diameter), that also bind with the neuron-specific monoclonal antibody (2D3) or tetanus toxin. Although neuronal clusters with neuritic processes were never observed on control cultures on plastic substrates, wells were considered positive if they contained any 2D3-staining cells. To establish whether the laminin and Matrigel promoted transdifferentiation to lens we also labelled the cells with γ-crystallin antibodies. We found that instead, these substrates suppressed lens formation. For [^{3}H] thymidine incorporation, cells were grown for two weeks *in vitro* and 1 μl ml^{-1} isotope (20 Ci mmol^{-1}) was added 48 h before trichloroacetic acid (TCA) precipitation. Cells were washed in normal frog Ringer's solution (three times), incubated for 10 min in 100 μl distilled water per well, and frozen and thawed to lyse the cells. Lysate (50 μl) was applied to Whatman No. 1 filter paper squares. The paper was washed three times in 200 ml 10% TCA (4 °C) and washed in ethanol and acetone. Scintillation fluid was added and the mixture was counted on an LKB β-counter. (Modified and reprinted with permission from Reh et al 1987.)

ineffective. This evidence further supported the hypothesis that retinal regeneration is initiated by changes in the composition of the extracellular matrix that is in contact with the pigment epithelial cells early in this process.

bFGF promotes regeneration of chick embryonic retina

Although a number of investigators have reported the transdifferentiation of chick embryo pigmented epithelial cells into neural epithelium after various perturbations of the normal developmental processes of the eye (see, for example, Orts-Llorca & Genis-Galvez 1960), Coulombre & Coulombre (1965) were the first to demonstrate that chick retina would regenerate after removal in early embryonic life (stages 21–26). Curiously, this response occurred only if a small piece of neural retina was reintroduced into the damaged eye, even though the regeneration that was observed was clearly from the remaining pigmented epithelial cells. Following up on these earlier studies, Tsunematsu & Coulombre (1981) found that the same regenerative process could occur in culture, independently of retinal cells. In fact, neural retinal tissue, added to the cultures, actually inhibited the process of transdifferentiation, instead of promoting it.

Since the neural retina contains considerable amounts of fibroblast growth factor (FGF) (Hicks & Courtois 1988), Park & Hollenberg (1989) tested whether this was the active substance released by the implanted retina in the Coulombres' experiment. By implanting slow-release gels which contained bFGF into the chick embryo optic cup after retinal removal, they demonstrated that only those eyes treated with bFGF showed regeneration, whereas eyes implanted with control gels failed to regenerate. Figure 1 shows the response observed when the eyes were examined seven days after retinectomy. Other growth factors, such as EGF and TGF-β, incorporated into similar slow-release gels, do not elicit this response (C. M. Park, personal communication).

Because these experiments were carried out in whole chick embryos, it is possible that the bFGF may have been stimulating some other cell type in the embryo to release other factor(s) that were responsible for the regeneration. In addition, since Tsunematsu & Coulombre (1981) had found that sensory retina was inhibitory for transdifferentiation of the pigment epithelium in culture, we were interested to see whether this process becomes independent of bFGF *in vitro*. We therefore investigated whether bFGF could induce the formation of neural retina in dissociated cell cultures of PE from embryonic chicks.

bFGF induces morphological change but not transdifferentiation in chick PE cells

Pigment epithelium was dissected from the stages of chick embryos that had been shown previously to be capable of regeneration. After dissociation (using trypsin EDTA) the cells were plated either at high density, or as small clusters of

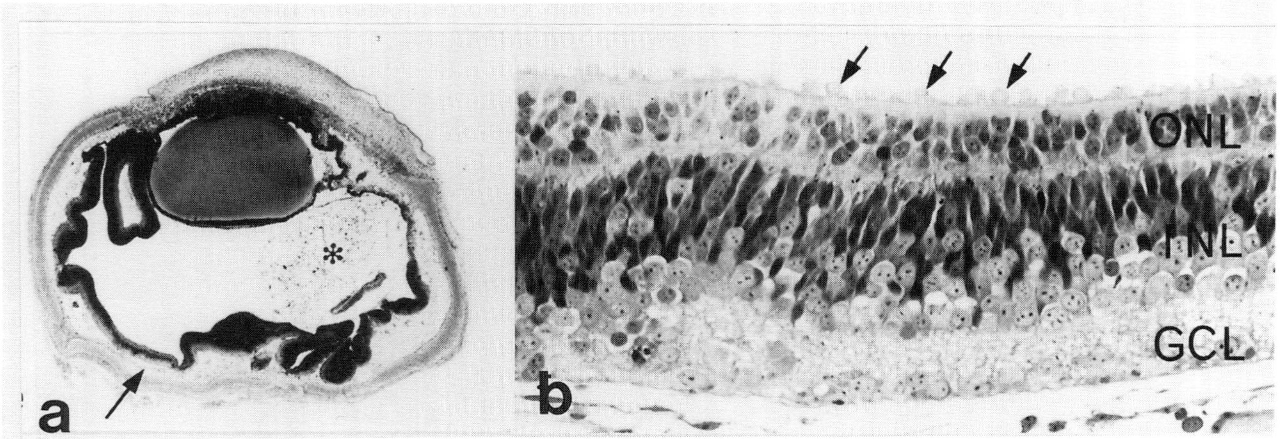

FIG. 1. Effects of bFGF on the regeneration of the embryonic chick retina *in ovo*, seven days after retinectomy at stages 22–24. The regenerated retina, shown in (a) and in (b) at a higher magnification, has already differentiated to the extent that photoreceptor outer segments are beginning to develop (arrows). *, denotes the slow-release gel in (a); ONL and INL, outer and inner nuclear layers; GCL, granule cell layer. (Reprinted from Park & Hollenberg 1989 with permission.)

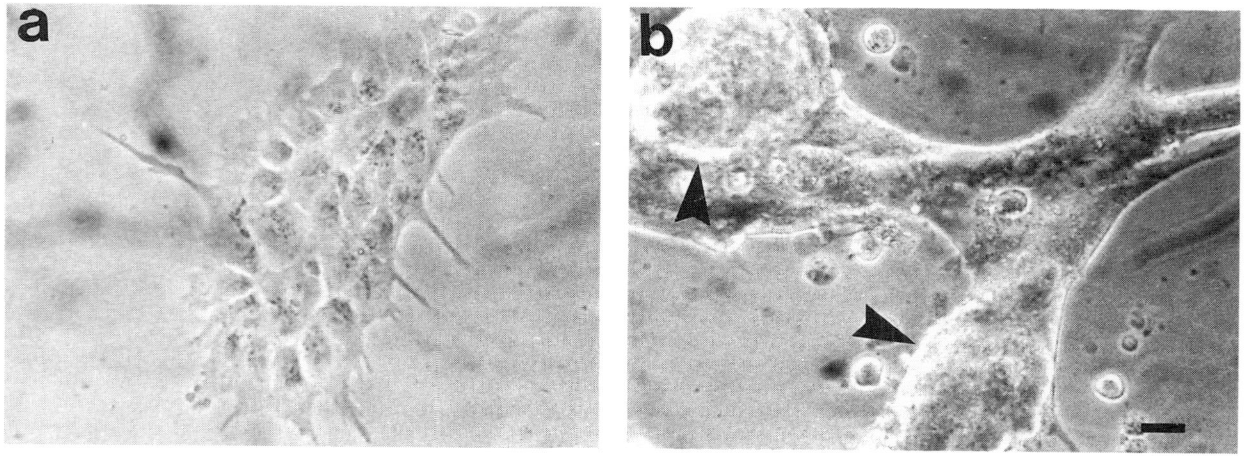

FIG. 2. Effects of bFGF on dissociated cell cultures of embryonic chick pigment epithelial cells. A small cluster of pigment epithelium cells (a) was placed into a Matrigel-coated well of a 24-well tissue culture plate and cultured with bFGF (10 ng/ml) for two weeks (b). Most of the cells had lost their pigment, and large aggregates or balls of cells (arrowheads) had developed in the cultures (b); however, these aggregates showed no immunohistochemical or morphological evidence of neuronal or neuroepithelial phenotypes. Scale bar, 30 μm.

<50 cells (often as few as 10), each in a well (Fig. 2a) on laminin, Matrigel or polylysine. In some wells, 10 ng/ml bFGF was added daily. Both bFGF-treated and control cultures actively proliferated; after 1–2 weeks, the few PE cells in the original culture could give rise to a confluent monolayer, regardless of the substrate onto which the cells were plated. Whereas the cells in the control cultures retained their pigmented phenotype, the cultures treated with bFGF began to show several morphological changes; most of the cells in these cultures lost their pigment and large aggregates had formed after 2 weeks (Fig. 2b). These aggregates resembled those found in the cultures of Tsunematsu & Coulombre (1981), so we examined them for the presence of neuroepithelial cells or differentiated neurons with immunohistochemical methods, in both whole cultures and cryostat sections of the aggregates. In over 20 bFGF-treated cultures that we examined, no specific immunoreactivity was observed, using a variety of neuron-specific antibodies. In addition, the aggregates that formed in the bFGF-treated chick cultures never extended neuritic processes onto the substrate—something we typically observed in the *R. catesbeiana* cultures. In over 50 experiments with chick cells from a variety of different embryonic stages, we never found neuronal cells in the cultures, at either high or low density.

We proposed two hypotheses to explain why chick embryo retina can regenerate *in vivo*, but apparently not *in vitro*. (1) The observed regeneration arose from a few neuronal progenitors left behind after removal of the retina, and the FGF stimulated their proliferation. Thus, regeneration at the ciliary margin was taking place from marginal retinal progenitors, and the central zone of regeneration occurred from similar cells along the ventral fissure of the embryonic retina. (2) The phenotypic switching of PE cells occurs only if the cells are in an epithelium, but not when they are dissociated and then allowed to flatten. There is some suggestion from earlier transplant studies that this is in fact the case. In a review of the experimental embryological literature on ocular development, Lopashov & Stroeva (1964) point out that '. . . the retina develops only where the eye rudiment cells form aggregates, and the pigment epithelium only where they spread as a thin layer . . .'

Tests of these hypotheses revealed that PE cells regenerate retina only when cultured as explants

In order to determine whether the cells at or near the ventral fissure of the chick embryo retina are uniquely capable of transdifferentiation we dissected cells from this region, dissociated them as before, and cultured them with or without bFGF. These cells behaved essentially identically to PE from other retinal regions, so we do not consider it likely that *in vivo* regeneration arises only from the cells of the ventral fissure.

We next cultured chick PE as explants, without prior dissociation. Strips of PE from chick embryos of stages 27–30 were dissected and cultured in 24-well

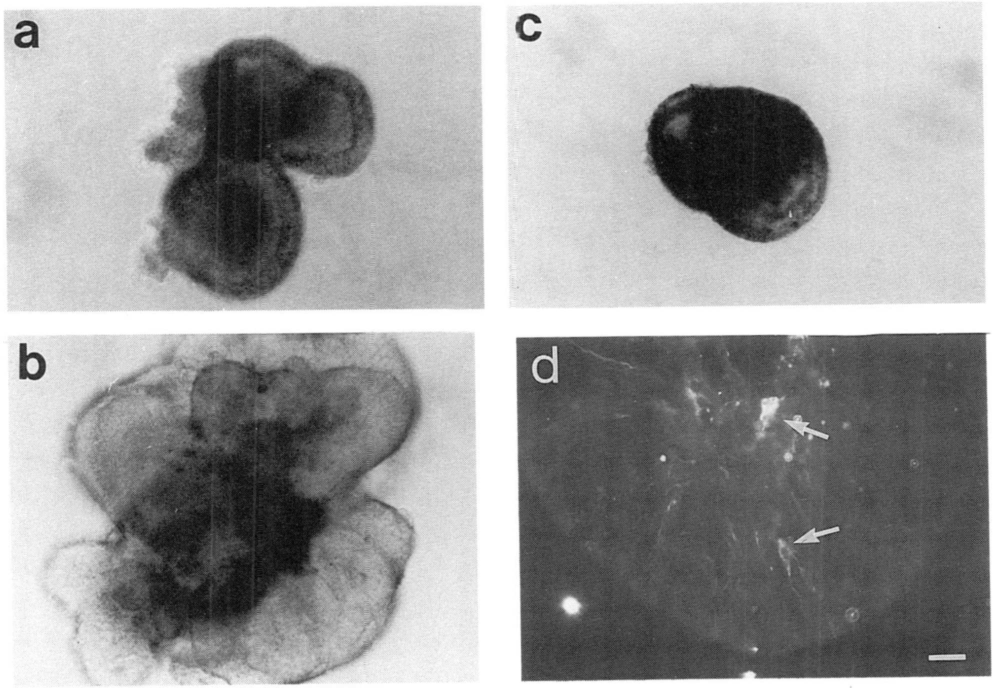

FIG. 3. Effects of bFGF on embryonic chick PE explants, stages 27–30. The cultures treated with bFGF showed a distinct morphological response within 24 h after the addition of bFGF to the medium; the cells began to lose some of their pigmentation (a). In (b), the degree of retinal transdifferentiation is quite extensive after seven days in culture with bFGF; (c) shows a similar PE explant cultured for seven days without bFGF. Sections of the bFGF-treated and control explants were processed for immunohistochemical staining for several neuron-specific antigens. In the micrograph shown in (d), neurofilament protein-immunoreactive cells in a bFGF-treated aggregate (arrows) appear much as they would look in the normal embryonic chick retina (Bennett & DiLullo 1985). The antiserum, raised in rabbit against the middle molecular weight subunit (160 kDa) of the chicken neurofilament protein, was generously provided by Dr G. Bennett.

plates and bFGF was added to some of the wells. Within the first day of culture, some of the bFGF-treated cultures began to lighten up (Fig. 3a) and more closely resemble a neuroepithelium. After seven days, in approximately 40% of the bFGF-treated wells the explants showed definite characteristics of neuroepithelial cells (Fig. 3b); however, this response was never observed in the control cultures (Fig. 3c). Some of the explants were fixed and sectioned for immunohistochemistry. The cells within the explant now appeared to be organized in a pseudostratified columnar epithelium, similar to that normally found in the embryonic retina. In addition, the cells within this neuroepithelium were immunoreactive for N-CAM, neuron-specific enolase, and neurofilament protein, in amounts and distribution that again were highly reminiscent of the neural retina of an embryonic chick (Fig. 3d). Cells with transitional morphology were also observed in sections of aggregates, yet these cells, as well as the PE cells that remained pigmented, did not label with any of the neuron-specific antisera. Therefore, aggregates of PE cells from chick embryos will transdifferentiate into neuronal epithelial cells and retinal neurons when cultured with bFGF (hypothesis 2).

An alternative explanation for these results is that a few neuroepithelial cells remained associated with the PE explants, and bFGF induced these remaining cells to proliferate rapidly. To determine whether our PE strips contained small pieces of neuroepithelium that were not removed in the dissection, we randomly selected a number of strips and sectioned them at 2 μm instead of culturing them. In addition, PE strips that were cultured without FGF were also sectioned and stained with neuron-specific antibodies. We examined the sections of both these control explants for the presence of neuronal progenitors. We found no evidence of neuronal epithelial cells in any of these control explants. We also added bFGF to cultures of chick neuroepithelial cells, to see whether it was mitogenic for such cells. The results of this experiment confirmed that bFGF is a mitogen for neuroepithelial cells: when bFGF was added to stage 27 and 28 retinal cells, they increased in number by two to three times over the control wells, after seven days *in vitro*. However, this degree of stimulation would not be enough to account for the amount of retinal epithelium present in the treated PE cultures.

It therefore appears that bFGF promotes the transdifferentiation of chick PE cells into neuronal precursors in cell culture when the PE is cultured as an explant. Recent *in situ* localization of FGF receptors shows that they are present both on the pigment epithelial cells and on the neuroepithelial cells of the Day 3–4 chick embryo (Fig. 4; Heuer et al 1990), so the observed effects of bFGF may be direct. However, FGF is also a mitogen for virtually all of the cells within the PE aggregate (Fig. 5); is it possible that this is why it has the property of shifting the phenotype of the cells? It is unlikely that mitogenic stimulation alone is responsible for the transdifferentiation of PE cells to the neuronal phenotype, because other growth factors (such as EGF and TGF-β) do not induce a similar

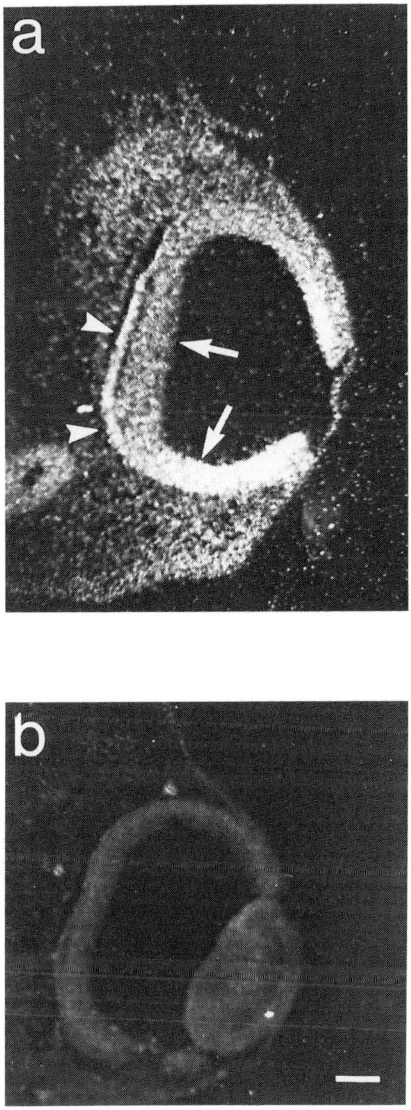

FIG. 4. *In situ* localization of bFGF receptors on both chick retinal neuroepithelial and pigment epithelial cells (at Day 3.5). Chicken FGF receptor riboprobes were used to probe paraffin-embedded and sectioned whole chick embryos at various stages (see Heuer et al 1990 for details of the methods). High levels of labelling in both the PE (arrowheads) and neuroepithelial cell layers (arrows) are observed with the antisense riboprobe (a), while only background labelling is observed with the sense strand (b). (Slides were provided by Dr M. Bothwell.)

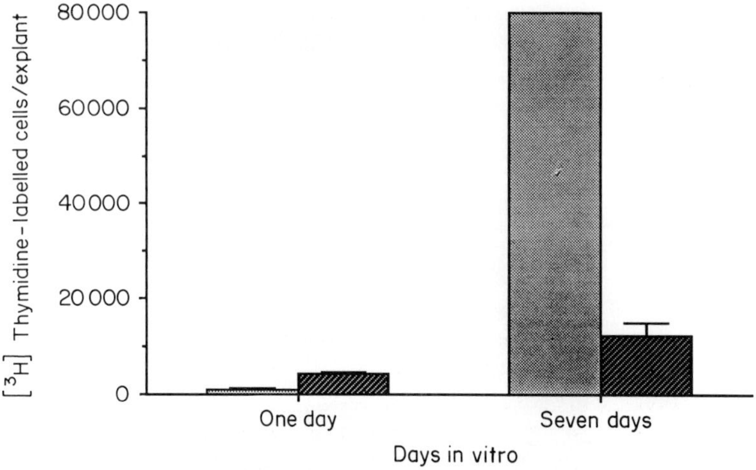

FIG. 5. bFGF is a potent mitogen for pigmented epithelial cells from the embryonic chick retina. PE explants were cultured with or without bFGF for one or seven days in the presence of [^{3}H]thymidine. The explants were then sectioned at 2 µm and processed for autoradiography. Three sections were chosen at random from each explant and the number of [^{3}H]thymidine-labelled cells was counted; the total number of labelled cells in the explant was then estimated as the product of the labelled cells per section and the total number of sections occupied by the explant; this number (ordinate) is illustrated as the mean ± SEM. The shaded bars are the values from cultures containing bFGF; the hatched bars represent cultures without bFGF added. After seven days there is a dramatic increase in the number of labelled cells in the explants cultured with bFGF, as compared to the controls.

regenerative response (C. M. Park, personal communication); however, we hope to further dissociate the mitogenic effects of bFGF from its effects on transdifferentiation by adding other peptide mitogens to our cultures.

Summary: how do laminin and FGF act to promote retinal regeneration?

There is heterogeneity in the composition of the basement membranes with which PE and neural retinal cells are in contact. Laminin is highly represented in the vitreal basement membrane (Reh et al 1987), and perhaps FGF and heparan sulphate proteoglycan are also present in the vitreal basement membrane to a greater degree than in Bruch's membrane. During eye development, differences in position within the optic primordia may be translated into differences in local inductive molecules; that is to say, cells at the interface with ectoderm could be exposed to higher concentrations of laminin and/or bFGF than cells surrounded by mesenchyme. In this scheme, the pigmented phenotype might require a matrix low in FGF and laminin in order to develop, while the neuro-epithelial phenotype would require contact with an environment rich in laminin and bFGF. As these two cell types become differentiated, they reinforce the initial matrix differences, leading to further heterogeneity in the basement membranes.

Stabilization of phenotype leads to the independence of the cell's state of commitment from microenvironmental factors in both chicks and mammals; urodeles and larval anurans retain their dependence on these local environmental factors for the stability of their phenotype. During regeneration of the retina in the frog, the PE cells migrate to the vitreal vascular membrane and respond to the laminin and possibly also its associated bFGF by proliferating and forming neuroepithelial cells; whereas in the early embryonic chick, after removal of the retina, no vitreal vasculature remains, so either a little piece of retina or bFGF is required to stimulate neuroepithelial cell formation.

Why do pigment epithelial cells only transdifferentiate when in an epithelial form rather than when flattened out? We can think of three possibilities. First, it could be that some other cell type within the explant is essential for the transdifferentiation of PE to neuronal epithelial cells, and these are removed in dissociated cell cultures. We are currently investigating this possibility by dissociating PE cells and then allowing them to reaggregate instead of flattening onto the substrate. In these experiments the cells have been dissociated from their underlying basement membrane and the scleral cells, but they should still retain their epithelial (as opposed to flat) morphology. Second, a critical number of cells may be required for the transdifferentiation to occur; Edelman (1988) has proposed that only aggregates can respond to inductive signals, and that the role of CAMs (cell adhesion molecules) or morphoregulatory molecules is to force cells into these aggregates, prior to the appearance of an inductive signal. However, it is unlikely that a critical number of PE cells is all that is required for bFGF to exert its effects, because in high density dissociated cultures, the cells do not transdifferentiate. Third, cytoskeletal changes associated with flattening may influence the expression of particular genes necessary for transdifferentiation to occur. Cell morphology has been shown to affect gene expression in many systems; mesenchymal–epithelial cell transformations during development dramatically influence the pattern of expression of extracellular matrix proteins (see Hay 1981 for a review).

Whichever of these possibilities turns out to be the case, it is clear from the experiments described here that the pigment epithelium of the chick can be induced to undergo a dramatic transdifferentiation *in vitro* when bFGF is added to the culture medium. This process of phenotypic switching should enable us to reach a further understanding of the molecular basis for neuronal determination.

Acknowledgements

The authors gratefully acknowledge the financial support of the RP Eye Research Foundation (Canada) for their financial support to T. A. R. in whose laboratory this work was carried out, at the University of Calgary, Alberta, Canada, and the University of Washington, Seattle. We also thank the University of Washington, School of Medicine Isthmus Program, for their support of M. J. We thank Macie Walker for excellent technical assistance throughout the studies on embryonic chick retinal transdifferentiation.

References

Bennett GS, DiLullo C 1985 Transient expression of a neurofilament protein by replicating neuroepithelial cells of the embryonic chick brain. Dev Biol 107:107–127

Clarke LF 1936 Regional differences in eye-forming capacity of the early chick blastoderm as studied in chorio-allantoic grafts. Physiol Zool 9:102–128

Coulombre JL, Coulombre AJ 1965 Regeneration of neural retina from the pigmented epithelium in the chick embryo. Dev Biol 12:79–92

Edelman GM 1988 Topobiology: an introduction to molecular embryology. Basic Books, New York

Hay E 1981 Cell biology of the extracellular matrix. Plenum Publishing Corporation, New York

Heuer JG, Nainie SF, Wheeler EF, Bothwell M 1990 Structure and developmental expression of the chicken NGF receptor. Dev Biol 137:287–304

Hicks D, Courtois Y 1988 Acidic fibroblast growth factor stimulates opsin levels in retinal photoreceptor cells in vitro. FEBS (Fed Eur Biochem Soc) Lett 234:475–479

Lassar AB, Paterson BM, Weintraub H 1986 Transfection of a DNA locus that mediates conversion of 10T½ fibroblasts to myoblasts. Cell 47:649–656

Lopashov GV, Stroeva OG 1964 Development of the eye. Experimental studies (translated by B Meytar). S Monson, Jerusalem

Okada TS 1980 Cellular metaplasia or transdifferentiation as a model for retinal cell differentiation. Curr Top Dev Biol 16:349–380

Orts-Llorca F, Genis-Galvez JM 1960 Experimental production of retinal septa in the chick embryo. Differentiation of pigment epithelium into neural retina. Acta Anat (Basel) 42:31–70

Park CM, Hollenberg MJ 1989 Basic fibroblast growth factor induces retinal regeneration *in vivo*. Dev Biol 134:201–205

Reh TA, Nagy T 1987 A possible role for the vascular membrane in retinal regeneration in *Rana catesbienna* tadpoles. Dev Biol 122:471–482

Reh TA, Nagy T, Gretton H 1987 Retinal pigmented epithelial cells induced to differentiate to neurons by laminin. Nature (Lond) 330:68–71

Reyer RW 1977 The visual system in vertebrates. In: Crescitelli F (ed) The amphibian eye: development and regeneration. Springer-Verlag, Berlin, p 309–390

Stone LS 1960 Regeneration of the lens, iris and neural retina in a vertebrate eye. Yale J Biol Med 32:464–473

Tsunematsu Y, Coulombre AJ 1981 Demonstration of transdifferentiation of neural retina from pigmented retina in culture. Dev Growth & Differ 23:297–311

DISCUSSION

Watt: In the experiments on the effect of basic FGF on chick embryo retinal pigment epithelium (PE), was the nature of the culture substrate important?

Reh: When the dissociated chick PE cells settled down on a substrate, bFGF had very much the same effects whatever substrate we used, Matrigel, laminin or plastic. In each case FGF made the PE cells form into balls in the culture, but they did not become neuroepithelial cells. In order for the explants to form a neuroepithelium we had to put them in a shaker culture; the cells had to remain suspended. If they attached to the substrate (any substrate) they would start to spread and would not undergo the process of differentiation.

Watt: Have you begun to look at extracellular matrix receptors of the integrin family in this system (see Hynes 1987, Ruoslahti & Pierschbacher 1987)?

Reh: We haven't looked at integrins in the frog, or the chick. We have examined the distribution of the extracellular matrix molecules that interact with integrins. Fibronectin is present in Bruch's membrane but not at the vitreal surface of the sensory retina. At the ventral fissure, the fibronectin immunoreactivity stops, at the transition from pigment epithelium to neuro-epithelium.

Laminin's distribution is intriguing; the immunoreactivity is most intense at the transition between PE and retinal neuroepithelium, both at the ventral fissure and at the peripheral margin of the eye.

Lopashov: Dr Reh, in your earlier paper (Reh & Nagy 1987) you observed regeneration of retina in tadpoles of *Rana catesbeiana*. I did some comparable experiments with the retina in tadpoles (*Rana temporaria* and *R. arvalis*). If I removed the retina and the iris, there was no regeneration at all. But if I retained the iris (both its layers), a perfect regeneration of the retina took place. I divided the inner, semi-transparent layer of the iris into eight parts, and removed them; if one part was retained, regeneration of the retina proceeded, although very rarely. So it seems that some remnants of the inner layer of the iris are necessary for regeneration of the retina (Lopashov 1955).

In your experiments, Dr Reh, you have kept the old retina in place. Perhaps it is not a passive, slowly degenerating structure, but an activator of regeneration from pigmented epithelium? This could lead to this difference, because in my experiments with tadpoles, there is no regeneration if the iris is removed, or its inner layer is fully eliminated, together with retina. There could be an explanation, for your and my results, in the experiments of Coulombre & Columbre (1965, 1970), in which they eliminated the retina from the chick embryo eye and then put a small piece of neural retina inside the empty eye cavity lined with pigmented epithelium. This piece activated the production of the retinal regenerate, which is otherwise always absent. Naturally, in experiments when a piece of pigmented epithelium is put in the posterior cavity of a lensless eye, all conditions for its conversion into retina are fully satisfied.

Reh: So you suggest that in all our experiments in frogs, we had the growth zone at the margin of the retina still intact, and that would have given rise to new neural retina, because then that would presumably have stimulated the regeneration, as you say, of the pigmented cells?

Lopashov: Yes, this is what I am suggesting.

Reh: That is certainly a possibility. We never removed the iris, so one may need something from the neural retina, as Coulombre suggested, and maybe that 'something' is FGF. It appears that FGF can replace this unknown factor, from the experiments of Park & Hollenberg (1989) *in vivo*, but again, we don't know whether they had removed all the neural retina. Some marginal zone may have remained. In culture, FGF can substitute for retina, but only if these cells are cultured as an explant; dissociation seems to stop these cells from undergoing the process of transdifferentiation. If the explants settle down, and flatten out on the substrate, they don't undergo transdifferentiation, even in the presence of FGF.

Lopashov: Naturally Bruch's membrane did not prevent the conversion of pigmented epithelium, because after implantation into the eye cavity it soon detaches from the pigmented epithelium.

Corwin: If one considers regeneration in general, not just of sensory organs but of limb buds and other organs—for instance, Marcus Singer's (1954) experiments which demonstrated that limb regeneration could be evoked in the forelimb in adult anurans, which normally do not regenerate, if a nerve was transplanted into the area—one general conclusion is that the limitation, where we don't see regeneration, is usually the failure to produce undifferentiated cells. If such cells are formed in a natural situation, the embryonic 'programme' that gives rise to the structure of the organ or tissue is present and goes on and forms that organ or tissue as a replacement. It sounds as if the difference between the outcome of supplying FGF when you have dissociated cells and the outcome from explanted cells may relate to its mitogenic property; maybe you have stimulated proliferation in the explant by adding FGF and then, because it is an explant in a normal tissue environment, the embryonic programme that allows differentiation to proceed, allows it to regenerate. Perhaps the basic FGF treatment is having the two effects, but mediated mostly through the proliferative effect.

Reh: We tried to distinguish between those two possibilities with epidermal growth factor (EGF). This growth factor appears to be as mitogenic for these PE cells as FGF; by contrast with FGF, EGF produces an expansion of the numbers of pigment cells, but they remain pigmented and do not form a neuroepithelium. Moreover, dissociated cell cultures also proliferate under exposure to FGF, yet when they proliferate their response is not to form an organized epithelium, or to form neuroepithelial cells at all.

Corwin: But after dissociation, cells are not in the normal, organized pattern of the embryo.

Reh: That is the point I think I am trying to make, that there is something about the normal organization of the epithelium that is important. What is the difference between an epithelium that has its basement membrane associated with it and is floating in the culture medium, and an epithelium that also has the basement membrane and is exactly the same explant, but settles down on the dish and begins to flatten and spread? Some of those cells, within 2–3 days, are still associated with the basement membrane, so far as we can tell, and yet are unable to produce a neuroepithelium.

Steinberg: When the cell sits down on the substrate and flattens, doesn't it becomes a cultured cell of a different type?

Reh: That's what we are seeing here; it's not the presence or absence of the extracellular matrix but a change in the cell state, by just flattening onto the substrate, or whatever occurs within the first couple of days in culture.

Rubel: Have you tried to look at cross-sections of the tissue, in terms of the interactions of the basement membrane with your other cells, to see what kinds of contacts occur between the tissues in the two conditions?

Reh: No. We haven't done any electron microscopy.

Lewis: There is a simple hypothesis that could explain the different behaviour. When the epithelial cells are stuck down on the substrate, nutrients and other molecules don't have access from the basal side, especially if there is a tight-junctional seal at the apical surface of the epithelium. This will isolate the basolateral surfaces of the cells from factors present in the culture medium.

Steinberg: If that's true, couldn't you culture the cells on a porous filter, to determine sidedness?

Reh: We thought about doing that, to see if we could flatten the cells out on a porous filter, to determine whether that would have an effect on their ability to transdifferentiate, once they were settled down on the filter.

Raymond: It would be interesting to see whether the PE cells are polarized; that is, to look at other markers for epithelial polarity that might change in the two cases. There are markers that could be used to distinguish the basolateral surface from the apical surface; for example, Na^{+}/K^{+}-ATPase and various ion channels and transporters involved in ion transport across the epithelium.

Steinberg: It would also be interesting to know the difference between cells in the eye that are in the form of a flattened epithelium and can grow into retina, and the same cells cultured as a flattened epithelium.

Reh: The PE that flattens around the lens, in the regeneration experiments by Park & Hollenberg (1989), remains as PE; yet it is exposed to plenty of FGF. This is the kind of result that had been noticed with transplantation; if you transplant a frog eye early in embryonic development, at a larval eyecup stage, to an ectopic location, it turns out that the ear is one of the best places to induce neural retina formation from the PE. But if the pigment epithelium sits next to a surface where it flattens out, it doesn't undergo this transdifferentiation into neural retina. So my overall feeling is that FGF is found all over the place, and its receptors are all over the place, and we can demonstrate inductive effects with these molecules; but what actually regulates the phenotype of these cells isn't the fact that that molecule is in the right place at the right time, but the cell having the right shape.

References

Coulombre JL, Coulombre AJ 1965 Regeneration of neural retina from the pigmented epithelium in the chick embryo. Dev Biol 12:79–92

Coulombre JL, Coulombre AJ 1970 Influence of mouse neural retina on regeneration of chick neural retina from chick embryonic pigmented epithelium. Nature (Lond) 228:559–560

Hynes RO 1987 Integrins: a family of cell surface receptors. Cell 48:549–554

Lopashov GV 1955 On the quantitative regularities in the regeneration of the retina. CR Acad Sci USSR 105:599–602

Park CM, Hollenberg MJ 1989 Basic fibroblast growth factor induces retinal regeneration *in vivo*. Dev Biol 134:201–205

Reh TA, Nagy T 1987 A possible role for the vascular membrane in retinal regeneration in *Rana catesbienna* tadpoles. Dev Biol 122:471–482

Ruoslahti E, Pierschbacher MD 1987 New perspectives in cell adhesion: RGD and integrins. Science (Wash DC) 238:491–497

Singer M 1954 Induction of regeneration of the forelimb of the postmetamorphic frog by augmentation of the nerve supply. J Exp Zool 126:419–472

Regenerative capacity of retinal cells and the maintenance of their differentiation

G. V. Lopashov

Institute of Gene Biology, USSR Academy of Sciences, Vavilov Street 34/5, Moscow 117334, USSR

Abstract. Mechanisms underlying cell type stability and the capacity of retinal cells for transdifferentiation are discussed. It is shown that cells of amphibian pigmented epithelium can be transformed into retina or lens cells depending on the inducing cell type: the influence of retina enables them to be transformed into retina, the influence of lens epithelium, to lens cells (lentoids or lenses). This led to an attempt to discover the molecular character of cell action by means of transfilter induction in early gastrula ectoderm of *Xenopus laevis*. The results show that the induced cell types correspond to the main inducing cell type, around which a range of neighbouring cell types is produced; this has been shown for five different cell types. The inducing factors involved seem to show qualitative differences. It is probable that they play a stabilizing role in the maintenance of the differentiated state of tissues, since temporary dissociation into cells leads eye tissues to transdifferentiate into other types. Such molecular factors can play a significant role in the maintenance of the type of differentiation and also in conversion into other cell types. These mechanisms of maintenance are not restricted to interactions between molecules and cells, since membranes on the surface of the retina and pigmented epithelium contribute to their shaping and consequently to the stability of the cell type.

1991 Regeneration of vertebrate sensory receptor cells. Wiley, Chichester (Ciba Foundation Symposium 160) p 209–218

The study of retinal regeneration has now moved to the molecular and cellular level. To understand the potential for retinal restoration we have to take into account mechanisms which lead to the appearance of new cell types through the conversion or transdifferentiation of existing ones. It is desirable also to take into account mechanisms leading to the formation of cell groups, particularly of mono- or multilayered structure, and their shaping, by which they approach the normal functional eye tissues. Since these problems cannot be treated separately, I shall also discuss the problem of lens formation from retinal cells. Because the retina proper (R) in many instances differs from the pigmented epithelium (PE), I shall treat them as different tissues.

Every approach to the restoration of the retina should take into account the outstanding work of Eguchi (1979, 1988) and Okada (1980, 1983, Okada et al 1983). It follows from their data that the R and PE cells are stable in terms of their cell type as tissues: R as a multilayered structure, PE as a monolayer (both covered by specific membranes). Their transdifferentiation can be brought about either by their long-term cultivation as isolated cells up to the formation of cell aggregates, or by their reaggregation after a brief dissociation into separate cells (Moscona & Degenstein 1981, Okada et al 1983). In both cases, they transform to clumps containing R, PE and L (lentoid) cells. This highlights the necessity of a continuous interaction of these cells for the maintenance of their differentiated type. A similar approach has been described by Eguchi (1988), who used PE to create a permanent semi-differentiated tissue by means of drugs which suppress the terminal steps of differentiation. As a result, L cells appear in addition to PE and R in a large variety of species, including newt, chick and man.

Induction of transdifferentiation

The next problem is the origin and function of factors that act between differentiated R cells to maintain the stability of their cell types. The PE of adult frogs, under the influence of retinal cells (R), consistently transforms into R with all its layers (Lopashov & Sologub 1972, Sologub 1977, Cioni et al 1986, Bosco 1988). This transdifferentiation appears to be due to molecular factors from the living retinal cells. But what is the actual role of these factors? It has been proposed that this retinal action leads to the stimulation of mitoses, whose number determines the course of transformation, whether into R or into L cells (Yamada 1982, Bosco 1988). Many studies show, however, that for the appearance of new cell types as a result of induction, cell division, and specifically the number of divisions, is not a deciding factor, and induction can take place in the absence of division (Cooke 1973, Gurdon & Fairman 1986, Blau 1989).

We chose to use lens epithelium (LE), instead of R, in an attempt to induce the transformation of adult PE. When adult frog (*Rana temporaria*) PE was wrapped in LE, in a sandwich, and transplanted into the tadpole orbit, the PE mostly transformed into R, after a series of mitoses. If such sandwiches were cultivated for three days in a protein-free medium, in which mitosis does not occur, and only then placed in the orbit, PE almost exclusively transformed into L. Mitoses appeared after implantation into the orbit in the same temporal sequence in both types of experiment. Thus, the type of inducing tissue can play a decisive role in determining the direction of transdifferentiation (Lopashov 1978, 1983). Similar results have also been obtained with adult frog iris, but no attention was paid to the significance of a delay in mitosis for the direction of transdifferentiation. It was also shown that embryonic eye vesicles, brought into contact through a filter with the same iris, led to its transformation into retina; in this case, possible contamination with inducer cells is ruled out (Hoperskaya & Zviadadze 1981).

It was important to use another method to exclude possible contamination of PE with LE cells, and to prove that the lenses were normal. Sandwiches of PE surrounded by LE in *Rana temporaria*, and in the combination of *Xenopus borealis* with *Xenopus laevis*, were cultivated for 3–5 days in protein-free medium. Then the PE was separated from LE and put into the pupils of tadpoles.

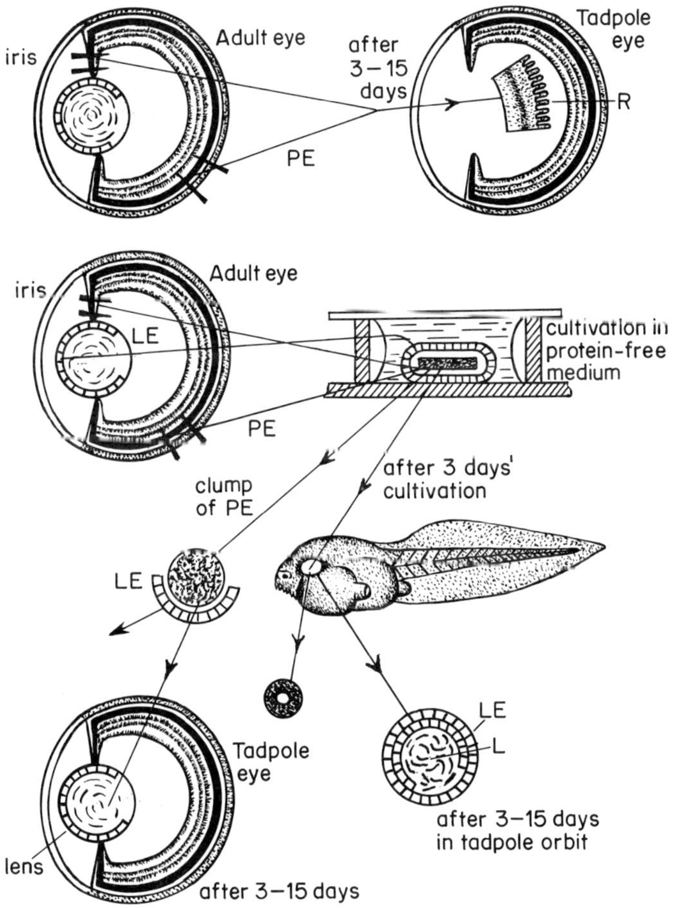

FIG. 1. The transformation of pigmented epithelium (PE) and iris under the influence of neural retina (R) and lens epithelium (LE).

It was clear that there was no tissue contamination. In nearly 40% of cases, normal lenses were formed. After transplantation they were subjected to the polarizing action of the pupil, which is necessary for the formation of normal lenses (Coulombre 1965). The relatively low percentage of normal lenses was probably due to a partial loss of the integrity of the PE clumps during the removal of the surrounding LE; this was confirmed by the structure of the resulting abnormal lenses (publication in preparation). Figure 1 summarizes these experiments.

Inducing factors in eye tissues

The experiments described above still do not clarify the problems of possible factors originating from R and LE and differing between them. We took a

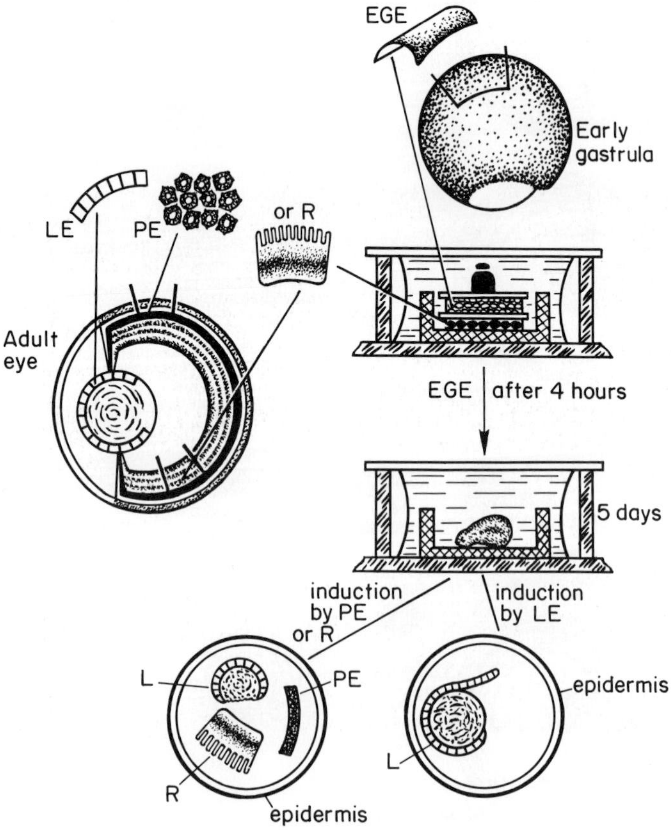

FIG. 2. Inductions by tissues of living adult eye in early gastrula ectoderm of *Xenopus laevis*.

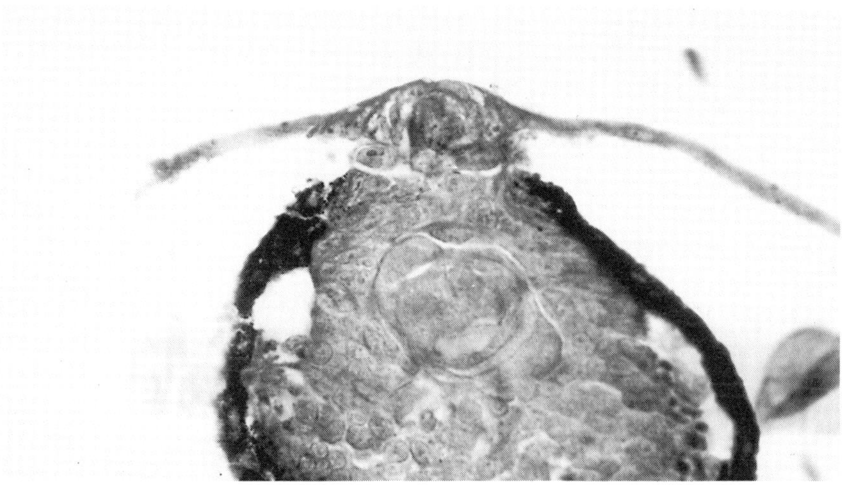

A

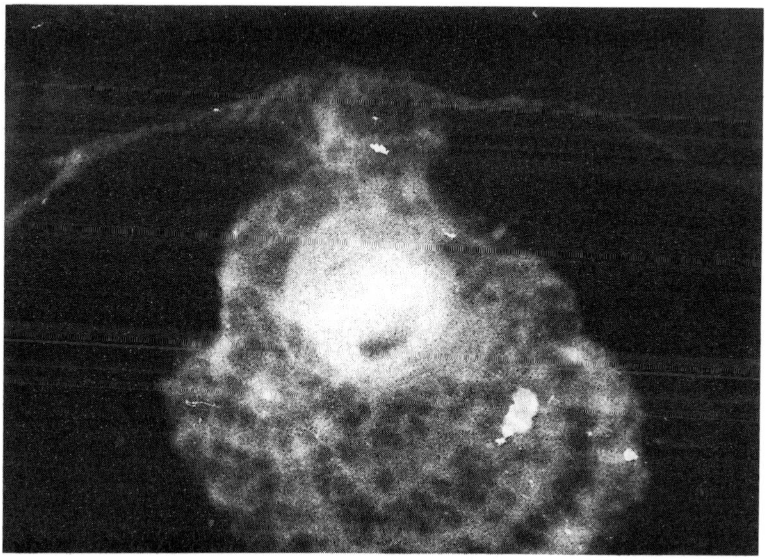

Б

FIG. 3. A. An example of the joint induction of L, PE and R in early gastrula ectoderm under the influence of living PE cells. B. Immunofluorescent reaction in L.

further step towards clarifying this question in a larger series of experiments. We exposed early gastrula ectoderm (EGE) of *Xenopus laevis* through a Millipore filter (25 μm thickness, pore size 0.45 μm) to a series of living adult tissues, mainly eye tissues: R, PE, LE and also forebrain. The results were impressive. R and PE induced R, PE, L and neural cells, as well as cells of a lower induction level, such as ectomesenchyme and melanophores. Brain induced only fragments of brain neural cells, ectomesenchyme and melanophores. But LE induced only L, with a very small number of mesenchymal cells. In all cases the nature of L was verified by indirect immunofluorescence with polyclonal antibodies (Lopashov et al 1986, Lopashov & Golubeva 1989). Figure 2 summarizes this set of experiments; Fig. 3 illustrates an example of this induction. Such specificity of action of adult living tissues is substantiated by the same kind of induction of beating hearts by living adult cardiomyocytes (also verified by indirect immunofluorescence) (Lopashov & Golubeva 1989, Golubeva & Lopashov 1991).

These results, together with the earlier ones, allow further conclusions to be drawn. Living R cells constantly produce substances inducing the formation of R from early gastrula ectoderm, and also from PE. Since dissociated R cells show a potential for PE and L formation, it is probable that the intrinsic inductive factors arise in R by a kind of homotypic induction (Lopashov 1961, Lopashov & Golubeva 1991). This type of induction by the mutual action of similar cells lasts throughout life; sufficient numbers of cells are required for such interactions. That such retinal inducing substances exist also follows from the isolation of a neuralizing factor, capable of inducing a range of cell types in EGE that corresponds to the cell types induced in it by living R cells. This neuralizing factor has been obtained by appropriate biochemical procedures from different sources, unrelated to eye tissue (Tiedemann 1984, Born et al 1989). Living cells differ from the killed tissues and their extracts by the selective production of only those substance(s) which induce(s) the same cell types in EGE, thus showing the advantage of using living cells in the search for active inducing factors. In the case of homotypic induction, the competence of eye cell types lasts for a long time, or throughout life, and in this respect clearly differs from the well-studied, strongly time-dependent competence of EGE (Holtfreter 1938, Gurdon 1987). On the basis of these data we can now approach an understanding of the mechanisms underlying the maintenance of cell type that are also involved in the capacity for retinal transdifferentiation which is essential for regeneration of the retina.

Maintenance of and variation in the shaping of the retina

Another set of mechanisms deals with the establishment and maintenance of the spatial structure of R. I shall mention them only briefly, in as far as they interact with the mechanisms already discussed. Extracellular matrices, covering the PE from the outer side and the retina from the inner side, play such a role. Isolated retinal vascular membrane stimulates the migration of various cells over its surface, on which fibroblasts leave a fibronectin track (Halfter et al 1988). This membrane, covering the inner surface of the retina (retinal vascular membrane or sheet of hyaloid vessels), consists of laminin (Halfter et al 1988). In contrast, the outer surface of PE is covered by Bruch's membrane, containing fibronectin and collagen.

Laminin as a substrate stimulates the formation of neurons from PE cells in culture (Reh et al 1987). The laminin membrane undoubtedly plays a part in the maintenance of retinal structure: when the membrane is conserved, R retains its structure; after its elimination, the retina rounds up; its substitution by another membrane leads to distinct changes in retinal structure (Reh & Radke 1988).

These data show the dependence of the structure of the retina and the pigmented epithelium on such extracellular matrices, but it is possible that they act on the cells' fate not directly, but through a modification of cell aggregation or the formation of cell sheets, which in turn can influence localized inductive interactions. This problem can only be mentioned here, but it merits attention if we are to reach a better understanding of the chain of phenomena leading to different kinds of cell and tissue differentiation.

Conclusions

From the results reported here we can conclude that the main mechanism responsible for both the maintenance and the transformation of eye cell types is the interaction between its component parts, such as R and PE, and their relationships with the surrounding membranes. But the roles of the two types of influence are probably different. Living eye cells continuously produce substances capable of repeatedly inducing such cells in the early gastrula ectoderm. They seem to be responsible for the transformations between different kinds of eye cell types, and probably also play a stabilizing role in the maintenance of the differentiated state.

Acknowledgements

The author is grateful to Dr George G. Gause (Institute of Developmental Biology of the USSR Academy of Sciences), to Dr Olga N. Golubeva (Institute of Gene Biology of the USSR Academy of Sciences), and to Dr Thomas Reh (University of Washington, Seattle) for their critical reading of the manuscript.

References

Blau HM 1989 How fixed is the differentiated state? Trends Genet 5:268–272

Born J, Janeczek J, Schwarz W, Tiedemann H, Tiedemann H 1989 Activation of masked neural determinants in amphibian eggs and embryos and their release from the inducing tissue. Cell Differ Dev 27:1–7

Bosco L 1988 Transdifferentiation of ocular tissues in larval *Xenopus laevis*. Differentiation 39:4–15

Cioni C, Filoni S, Aquila C, Bernardini S, Bosco L 1986 Transdifferentiation of eye tissues in anuran amphibians: analysis of the transdifferentiation capacity of the iris in *Xenopus laevis* larvae. Differentiation 32:315–320

Cooke J 1973 Properties of the primary organization field in the embryo of *Xenopus laevis*. IV. Pattern formation and regulation following early inhibition of mitosis. J Embryol Exp Morphol 30:49–62

Coulombre AJ 1965 Experimental embryology of the vertebrate eye. Invest Ophthalmol 4:411–419

Eguchi G 1979. Transdifferentiation in pigmented epithelial cells of vertebrate eyes *in vitro*. In: Ebert JD, Okada TS (eds) Mechanisms of cell change. Wiley, New York, p 241–258

Eguchi G 1988 Cellular and molecular background of Wolffian lens regeneration. Cell Differ Dev (suppl) 25:147–158

Golubeva ON, Lopashov GV 1991 Inducing capacity of cardiomyocytes. Ontogenez, submitted

Gurdon JB 1987 Embryonic induction—molecular prospects. Development 99:285–306

Gurdon JB, Fairman S 1986 Muscle gene activation by induction and the non-requirement for cell division. J Embryol Exp Morphol (suppl) 97:75–84

Halfter W, Diamantis I, Monard D 1988 Migratory behavior of cells on embryonic retina basal lamina. Dev Biol 130:259–275

Holtfreter J 1938 Veränderungen der Reaktionsweise in isolierten alternden Gastrula-ektoderm. Wilhelm Roux's Arch Dev Biol 138:163–196

Hoperskaya OA, Zviadadze KG 1981 Transdifferentiation of adult frog iris into retina or lens by exogenous influences. Dev Growth & Differ 23:201–213

Lopashov GV 1961 The development of the vertebrate eye in relation to the problem of embryonic induction. Ann Zool Soc Zool Bot Fenn Vanamo 22(6):1–17

Lopashov GV 1978 Developmental approaches to organ restoration. Med Biol (Helsinki) 56:344–348

Lopashov GV 1983 Transdifferentiation of pigmented epithelium induced by the influence of lens epithelium in frogs. Differentiation 24:27–32

Lopashov GV, Golubeva ON 1989 Assimilatory induction by living cells of adult tissues in gastrula ectoderm of *Xenopus*. Cell Differ Dev (suppl) 27:70

Lopashov GV, Golubeva ON 1991 Homotypic induction. Ontogenez, submitted

Lopashov GV, Sologub AA 1972 Artificial metaplasia of pigmented epithelium into retina in tadpoles and adult frogs. J Embryol Exp Morphol 28:521–546

Lopashov GV, Golubeva ON, Zviadadze KG 1986 Common determinants of embryonic induction and transdifferentiation. In Slavkin H (ed) Progress in developmental biology. Alan R Liss, New York, part A, p 43–46

Moscona A, Degenstein I 1981 Lentoids in aggregates of embryonic neural retina cells. Cell Differ 10:39–46

Okada TS 1980 Cellular metaplasia or transdifferentiation as a model for retinal cell differentiation. Curr Top Dev Biol 16:349–380

Okada TS 1983 Recent progress in studies of the transdifferentiation of eye tissue in vitro. Cell Differ 13:177–183

Okada TS, Nomura K, Yasuda K 1983 Commitment to transdifferentiation into lens occurs in neural retina cells after brief spreading culture of dissociated cells. Cell Differ 12:85–92

Reh TA, Radke K 1988 A role for the extracellular matrix in retinal neurogenesis *in vitro*. Dev Biol 129:283–293
Reh TA, Nagy T, Gretton H 1987 Retinal pigmented epithelial cells induced to transdifferentiate to neurons by laminin. Nature (Lond) 330:68–71
Sologub AA 1977 Mechanisms of repression and derepression of artificial transformation of pigmented epithelium into retina in *Xenopus laevis*. Wilhelm Roux's Arch Dev Biol 182:277–291
Tiedemann H 1984 Neural embryonic induction. In: Duprat AM, Kato AC, Weber M (eds) The role of cell interactions in early embryogenesis. Plenum Press, New York, p 89–105
Yamada T 1982 Transdifferentiation of lens cells and its regulation. In: McDevitt DS (ed) Cell biology of the eye. Academic Press, New York, p 193–242

DISCUSSION

Rubel: Can you do these experiments at any age of the frog? Can you induce differentiated pigment epithelium to transform into other eye tissues?

Lopashov: Yes, we use only adult tissues from the frogs (*Rana temporaria* and *Xenopus laevis*), and expose them to the action of retina and pigmented epithelium. These tissues are very active as inducers, as is clear from the experiments with gastrula ectoderm that I reported. Recently (unpublished results) we found a means of collecting the factors produced by living eye tissues and of testing them on ectodermal explants.

Raymond: When you say that you use *adult* tissue from these frog species (*Rana* and *Xenopus*), do you use frogs which have *just* passed through metamorphosis? How old are they?

Lopashov: We used fully grown adult animals (*Rana* and *Xenopus*) as the source of the test tissues; for induction we used also tadpole retina, but lens epithelium only from adult animals.

Costanzo: You mentioned experiments with cardiac myocytes. Can you say more about those experiments?

Lopashov: We place living myocytes from frogs (*Xenopus laevis*) in culture, then a Millipore filter, and then gastrula ectoderm. We leave this for four hours of exposure, similar to the arrangement shown in Fig. 2 (p 213). After this, we free the gastrula ectoderm from the filters and myocytes and cultivate it for three or five days. We observe that nearly half of the explants beat regularly. We have histological and immunochemical evidence of muscle fibres in the explants.

Lewis: I want to be sure that I understand your results. You take gastrula ectoderm and you put it with lens tissue and the ectoderm differentiates into lens; if you put gastrula ectoderm with PE, it differentiates into PE. Is that right?

Lopashov: It is not so simple! With PE or with retina, we obtain more complex results. In most cases we obtain retina, PE and lens. But with lens epithelium we obtain only lenses from gastrula ectoderm, with only rare mesenchymal cells.

Lewis: But you do specifically get eye tissues when you put a fragment of eye with gastrula ectoderm?

Lopashov: That is correct. If you recall the papers of Okada (1980, 1983), the pigmented epithelium and retina have a high capacity for transdifferentiation, but lens epithelium has no such capacity. Correspondingly, the action of LE is 'selective', whereas the action of retinal tissues is less selective. It is not a question of selectivity or not; it is a question of the *degree* of selectivity.

Rubel: How many types of retinal cells do you obtain when gastrula ectoderm differentiates into retina? Do you see all the different types of retinal cells, including round cells, horizontal cells and amacrine cells?

Lopashov: In different cultures the completeness of the structure of the retina is not the same, but there are cases in which all nuclear and reticular layers are well represented, as also the photoreceptors. We cannot do a very fine analysis on these small explants, which show different degrees of complexity.

Reh: What cell type in the differentiated retina do you suggest can give rise to lens and pigment epithelium?

Lopashov: I cannot answer this question on the basis of our own experiments, but the experiments of Okada (1980, 1983) and Moscona (1986) are suggestive. The latter author supposes that they are glial cells, but as I understand it, the question is not fully decided.

Lewis: Have you done similar experiments with ear tissue? That is to say, if you put otic vesicle with gastrula ectoderm, can you induce ear development?

Lopashov: No. Unfortunately, it has not been possible to do such experiments yet. Your suggestion is valuable, and could be followed up in further experiments.

References

Moscona AA 1986 Conversion of retina glia cells into lenslike phenotype following disruption of normal cell contacts. Curr Top Dev Biol 20:1–19

Okada TS 1980 Cellular metaplasia or transdifferentiation as a model for retinal cell differentiation. Curr Top Dev Biol 16:349–380

Okada TS 1983 Recent progress in studies of the transdifferentiation of eye tissue *in vitro*. Cell Differ 13:177–183

General discussion

bFGF and photoreceptor regeneration

Steinberg: In addition to Müller cells in the mammalian retina there is another support cell, the retinal pigment epithelial cell or RPE cell, about which we have heard from Professor Lopashov. This is a multipotential cell and it is also a multifunctional cell, in the mature retina. The RPE cell is a transporting cell, carrying metabolites to and from the subretinal space, which is the primitive third ventricle space. The RPE cell recycles vitamin A and isomerizes it during dark adaptation. This cell also functions as a macrophage, phagocytosing the shed tips of the outer segments from rods and cones, which replenish these discs on a daily basis.

That the outer segment grows every day also provides for its ability to regenerate. This is not regeneration of the whole photoreceptor, just of the outer segment. We can lose the outer segment, as occurs in retinal detachment when the neural retina becomes displaced from the RPE. The outer segment can regrow, as long as the inner segment and soma region survive. We don't know the physiological characteristics of this survival, or the degree of injury that will prevent this regeneration.

I want now to tell you about some recent research that I have been engaged in on the role of the basic fibroblast growth factor in photoreceptor regeneration and survival in the rat. This research was done in full collaboration with Dr Matthew M. LaVail and with a medical student at UCSF, Ella G. Faktorovich. Douglas Yasumura and Michael T. Matthes also contributed to these experiments*.

The retinal pigment epithelium contains basic fibroblast growth factor (bFGF) (Schweigerer et al 1987, Sternfeld et al 1989), which has been shown to act as a mitogen and differentiation-promoting factor in many systems. For instance, bFGF induces retinal regeneration from the RPE in the chick embryo (Park & Hollenberg 1989). Moreover, bFGF has been shown to act as a neurotrophic agent in several regions of the central nervous system (Sievers et al 1987, Anderson et al 1988, Otto et al 1989). To investigate its possible role as a survival factor in the vertebrate retina, we studied its effect in two conditions that lead to profound degeneration of photoreceptors in the rat—inherited retinal

*We thank D. Gospodarowicz for helpful discussions and the gift of aFGF and bFGF, and N. Lawson and G. Riggs for technical and secretarial assistance. This work was supported by the National Eye Institute, the National Retinitis Pigmentosa Foundation, Research to Prevent Blindness and That Man May See, Inc. M. M. LaVail is an RPB Senior Scientific Investigator.

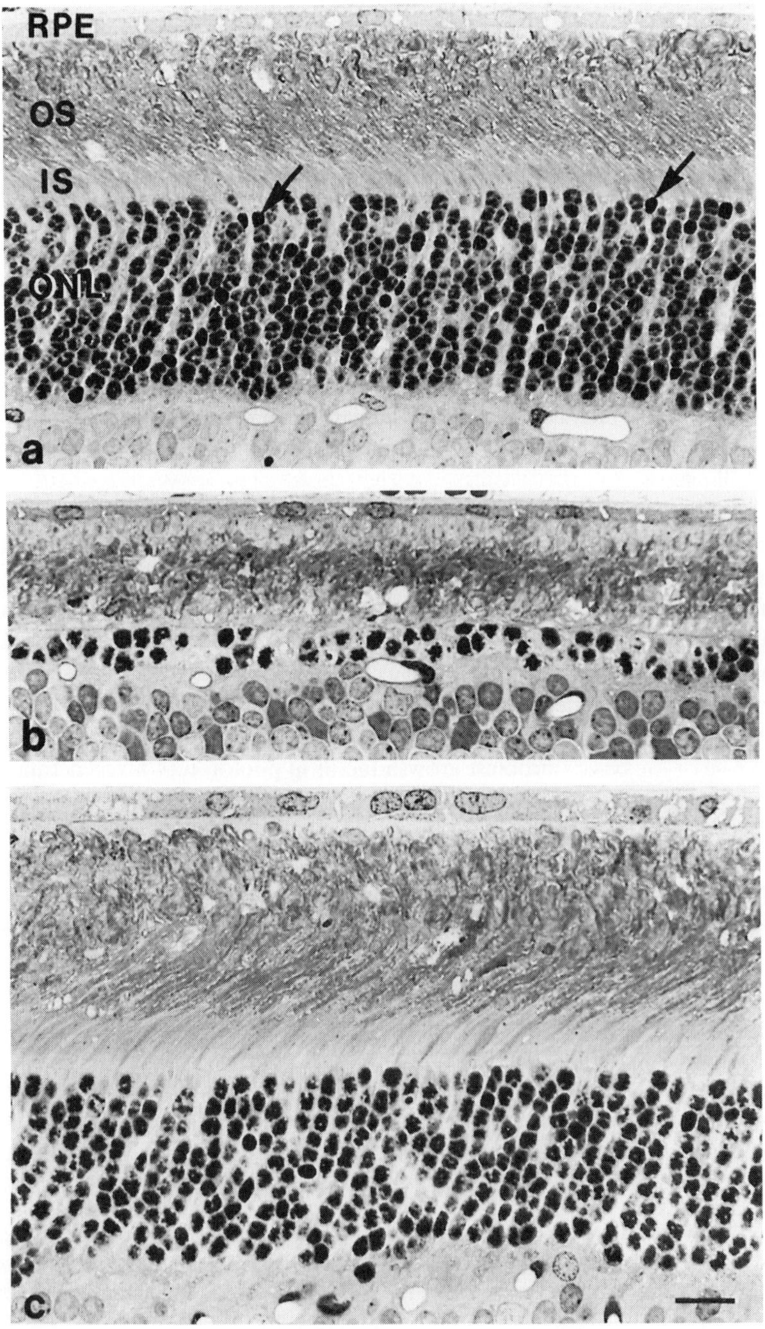
RPE
OS
IS
ONL
a
b
c

dystrophy in the Royal College of Surgeons (RCS) rat (Mullen & LaVail 1976), and exposure to constant light in the albino rat (Noell et al 1966).

In the RCS rat (Faktorovich et al 1990) we transsclerally injected bFGF dissolved in phosphate-buffered saline (PBS) (500 or 820 ng in 1 μl) into the interphotoreceptor space or subretinal space of the superior hemisphere of the eye at postnatal day 23 (P23), when photoreceptors have just started to degenerate. We removed the eyes after one or two months and and examined them histologically in plastic sections or polyester wax serial sections. For comparison and as controls we injected acidic fibroblast growth factor (aFGF) (500 ng in 1 μl) dissolved in PBS, or PBS alone. We also did the 'mechanical' control of penetrating the interphotoreceptor space with a needle with no injection. To obtain wider dissemination of bFGF, we also injected it into the vitreous body and did intravitreal controls.

When bFGF was injected into the interphotoreceptor space we saw remarkable preservation of many photoreceptors in the superior hemisphere of the injected eye. After one month (P53), the retina appeared practically intact compared to the extensive photoreceptor degeneration observed in uninjected controls, as judged by the thickness of the outer nuclear layer (ONL) and the presence of inner and outer segments (Fig. 1). After two months (P83), bFGF-injected retinas still showed substantial photoreceptor rescue, although less than at one month. Remarkably, all the subretinal controls, including PBS, aFGF and the dry needle, showed rescue that was *qualitatively* similar to the rescue produced by bFGF, in terms of preservation of photoreceptors. Rescue in the controls, however, was much more limited in extent, usually being localized to the site of injection. Intravitreal injection of bFGF produced a greater extent of rescue than subretinal bFGF, encompassing almost the entire retina. It should be noted that bFGF did not reverse the phagocytosis defect known to occur in the RCS rat, from the absence of phagosomes in the RPE and the presence of a debris zone in the rescued areas (see Faktorovich et al 1990).

Similar experiments were repeated in albino rats exposed to constant light. This causes photoreceptor degeneration (Noell et al 1966). We injected bFGF

FIG. 1 (*Steinberg*) Plastic-embedded sections of RCS rat retinas. (a) P23, at the time of injections; cell death had just begun and some photoreceptor nuclei in the ONL were pyknotic (arrows). Photoreceptor inner segments (IS) and outer segments (OS) are present, as well as some accumulated outer segment debris membranes near the apical surface of the RPE. (b) P53, an uninjected control animal; the ONL is reduced to 1–2 rows in thickness, no photoreceptor inner or outer segments are present, and the outer segment zone is composed only of debris membranes. (c) P53, a region of optimal photoreceptor rescue one month after subretinal bFGF injection; photoreceptor inner and outer segments are present and the retina generally appears unchanged from the time of injection. The retina in the opposite hemisphere of the same eye as in (c) was fully degenerated and indistinguishable from that shown in (b). Toluidine blue stain. Scale bar, 10 μm (From Faktorovich et al 1990, with the permission of Macmillan Magazines Ltd 1990.)

intravitreally 48 hours before placing 3–4-month-old Fischer 344 or Sprague Dawley albino rats in constant light for seven days (LaVail et al 1991). One week of exposure to fluorescent lights (115–130 ft-c) reduced the thickness of the ONL from the normal 8–10 rows of photoreceptor nuclei to 1–2 rows in the most degenerated region of the eye in uninjected or PBS-injected controls

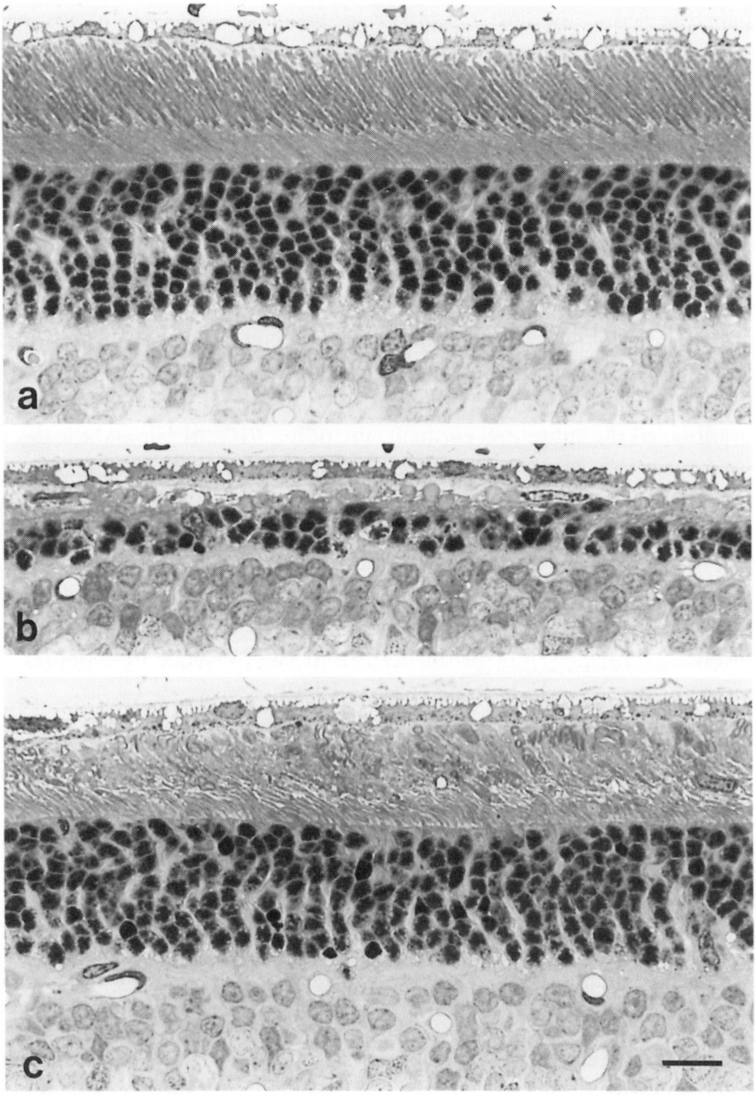

(Fig. 2). There was also a profound loss of most photoreceptor inner and outer segments in this region.

We saw extensive photoreceptor rescue in the bFGF-injected rats: the most degenerated regions of the ONL had 4–5 (Fischer 344) or 6–8 (Sprague Dawley, Fig. 2) rows of nuclei, and many photoreceptor inner and outer segments remained. When bFGF-injected and control rats were allowed to recover in cyclic light for 10 days, after the week of exposure to constant light, the bFGF-injected retinas showed much greater regeneration of photoreceptor inner and outer segments than did the control retinas. The insertion of a dry needle in the subretinal space, 48 hours before exposure to constant light, also greatly reduced the degree of photoreceptor degeneration.

These results show that bFGF can *delay* photoreceptor degeneration in an inherited retinal degeneration and can *prevent* the degeneration of photoreceptors that occurs in light damage. The mechanism for these rescue effects is not known, nor is it known whether bFGF acts directly on photoreceptors or through other cells or other factors. The rescue produced in the subretinally injected controls may result from the release of endogenous bFGF when RPE cells or other cells are damaged during injection. We caution about generalizing these results to humans, especially because of the potentially harmful side-effects of bFGF, given its mitogenic and angiogenic properties.

Farbman: You conclude that the lesion in the RCS rats is in the pigment epithelium; but is it also in the photoreceptors?

Steinberg: So far as we know, there's no lesion in the photoreceptor. One defect in the RCS rat is the failure of phagocytosis. By giving this growth factor, we can delay the death of the photoreceptors. I don't know whether there is a defect in the production of growth factor by the RPE or a problem in access of that growth factor to the photoreceptors through the debris zone. The interphotoreceptor matrix might be sufficiently disturbed in this region that FGF, which might reach the photoreceptors normally, as a survival factor, is lost.

Farbman: When you inject FGF into the vitreous and get survival, how does this large molecule, which is 16 000 Da in molecular mass, get to the photoreceptor?

FIG. 2 (*Steinberg*) Plastic-embedded sections of retinas from three-month-old Sprague Dawley albino rats. (a) Normal retina from a rat maintained in a 12 hour on:12 hour off cyclic light environment (15 ft-c). Photoreceptor inner segments and outer segments are present and the ONL is 8–10 rows of photoreceptor nuclei in thickness. (b) Retina from an uninjected control, killed at the end of a seven-day period in a constant light environment (115–130 ft-c). The ONL is reduced to 1–2 rows in thickness, and there are no inner segments and no normally appearing photoreceptor outer segments. The RPE appears intact. (c) A region of photoreceptor rescue from an animal injected intravitreally with 1000 ng of bFGF in 1 μl of PBS 48 hours before being placed in the constant light environment for seven days. The ONL is 6–8 rows of photoreceptor nuclei in thickness. Many inner segments are present, and the outer segment zone is filled mostly with swollen disorganized outer segment profiles. Toluidine blue stain. Scale bar, 10 μm.

Steinberg: It could get there by diffusion, if it works directly on the photoreceptor; we don't know that for sure.

Calof: FGF is known for its ability to bind to molecules of the extracellular matrix, such as glycosaminoglycans; it's a heparin-binding molecule. Since it binds well to extracellular matrix, how do you know it can diffuse rapidly in the eye? Also, do you know anything about the half-life of FGF in this situation?

Steinberg: We can work out the answer to many of these questions in the light-damage model. We are looking at the half-life by giving it various times before the continuous light period. The effect of FGF has decreased considerably after about eight days.

Calof: So is there an immediate effect of FGF that results in the rescue effect?

Steinberg: To investigate the immediacy of its effect we would need to give FGF, then put the light on and look within hours; we are doing that.

Calof: Do you see any effect of FGF on the pigment epithelium itself, in the RCS rat degenerative model?

Steinberg: There's no evidence yet of any change in the RPE due to the growth factor, itself.

Calof: So histologically it looks normal, but there could be functional differences?

Steinberg: Yes.

Reh: You compared the effect of injecting FGF with a saline injection and inserting a dry needle. It is known that certain growth factors, TGF-β and FGF for example, injected under the skin, cause excessive inflammation. Therefore one explanation for your results is that if a small, sham injection causes a little bit of inflammation, injecting a growth factor into that same space causes a lot of inflammation. Could that be an explanation for the long-term rescue?

Steinberg: One hypothesis for the rescue seen in the dry-needle controls would be release of FGF from the pigment epithelium by the insertion of the needle through it. Regarding inflammation, TGF-β, alone, does not rescue the photoreceptors, nor do any of the other growth factors that we have tested that are involved in the inflammatory response. With vitreous injections, we see an increase in macrophages in the retina, so macrophages coming in as part of an inflammatory response may be very intimately related to the rescue. Ultimately, we are interested in what molecules and cells cause the rescue, and how they are part of a sequence that begins with FGF.

Raymond: You are injecting FGF shortly after the rods have been born, postnatally, in rats. We (unpublished work) have seen in goldfish that after stab wounds of the eye there is a massive up-regulation in the rate of rod precursor proliferation, so potentially instead of rescuing cells, you may be augmenting proliferation of the rod progenitor cells so that new rods are being generated.

Steinberg: One way of dealing with that question was to try the constant light model, in an adult rat, where there are no precursor cells.

Rubel: Is there any differential effect on the outer segments? That is, if you wait until the outer segments have pretty much disappeared, and inject FGF, do you get regrowth of the outer segments?

Steinberg: That is a very good question, and we don't have the results. It's a difficult experiment to do in terms of the timing, because while the outer segments are degenerating, some photoreceptor cells are dying as well. It is hard to define a time point when only outer segments have been injured.

Powers: In the RCS rats, how long does the FGF effect last, once you have given the injection?

Steinberg: After the one injection of growth factor, by three months the rescue effects have diminished. So we are delaying the degeneration; whereas in the light-damaged retina we are preventing it.

Reh: Does the small amount of 'saving' that you get in the 'dry needle' sham control show the same time course as FGF, in the RCS rat or in the other model?

Steinberg: We have not analysed the data for that, but I would expect it to be the same.

Powers: You said phagocytosis is not normal, but you didn't say how normal the rod outer segments are.

Steinberg: They look normal; as to whether they are functioning or not, we are looking at that with electroretinograms.

Powers: Are they shed normally?

Steinberg: Yes, because the debris zone builds up.

Pujol: What causes the degeneration in the light damage?

Steinberg: One hypothesis is lipid peroxidation due to a build-up of superoxide radicals. Ascorbate injections, before the continuous light, also retarded degeneration in this model (Organisciak et al 1985, Li et al 1985).

Pujol: Do you see swelling of the neurites?

Steinberg: In the light-damage model the photoreceptor outer segments swell a great deal.

Fernald: If the process that you have uncovered is part of a continuum that goes from overexposure to light to real damage, is it possible to find an up-regulation of something like FGF in slightly brighter light, if what you have found is a natural response of the eye, designed to protect it?

Steinberg: We do not have data, yet, on whether there's an up-regulation of FGF in the light-damaged retina. That would be a good experiment.

Fernald: The question of the time course becomes interesting, because one might wonder why you get a 'protective' response with FGF given before photic damage.

Steinberg: We think it takes time for bFGF to affect the photoreceptor cell and turn genes on or off and produce new proteins. The exact timing of the effect is still under investigation.

Fernald: Is there any chance that the RPE is itself responsive to light? If there is photoreceptive capability in the RPE, then very bright lights might activate RPE cells directly, to produce growth factors.

Steinberg: Dr K. T. Brown obtained electrical responses of the RPE cells of frogs when driven by very intense light. This is a response coming from a direct effect of light absorbed by melanin granules. The light used to produce damage in our second model is not very intense light; it's normal room lighting, and that intensity has never been shown to produce a direct electrical response from the RPE.

Reh: We have been studying the neurogenesis of the retina in the rat, and the response of neuroepithelial cells to various growth factors. We have looked at FGF in terms of the question that Pamela Raymond addressed earlier—whether it would stimulate the production of new rods in cultures and what effects it has on the rods. Others have looked in culture models to see if FGF has survival effects on rods.

There is evidence that the retina has many growth factors in it, during its development: mRNA for FGF and its receptor; for EGF and TGF-α, message and receptor; and NGF and the NGF receptor. So there are enough growth factors to regulate the whole process of neurogenesis. But in our culture models, so far, none of these factors selectively influences the production of one type or another of retinal cell; but the factors can, in combination or alone, promote the proliferation of neuroepithelial cells to produce more neurons and glia overall. So from our experiments we would not conclude that any of these factors is a controlling element that regulates the production of different types of cells, but rather that they may act together, in concert—in survival, in axon and dendrite development, and in proliferation—to regulate those processes. They don't seem to be (at least in culture models) what determines what a cell becomes.

Rubel: To follow that up in the ear, I can't talk about normal development because we don't fully understand the expression of the receptors for these factors, but during regeneration in the chick we have looked at the expression of RNA for receptors for FGF, EGF and NGF. The results with at least two of these probes are negative; we don't see an increase in the expression of these receptors during regeneration.

Lewis: Dr Steinberg, can you show any defect in FGF production in any of the retinal degenerative disorders?

Steinberg: We are in the process of investigating this, in the RCS rat. We have looked at some other rodent mutants, but that study is not completed yet.

Rubel: Is there any evidence that retinal precursor cell types such as the rod precursor cell remain after birth, and has anybody tried to induce light damage or any other kind of damage to see if anything happens?

Raymond: We (Raymond et al 1988) tried to produce light damage in goldfish. Unless we use very high, photically damaging, levels of light, we don't get the loss of photoreceptors that Roy Steinberg sees in his albino rat model. Potentially this is because in these retinas there is a continual replenishment of rods. We did not assess whether rods were being killed by light, and subsequently replaced; we only scored the number of rods that were present after various intervals. It's possible that there was actual turnover—that is, loss and replacement. After one year of constant light exposure, from hatching, there was a 30% loss of rods and a concominant loss in sensitivity, and no loss of cones (Raymond et al 1988, Powers et al 1988).

With regard to the persistence of neuroepithelial cell precursors in retinas that have differentiated, there are reports that in frog retinas, in *Xenopus* for example, cells that incorporate thymidine persist in the differentiated retina (Taylor et al 1989). Some of these cells are probably analogous to the cells that we have shown in the fish retina to give rise to rod precursors. There are no rod precursors in amphibians, because there is not continuous addition of rods in these retinas, but there is evidence of a population of resting stem cells, potential precursor cells, that do not normally divide, or do so very slowly. Under certain conditions perhaps they could be induced to proliferate. In my paper I described only rod precursors in the fish retina because that is what we understand best, but even in the teleost retina we suspect that there might be other cells in the inner nuclear layer in differentiated regions of the retina that do not divide under normal circumstances but that can be induced to proliferate under conditions of damage.

Rubel: Are you suggesting that, unlike normal development, different precursors exist for different cell types?

Raymond: No; I am suggesting that it is possible that there are small numbers of neuroepithelial cells, which do not differentiate but remain in a resting phase, undifferentiated, and have the potential to proliferate. That doesn't mean that they are a different kind of cell from the primitive neuroepithelial cells of the embryonic retina. We have no markers, however, for recognizing an undifferentiated cell that is not dividing. So there could be lurking in all of these differentiated sensory tissues, particularly those tissues that are continuing to grow, potential proliferative cells that we are not aware of.

Reh: In the rat, studies have been made of the regenerative capability of the retina in the neonatal period, after a crush lesion of the optic nerve to eliminate all the ganglion cells, or earlier, in embryogenesis, using X-irradiation to get rid of developing ganglion cells. After X-irradiation at a stage in embryogenesis where neuroepithelial cells are still present and generating all cell types, histological examination of the adult showed that the retinas are fairly normal (Rugh & Wolf 1955).

By contrast, after a neonatal optic nerve lesion which destroys virtually all the ganglion cells, there is no recovery of ganglion cells, even though neurogenesis is still going on in the retina. The neuroepithelial cells in neonatal rats are still producing rods, bipolar cells and Müller cells, yet when ganglion cells are deleted they don't make new ganglion cells.

Rubel: As if those precursors have lost the ability to make ganglion cells?

Reh: Or they are just too far away to receive the signal that the ganglion cells are gone. We could think of many other explanations.

A more recent study with kainate in the rabbit shows that there appears to be some reorganization of postnatal rabbit retina after damage, although it is not clear what the final outcome of the recovery process is (Messersmith & Redburn 1990).

Raymond: Those authors saw a change in the ratio of photoreceptors, but they have not demonstrated any change in overall cell number.

Pujol: The effect of light damage is very surprising. Is there any other mammalian species besides the albino rat in which normal light (continuous) causes complete degeneration of the retina?

Calof: In albino humans, there is retinal degeneration.

Rubel: There are medical reports of light blindness in humans. It was common in the early explorers of the West.

Steinberg: One can damage the retina in man with light, but not with normal intensities, as far as I know. Light damage has been produced during the removal of cataracts, but at levels above the normal light level.

Pujol: Is there another experimental animal?

Steinberg: There are some light-damage models, but usually with much more intense light.

Raymond: Premature babies exposed to ultraviolet light to treat jaundice were found to have damaged retinas (Fulton et al 1986). Now their eyes are shielded during these treatments.

Steinberg: The light-damage model in the rat has been extensively described (Noell 1980). There's an explanation for light damage that involves the rat regulating the amount of rhodopsin in its eyes, depending upon the lighting environment in which it lives, and age is also a factor (Penn et al 1985, Rapp et al 1990). It will down-regulate when it goes into brighter light, and increased rhodopsin levels are found when rats are kept in the dark.

Margolis: Is it a possible extension of your hypothesis that FGF protects against normal damage in these animals?

Steinberg: There's no evidence yet that FGF protects normally against damage.

Reasner: Dr Raymond, in your light experiment with goldfish, did you use a full spectrum light?

Raymond: It was daylight fluorescent light, at levels typical of normal room illumination.

Jørgensen: In your fish with normal retinas, do you see cells similar to the dark cells that we find in the inner ear?

Raymond: Yes, we see pyknotic (dying) cells during normal development.

Jørgensen: Do you have any idea of their function? Are you sure they are dying?

Raymond: We presume so, from their appearance in the electron microscope; they look similar to those you are showing, with very condensed chromatin. We have seen such cells mostly in developing retinas and at the margins of the adult retina, so we believe that cell death is associated with cell proliferation. We presume that the cells that die are the ones that have failed to act on the differentiation signals that they have received, or those in which a fatal error has occurred in the mitotic process.

Jørgensen: How about macrophages in the retina? They have been described by Braekevelt (1980). Do you see them?

Raymond: There are macrophages, both within the neural retina itself and also in the subretinal space (between the neural retina and pigmented epithelium). In goldfish, in addition to the involvement of RPE cells in the phagocytosis of photoreceptor outer segments, there are wandering phatocytic cells in the subretinal space that also serve this function.

Rubel: One of the major differences between what has been presented in the retina and what we presented in the ear is that in the retina you have a fairly good idea at least of the precursors for the rod cells, and, in the case of pigment epithelium, where the regeneration is occurring.

We have been interested in the nature of the signals that stimulate cells to enter into mitotic activity. Do any of the those who study retina want to comment?

Raymond: I have tried hard to make the fish like a frog! In other words, I have devascularized the fish retina, thinking that our failure to get the RPE to take part in retinal regeneration might be due to differences in the way we produce degeneration of the retina. We typically use neurotoxins, ouabain or 6-hydroxydopamine, to destroy the retina, but I have also ligated the ophthalmic artery, and then released the ligature after three to 24 hours to allow blood back into the retina. Under those conditions the goldfish retina degenerates as it does in the amphibian, and it regenerates, but with the same mechanism that we showed with ouabain. Again the RPE cells do not seem to be directly involved. Tom Reh has looked at some of my retinal sections and confirms that they look different from the frog, in terms of the pattern of cell proliferation. We did not see RPE cells detaching and migrating into the retina, touching the vascular membrane and initiating the whole sequence of events that he has described for the frog.

Reh: It just appears that frogs are a lot better than goldfish at this!

Raymond: Or that the frog RPE is more able to regenerate a totally intact, nicely laminated structure than are the foci of proliferating cells which originate from rod precursors in the goldfish.

Reh: Also, the frog RPE often appears to spread out on the vitreal vascular membrane, before undergoing the switch to neuroepithelial cells. So there seems to be a continuous layer of cells of approximately the same developmental stage. Whereas with the foci generating from rod precursors in the fish, the cells are at various developmental stages. In the frog there is always a discontinuity at the point where the retina regenerated from the marginal zone meets retina regenerated from the pigmented epithelium.

Powers: Can you obtain regeneration in adult frogs?

Reh: We never have, but Lopashov & Sologub (1972) transplanted the pigment epithelium of adult frogs into the eye of a tadpole. It appeared to give rise to new retina and pigment epithelium; so the potential seems to be there, though we have never observed it in the adult.

Powers: It strikes me that, comparing these two sets of studies (tadpole and goldfish), you have in the frog a larval tissue which may be able to produce new cells in a more orderly fashion than an adult tissue, which has to go back and do things over again, as it were.

Raymond: I think it may be more a function of the mechanics of the situation, and the fact that in the amphibian case there is a continuous sheet of dividing cells, more or less synchronized, whereas in the fish retina regenerating from rod precursors there are isolated foci. As Dr Reh suggests, the less-than-perfect result in the regenerated fish retina may simply be a function of the asynchrony of the process.

Rubel: How good is the mosaic within one of your foci? How normal is it?

Raymond: When you talk about a mosaic, retinal aficionados think of a cone mosaic, because the most regularly arrayed cells in the fish retina (and most other retinas as well) are the cone photoreceptors. The cone mosaic in the regenerated retina appears to be quite regular, but the inner layers of the retina are much more disorganized.

Rubel: Is the whole retina in the frog pluripotent in terms of its ability to regenerate?

Reh: Nobody has looked carefully to see whether one area of frog retina is particularly good at regenerating, but we have never seen any evidence of that. Generally one classifies the regeneration as coming either from the marginal zone or from the back of the eye.

Fernald: On the question of what might be regulating cell proliferation, we have shown a robust circadian rhythm in the teleost *Haplochromis burtoni*; we find that cell generation in the temporal region of the eye can be enhanced when that in central retina is not, so the rod progenitors are out of phase with the marginal zone (Fernald 1991). So I would suspect that there may be a circadian regulation of a number of these factors, and, when we look at regeneration, circadian features may play distinct roles at different times of day.

Rubel: Do we know anything about the signals that up-regulate the amount of mitotic activity in the retina after damage?

Raymond: We are now starting to learn what growth factors regulate proliferation *in vitro*, but *in vivo* we do not know.

Powers: We are beginning to investigate what happens to visual function during retinal regeneration, using the model of ouabain destruction that Dr Raymond presented earlier. In the ouabain-treated fish (*Carassius auratus*) the retina degenerates, and connections to the brain also degenerate. We have to consider therefore not only the regeneration of the receptors themselves; regeneration of the connections has to be appropriate to mediate the behaviour.

I have done two preliminary experiments. In one we use the optokinetic nystagmus response, which consists of the smooth pursuit response of the eye to a moving object, followed by a rapid saccadic re-set in the direction opposite to the movement that is being followed. We put a fish in a drum and rotate stripes around the fish. We then record eye movements. We have tried to measure visual acuity. As shown in Pamela Raymond's results, the photoreceptors come back; the question was when vision comes back. It returns shortly after the photoreceptors are regenerated, which suggests that some neural connections must be made.

There is a flaw in this experiment, namely that there is a ring of normally developed cells around the eye, so we don't know whether the fish were using that ring of normal cells or the regenerated cells to follow the stripes. We have therefore turned to an electrophysiological measure, recording electroretinograms in fish injected with ouabain and allowed to recover. We inject one eye with ouabain and leave the other eye uninjected; then periodically, after the injection, we record from injected and control eyes. Because ouabain causes the lens to become cloudy, we slice the cornea off, remove the lens, and have the fish lying on its side; then we stimulate with a tiny fibre optic probe that contains only seven fibres. I am pretty sure it is stimulating only the part of the retina that was destroyed and then is regenerating. In this initial experiment I am looking at rod function by examining b-wave responses in dark-adapted retinas. Although the results are incomplete, I find that at about the time that rods appear in the regenerating retina, I can record an ERG in the ouabain-treated eye. I can't yet say whether the amplitude is the same as in the control eye when the ERG first comes back, or what the spectral sensitivity of the response is. However, it looks as if at least the first levels of retinal function are wired up as soon as the rod photoreceptors come back, because the b-wave response is known to reflect the activity of bipolar cells.

We plan to look also at the output of the eye by recording the whole optic nerve response with a suction electrode, to see whether signals are getting out of the retina, and, if so, how they compare to normals. Then we want to address the issue of spatial vision; we plan to record at the level of the optic tectum from single ganglion cell fibres. There is a map of the visual world—the retinal field—onto the tectum. By looking at the nature of this map and how it changes during retinal regeneration, we can determine what connectivities exist after regeneration and develop hypotheses about the nature of the animal's spatial vision which we can ultimately test psychophysically.

References

Anderson KJ, Dam D, Lee S, Cotman CW 1988 Basic fibroblast growth factor prevents death of lesioned cholinergic neurons *in vivo*. Nature (Lond) 332:360–361

Braekevelt CR 1980 Wandering phagocytes at the retinal epithelium–photoreceptor interface in the teleost retina. Vision Res 20:495–499

Faktorovich EG, Steinberg RH, Yasumura D, Matthes MT, LaVail MM 1990 Photoreceptor degeneration in inherited retinal dystrophy delayed by basic fibroblast growth factor. Nature (Lond) 347:83–86

Fernald RD 1991 Diurnal variation in cell production in the retina of a teleost. Submitted

Fulton A, Abramov I, Allen J et al 1986 Visual development. In: Waxler M, Hitchins V (eds) Light toxicity. CRC Press, Boca Raton, FL (Report of the FDA-NCRDH workshop on long-term visual health risks of optical radiation) p 137–146

LaVail MM, Faktorovich EG, Matthes MT et al 1991 bFGF protects photoreceptor cells from light damage in the rat. Invest Ophthalmol & Visual Sci 32 (suppl): 1094

Li Z-Y, Tso MOM, Wang H, Organisciak DT 1985 Amelioration of photic injury in rat retina by ascorbic acid: a histopathologic study. Invest Ophthalmol & Visual Sci 26:1589–1598

Lopashov GV, Sologub AA 1972 Artificial metaplasia of pigmented epithelium into retina in tadpoles and adult frogs. J Embryol Exp Morphol 28:521–546

Messersmith EK, Redburn DA 1990 Kainic acid lesioning alters development of the outer plexiform layer in neonatal rabbit retina. Int J Dev Neurosci 8:447–461

Mullen RJ, LaVail MM 1976 Inherited retinal dystrophy: primary defect in pigment epithelium determined with experimental rat chimeras. Science (Wash DC) 192:799–801

Noell WK 1980 There are different kinds of retinal light damage. In: William TP, Baker BN (eds) The effects of constant light on visual processes. Plenum Press, New York, p 3–28

Noell WK, Walker, Kang BS, Berman S 1966 Retinal damage by light in rats. Invest Ophthalmol 5:450–473

Organisciak DT, Wang H, Li Z-Y, Tso MOM 1985 The protective effect of ascorbate in retinal light damage of rats. Invest Ophthalmol & Visual Sci 26:1580–1588

Otto D, Frotscher M, Unsicker K 1989 Basic fibroblast growth factor and nerve growth factor administered in gel foam rescue medial septal neurons after fimbria fornix transection. J Neurosci Res 22:83–91

Park CM, Hollenberg MJ 1989 Basic fibroblast growth factor induces retinal regeneration *in vivo*. Dev Biol 134:201–205

Penn JS, Baker BN, Howard AG, Williams TP 1985 Retinal light damage in albino rats: lysosomal enzymes, rhodopsin, and age. Exp Eye Res 41:275–284

Powers MK, Bassi CJ, Raymond PA 1988 Lighting conditions and retinal development in goldfish: absolute visual sensitivity. Invest Ophthalmol & Visual Sci 29:37–43

Rapp LM, Tolman BL, Koutz CA, Thum LA 1990 Predisposing factors to light-induced photoreceptor cell damage: retinal changes in maturing rats. Exp Eye Res 51:177–184

Raymond PA, Bassi CJ, Powers MK 1988 Lighting conditions and retinal development in goldfish: photoreceptor number and structure. Invest Ophthalmol & Visual Sci 29:27–36

Rugh R, Wolff J 1955 Resilience of fetal eye following radiation insult. Proc Soc Exp Biol Med 89:248–253

Schweigerer L, Malerstein B, Neufeld G, Gospodarowicz D 1987 Basic fibroblast growth factor is synthesized in cultured retinal pigment epithelial cells. Biochem Biophys Res Commun 143:934–940

Sievers J, Hausmann B, Unsicker K, Berry M 1987 Fibroblast growth factors promote the survival of adult rat retinal ganglion cells after transection of the optic nerve. Neurosci Lett 76:157–162

Sternfeld MD, Robertson JE, Shipley GD, Tsai J, Rosenbaum JT 1989 Cultured human retinal pigment epithelial cells express basic fibroblast growth factor and its receptor. Curr Eye Res 8:1029–1037

Taylor JSH, Jack JL, Easter SS 1989 Is the capacity for optic nerve regeneration related to continued retinal ganglion cell production in the frog? A test of the hypothesis that neurogenesis and axon regeneration are obligatorily linked. Eur J Neurosci 1:626–638

Regeneration of olfactory receptor cells

Richard M. Costanzo

Department of Physiology, Medical College of Virginia, Virginia Commonwealth University, Richmond, VA 23298-0551, USA

Abstract. The vertebrate olfactory system has become an important model for the study of neural regeneration. The most remarkable feature of this system is its unique capacity for neurogenesis and replacement of degenerating receptor neurons. This replacement is made possible by a persistent neurogenesis among basal cells. Basal cells differentiate, develop into sensory neurons and grow axon processes. Receptor cell axons project back to the olfactory bulb where they re-establish connections with the central nervous system. When mature receptors reach a critical age, are damaged by nerve injury, or are exposed to environmental agents that enter the nasal cavity, they degenerate and are subsequently replaced by newly regenerated receptor cells. Recent experiments demonstrate that olfactory neurogenesis is not simply an extension of growth and development but is a unique capacity for cell replacement that persists beyond maturity and well into old age. Even more remarkable is the finding that replacement receptor cells re-establish connections with the CNS and restore sensory function. It is expected that further studies of olfactory neurogenesis using cell and tissue culture methods will provide important advances for the field of neural regeneration.

1991 Regeneration of vertebrate sensory receptor cells. Wiley, Chichester (Ciba Foundation Symposium 160) p 233–248

Characteristics of the olfactory epithelium

The vertebrate olfactory epithelium is a pseudo-stratified columnar epithelium and, like many other sensory systems, is derived from a thickening (placode) of the superficial ectoderm (Costanzo & Graziadei 1987). Olfactory placodes appear early in development as paired structures adjacent to the neural tube. The olfactory placodes invaginate to form olfactory pits and later the olfactory nasal cavity. Cells in the placode differentiate into receptor neurons and grow axons. These axons penetrate the primitive wall of the neural tube, make contact with cells in the telencephalon, and give rise to the olfactory bulb.

The fully developed olfactory epithelium consists of three cell types: supporting cells, olfactory receptor cells, and basal (stem) cells (Fig. 1).

The function of supporting cells is not fully understood; they have been shown to have secretory granules and thus may play a role in mucus formation. Costanzo & Morrison (1989) have recently shown that there is a close structural relationship between supporting cells and receptor cell dendrites. It has been suggested that supporting cells derived from the non-nervous ectoderm may have properties similar to those of glial cells in the CNS. The olfactory receptor cells make up most of the sensory neuroepithelium. These cells are responsible for

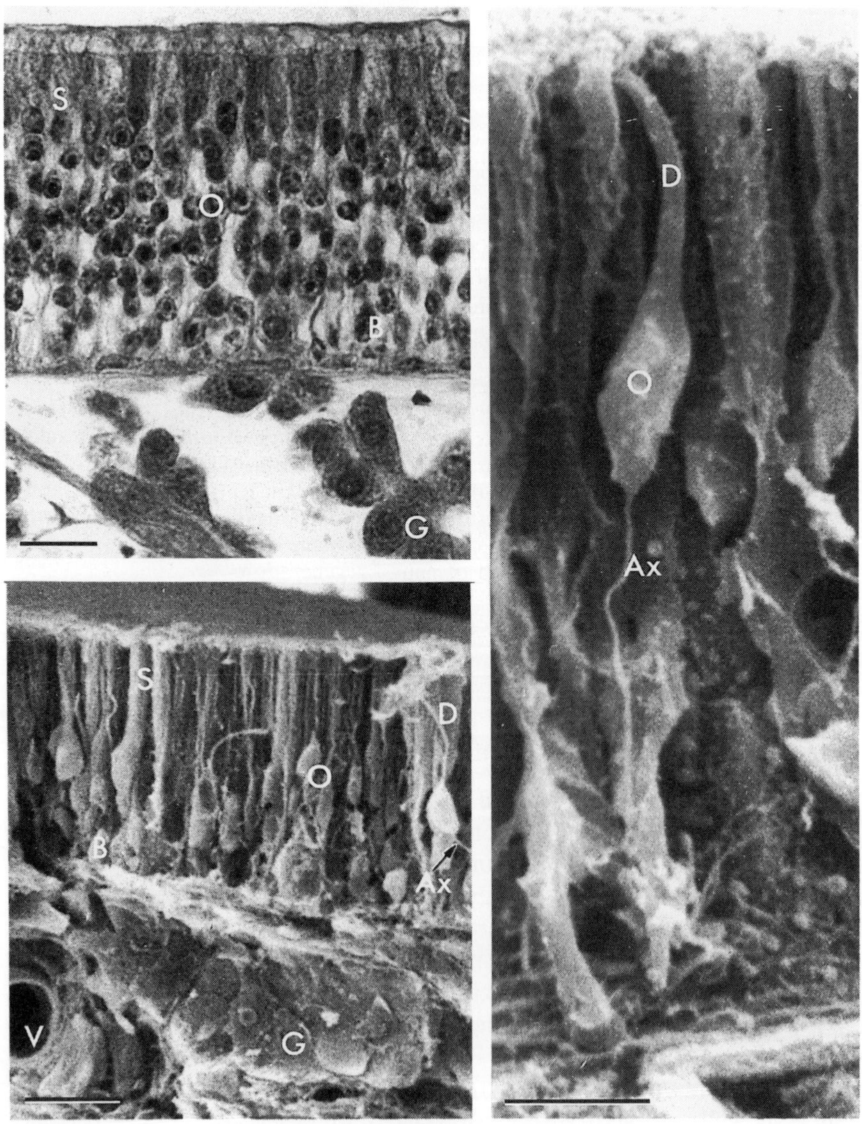

the detection of chemical signals in the environment and transmit sensory information to the CNS. Olfactory receptors are true neurons. They have dendrites that extend to the epithelial surface where they give rise to sensory cilia, and they grow axons that project directly to the olfactory bulb. Most sensory receptors are specialized epithelial cells, do not have axons, and do not have direct access to the CNS. In this respect, olfactory receptors are unique among sensory receptors. In addition, these sensory neurons are unique in that they can be replaced by a continuous neurogenesis among basal cells. These basal stem cells provide a constant source of replacement cells that differentiate into sensory neurons (Costanzo & Graziadei 1987).

Neurogenesis and replacement of sensory cells

The capacity of the olfactory system to undergo continuous neurogenesis and replacement of sensory neurons is truly remarkable. Unlike most sensory receptors, olfactory dendrites and cilia are in direct contact with the external environment (Farbman 1990). This puts them at risk of exposure not only to odour stimuli but also to neurotoxins, chemical irritants, viruses and a number of other potentially harmful agents. It is therefore not surprising that they would need to be continuously replaced.

A recent study of the human olfactory epithelium using the scanning electron microscope revealed for the first time patches of non-olfactory cells located in the olfactory region of the nasal cavity (Morrison & Costanzo 1990). It has been suggested that such patches may represent areas that were exposed to environmental agents, and that the olfactory receptors may have degenerated in these areas. Further examination revealed a few olfactory cells among the non-sensory cells, suggesting that regeneration and replacement of sensory receptors may have occurred.

FIG. 1. Photomicrographs of the sensory epithelium lining the olfactory area of the nasal cavity of the hamster. *Upper left*. With light microscopy cell nuclei are observed in different layers of the olfactory epithelium. Oval supporting cell nuclei (S) are near the surface, olfactory receptor cell nuclei (O) occupy the lower two-thirds, and basal cell nuclei (B) are found at the base of the epithelium. Bowman's gland (G) cells located below the epithelium produce mucus secretions that cover the surface of the nasal cavity. Scale bar, 20 μm. *Lower left*. Image taken with the scanning electron microscope (SEM) provides a three-dimensional view of cell morphology from the surface to the base of the epithelium. Olfactory receptor cells (O) are seen having dendrites (D) and axon (Ax) processes. Blood vessel (V) in the lamina propria. Scale bar, 25 μm. *Right*. High magnification SEM shows features of olfactory receptors demonstrating that they are true neurons. The oval cell body (O) gives rise to an apical dendrite (D) that extends to the epithelial surface and a slender axon process (Ax) that is directed towards the lamina propria. Scale bar, 5 μm. (From Costanzo & Morrison 1989 by permission of Chapman & Hall.)

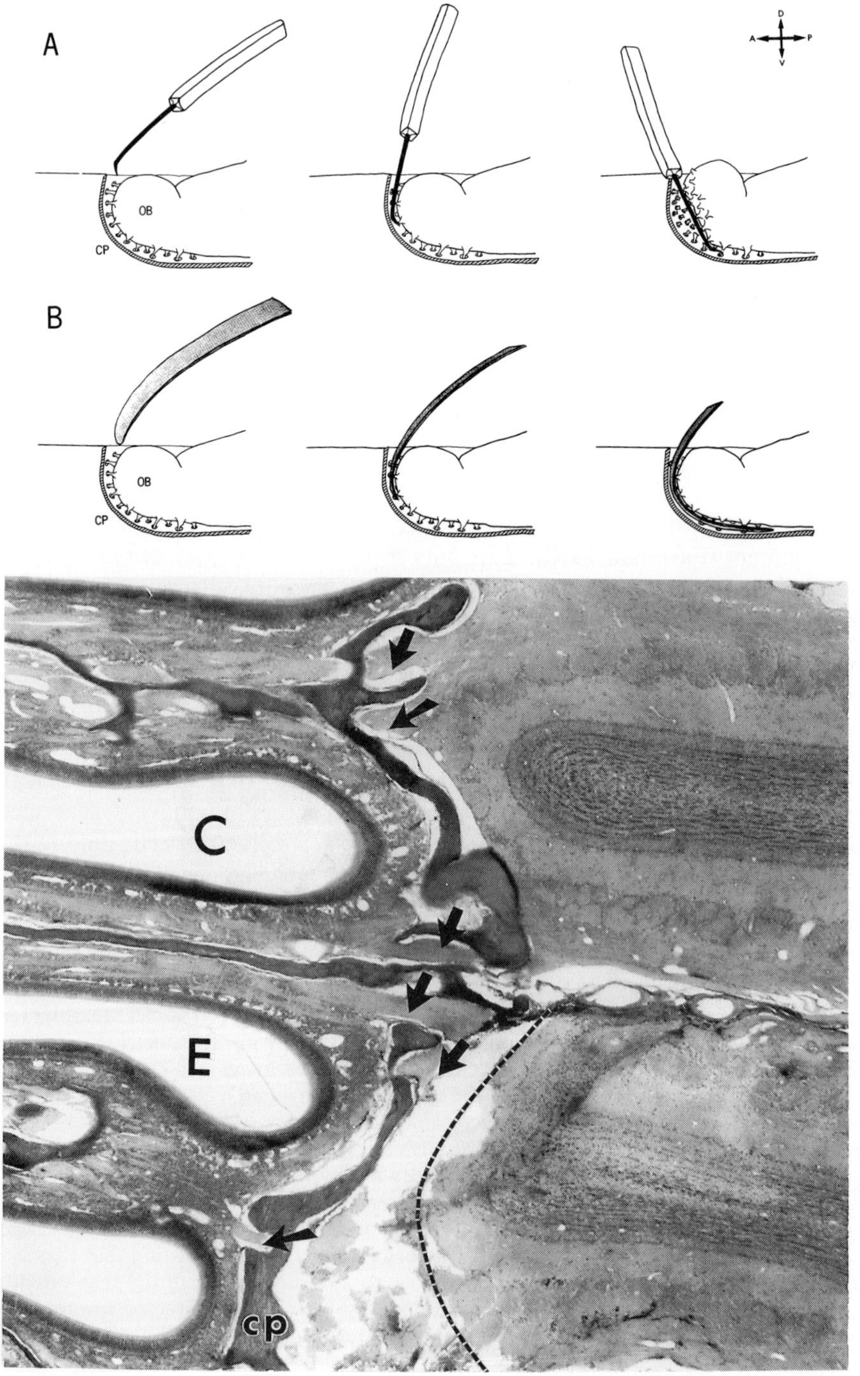
A
B
OB
CP
D
A
P
V
C
E
cp

What is most remarkable about the replacement of olfactory receptor cells is the fact that they are neurons. Not only do they function to transduce sensory signals, but they have axons that transmit sensory information back to the CNS. Until recently it was thought that replacement of neurons was impossible in the vertebrate nervous system. Because of its ability to replace neurons, the olfactory system has become a model for the study of regeneration and repair in the nervous system.

Model for regeneration and repair in the nervous system

There are two experimental designs most commonly used to study olfactory neurogenesis and replacement of degenerating neurons: chemical exposure, and axotomy. In chemical exposure models, chemical solutions are introduced into the nasal cavity, causing a degeneration of cells in the olfactory epithelium. Chemically induced lesions are usually non-specific and involve both supporting cells and sensory receptors. In axotomy, there is a selective degeneration of olfactory neurons. After transection of the axons of olfactory receptor cells, a retrograde degeneration of receptors in the sensory epithelium occurs. The supporting cells are left intact. Axotomy is the most common method used. This can be done in conjunction with a bulbectomy (removal of the target neurons), or by nerve transection where the olfactory bulb (target) is left intact (Costanzo & Graziadei 1983, Costanzo 1984). Recently, a new method for nerve transection has been established that reduces damage to the olfactory bulb, yet allows for complete transection of nerve fibres (Fig. 2).

After the degeneration of sensory receptor cells there is an increase in activity among basal cells and a neurogenesis of newly differentiating olfactory receptors. These cells develop into sensory neurons, send a dendrite to the surface of the epithelium, and grow an axon process that projects back to the olfactory bulb. The replacement of olfactory receptors has been examined and quantified in mammals (Graziadei & Monti Graziadei 1979, Costanzo & Graziadei 1983, Hinds et al 1984). In several studies, young animals, some of which were still undergoing development and sexual maturation, have been used. This raises a question about whether the observed neurogenesis is a late stage of neural growth

FIG. 2. Experimental methods used in olfactory nerve transection studies. *Upper*. Diagram comparing the conventional method (A) that uses a stainless steel microdissecting knife and a new method (B) that uses a flexible teflon blade. The new method provides better access to ventral nerve fibres and causes less damage to the olfactory bulb (OB). *Lower*. Horizontal section through the nasal cavity and olfactory bulbs prepared immediately after nerve transection. Nerve fibres (arrows) on the control side (C) connect receptors in the nasal cavity to the olfactory bulb. Immediately after nerve transection (dashed line), nerve fibres on the experimental (E) side can be seen completely transected at the point where they emerge from the cribriform plate (CP) and project to the olfactory bulb.

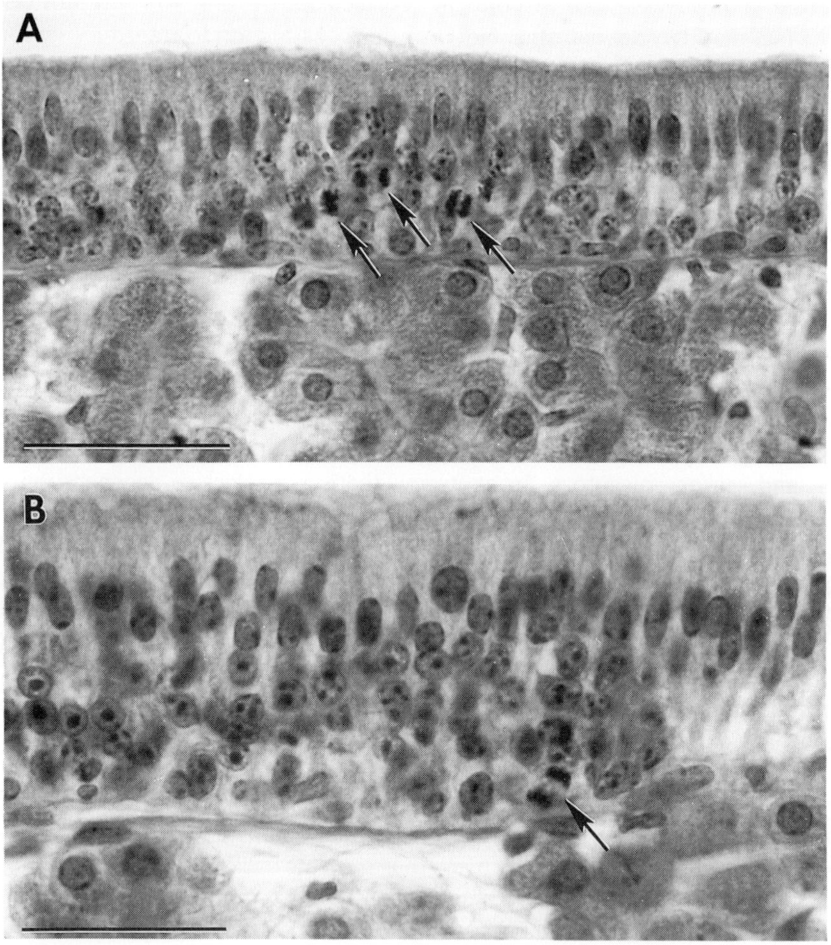

FIG. 3. Photomicrographs of the olfactory epithelium of old adult hamsters (>one year) after recovery from nerve transection. The presence of mitotic figures (arrows) at both 60 days recovery (A) and 120 days recovery (B) demonstrates that the capacity for neurogenesis and replacement of receptor cells persists beyond development and into old age. Scale bars, 50 μm.

and development, or a unique process that continues beyond development and into adulthood.

To examine this issue we studied three age groups: immature (<two months), sexually mature adult (2–4 months), and old adult (>one year) hamsters (*Mesocricetus auratus*). Methods similar to those used by Costanzo & Graziadei (1983) were employed to quantify the number of sensory receptor neurons during degeneration and recovery. One of the most significant findings was the persistence of neurogenesis beyond sexual maturity and well into old age. Histological examination of the sensory epithelium in the old adult group at both 60 and 120 days of recovery revealed mitotic figures among the basal cells of the epithelium (Fig. 3). This confirmed that the capacity for neurogenesis was not a process limited to epithelial growth and development. Epithelial cell counts were taken from the experimental side (nerve transection) and compared to the control side during recovery from nerve transection. Results were expressed as a percentage of the control side.

Figure 4 gives the recovery curves for the three age groups studied. Although the number of cells increased significantly above degeneration levels, recovery rarely reached control levels. During the initial degeneration period (0–5 days) there was a marked decrease in cell number on the lesioned side. Although the presence of supporting cells (not damaged by nerve transection) explains why

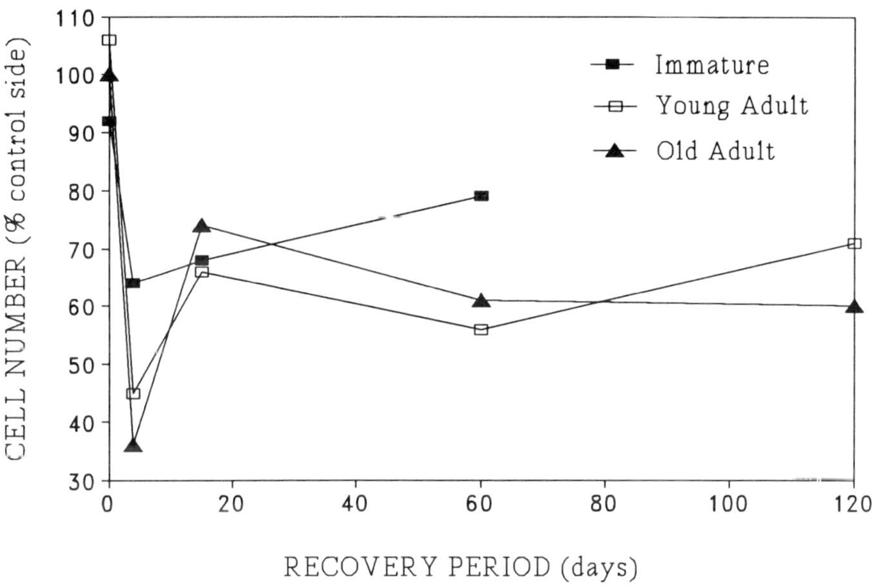

FIG. 4. Recovery curves plotting receptor cell replacement after nerve transection in hamsters in three age groups: Immature, <two months (filled squares); young adults, 2–4 months (open squares); and old adults, >one year (filled triangles). Transection occurred at Day 0. For each age group there was a marked increase in cell number, well above degeneration levels. This demonstrates that there was a replacement of receptor cells during recovery from nerve transection. Recovery in the old and young adults reached 60–70% of the control values by 120 days. Immature animals recovered to 80% of control levels by Day 60. This higher level of recovery could be the result of an increased number of immature receptors present in the epithelium during early stages of growth and maturation.

cell numbers did not decrease to 0%, some of the cells counted during this period may also be immature sensory neurons. Axons of newly developing neurons may not have grown far enough to reach the site of nerve transection and, therefore, would be spared from any retrograde degeneration resulting from nerve transection. This could explain the significantly higher cell numbers (>60%) for immature animals, where there is an increase in the number of young neurons as a result of growth and development. Immature animals also showed the highest recovery level (80% by Day 60). Data for Day 120 are not available. Although overall recovery was somewhat less in the sexually mature young adult and old adult animals, these data demonstrate significant neurogenesis and increase in cell number well above degeneration levels. The results show that the capacity for neurogenesis and nerve cell replacement continues beyond growth and development and into old age (1–2 years in hamsters). It seems likely that two mechanisms contribute to the neurogenesis of sensory receptor cells in the olfactory system: one associated with normal growth and development, and a second involved with the replacement of degenerating receptors lost through injury, exposure, or the normal ageing process.

Functional replacement of damaged pathways

Olfactory neurogenesis results not only in the replacement of sensory receptor neurons in the epithelium, but also in the growth and reconnection of severed axons that project to cells in the olfactory bulb. The observation that replacement neurons are capable of re-establishing connection with target cells is an important finding. Not only is there regeneration of sensory receptors; there is also a rewiring of damaged pathways. Reconnection of axon projections has been observed in histological sections and confirmed by horseradish peroxidase tracing of axons from the epithelium to the olfactory bulb. The extent of these connections and the spatial patterns of projections remain to be investigated. Can these reconnected axons transmit sensory information from receptors back to the olfactory bulb? There have been a few studies of receptor cell function after recovery from nerve transection showing that receptors do respond to odour stimuli (Simmons & Getchell 1981). There have also been behavioural studies in birds that suggest recovery of olfactory function (Oley et al 1975, Kiyohara & Tucker 1978). The findings reported here suggest that axon connections are re-established with cells in the olfactory bulb of hamster and that those connections are capable of transmitting sensory information to the olfactory bulb.

Electrophysiological recordings from cells in the hamster olfactory bulb were made to determine whether reconnected axons from replacement receptor cells could provide sufficient input to the bulb for the detection of odour stimuli. Recordings were made at recovery times of 60, 120 and 180 days

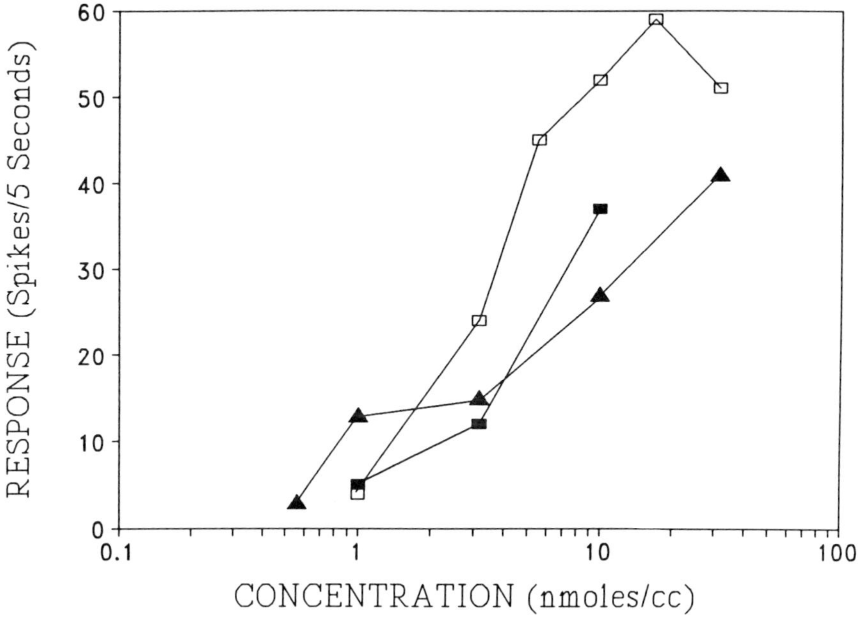

FIG. 5. Concentration–response relationships recorded from hamster olfactory bulb cells after recovery from nerve transection. The stimulus illustrated was amyl acetate in air, diluted using a flow dilution olfactometer. The response of a cell from the control side (open squares) where olfactory nerve projections were left intact illustrates a typical odour response profile. The response profiles of two cells from the experimental side (filled symbols) at 120 days after nerve transection demonstrates functional recovery. Replacement receptor cells re-established axon connections with the olfactory bulb and the new connections transmitted sensory information (stimulus intensity) from the receptors to cells in the olfactory bulb.

after nerve transection. Stimulus concentration was controlled by a flow dilution olfactometer and concentration–response functions were obtained for units that responded to odour stimulation. The concentration–response functions of three units are given in Fig. 5. The fact that the units responded at all to odour stimulation demonstrated that replacement receptor cell axons had re-established synaptic connections with the olfactory bulb. The observed stimulus concentration–response relationship is further evidence that there was a sufficient number of connections to restore sensory function.

Thus, there is anatomical, behavioural and electrophysiological evidence that olfactory receptors have the ability to replace damaged pathways and to restore sensory function. The underlying mechanisms have yet to be uncovered and present important challenges for future research.

Summary/conclusion

Until recently, regeneration and repair of the nervous system was thought to occur only when nerve cells were partially damaged. If the cell body was spared, then regeneration of an axon or dendritic process was possible. Regeneration studies in the olfactory system suggest not only that neurons can be replaced (neurogenesis) but that replacement neurons can re-establish connections with the CNS, and in some cases restore sensory function. What is it about olfactory receptors that make them unique? Is there a neurogenic factor residing within the sensory epithelium? Is regeneration a response to signals triggered by cell degeneration? Or is it some genetically programmed property of olfactory receptor cells? Further studies using cell and tissue culture methods, along with advances in molecular biology, may soon provide a better understanding of the mechanisms underlying olfactory neurogenesis and regeneration in these sensory cells.

Acknowledgement

This work has been supported by grant DC00165 from the National Institute on Deafness and Other Communication Disorders.

References

Costanzo RM 1984 Comparison of neurogenesis and cell replacement in the hamster olfactory system with and without a target (olfactory bulb). Brain Res 307:295–301

Costanzo RM, Graziadei PPC 1983 A quantitative analysis of changes in the olfactory epithelium following bulbectomy in hamster. J Comp Neurol 215:370–381

Costanzo RM, Graziadei PPC 1987 Development and plasticity of the olfactory system. In: Finger TE, Silver W (eds) Neurobiology of taste and smell. Wiley, New York, p 233

Costanzo RM, Morrison EE 1989 Three-dimensional scanning electron microscopic study of the normal hamster olfactory epithelium. J Neurocytol 18:381–391

Farbman AI 1990 Olfactory neurogenesis: genetic or environmental controls? Trends Neurosci 13:362–365

Graziadei PPC, Monti Graziadei GA 1979 Neurogenesis and neuron regeneration in the olfactory system of mammals. I. Morphological aspects of differentiation and structural organization of the olfactory sensory neurons. J Neurocytol 8:1–18

Hinds JW, Hinds PL, McNelly NA 1984 An autoradiographic study of the mouse olfactory epithelium: evidence for long-lived receptors. Anat Rec 210:375–383

Kiyohara S, Tucker D 1978 Activity of new receptors after transection of the primary olfactory nerve in pigeons. Physiol Behav 21:987–994

Morrison EE, Costanzo RM 1990 Morphology of the human olfactory epithelium. J Comp Neurol 297:1–13

Oley N, DeHan RS, Tucker D, Smith JC, Graziadei PPC 1975 Recovery of structure and function following transection of the primary olfactory nerves in pigeons. J Comp Physiol Psychol 88:477–495

Simmons PA, Getchell TV 1981 Neurogenesis in olfactory epithelium: loss and recovery of transepithelial voltage transients following olfactory nerve section. J Neurophysiol 45:516–528

DISCUSSION

Rubel: There have been enough data on the generation of olfactory neurons in mammals that I don't doubt it. But how do you know that some or all of the recovery you see, in terms of neural potentials or axons to the brain, is not due to the regrowth of axons that were cut, as opposed to axons from new olfactory cells?

Costanzo: In studies in which we label cells with horseradish peroxidase, we don't see any HRP in the nerve layers during this period of maximum degeneration.

Rubel: That doesn't resolve the question. The only way to do that is by a double-labelling study with tritiated thymidine, followed by back-labelling with HRP or some other tracer.

Calof: In other words, you should label with a compound such as [^{3}H]thymidine that will get into the progeny of proliferating cells, and then, after the axons have grown into the olfactory bulb, inject something like diI or HRP into the bulb and let it be taken up by the axons. You then look for a cell in the epithelium that contains both tritiated thymidine and the second label; this was therefore a cell that was newly generated and had made axonal connections with the olfactory bulb.

Margolis: The answer to this question will also come from consideration of the *numbers* of neurons. After deafferenting lesions or olfactory axotomy, the olfactory receptor cells present in the neuroepithelium are very few, relative to what you see, say six weeks after recovery, or before lesioning.

Reasner: Yes, the normal olfactory receptor neuron layer is quite thick, and after degeneration it thins down very dramatically. There are few remaining cells that would even need to be considered.

Rubel: Dr Costanzo found that 30% of the cells were left, however.

Costanzo: The cell counts I showed were the total number of cells after nerve transection, so they included all the supporting cells, the basal cells and all immature neurons that had not yet developed axon processes.

Farbman: The question of olfactory neuronal survival after axotomy has been studied in the garfish (*Lepisosteus osseus*) which has a long olfactory nerve (20–30 cm). It was shown that if the olfactory nerve was severed a long distance from the cell body, about 10 cm or more, the cell bodies of the olfactory neurons would survive and simply regenerate their axons. If the lesion was made less than this distance away, the cell bodies would die (Cancalon 1987).

Rubel: Those studies don't answer the question about axonal regeneration for the hamster, where the olfactory nerve is much less than 10 cm.

Calof: Dr Pujol has asked me why, if you take a sensory or motor neuron and cut its axon in the periphery, the axon will regenerate, whereas if you axotomize an olfactory receptor neuron, you get death of the neuron rather than regeneration of the axon. I didn't know about the garfish experiment, however.

Farbman: If you cut the axon close to the cell body of any neuron, that neuron will die. In this respect, the olfactory neuron does not differ from other neurons.

Calof: Bulbectomy, which is the most common way to do that sort of axotomy, is about as far away as you can get from the cell bodies of the olfactory receptor neurons.

Rubel: Is Dr Farbman's point universally true, that all olfactory neurons go through retrograde degeneration? Does it depend on the structure of the neuron?

Palay: It is correct that every nerve cell will die if you cut very close to the perikaryon.

Farbman: Yes, and all the lesions were made very close.

Rubel: I am satisfied, if you say that all the *neurons* are destroyed, Dr Costanzo.

Costanzo: It is all the 'mature' neurons that are destroyed by nerve transection. Those immature neurons that haven't grown axons are spared.

Cotanche: Dr Costanzo, you have mitosis occurring in what appears to be the base of the epithelium, rather than at the lumen, despite what was said earlier—namely that mitoses occur near the lumen of the epithelium.

Costanzo: It is definitely at the base, not the lumen; that's where mitosis is observed. The basal cells are at the base of the epithelium; they divide there and then differentiate as the cell migrates toward the lumen, sending one process, the dendrite, up and another process, the axon, down (see Fig. 1, p 234).

Ryals: So it's very different from the lateral line organ in fish, where the cell migrates from the basal portion of the epithelium toward the lumen, as Dr Presson described?

Costanzo: Yes. The basal cell doesn't migrate, in the adult hamster olfactory system.

Farbman: In the rodent embryo, however, it happens that way. In prenatal development in rodents, all the mitoses are at the surface of the olfactory epithelium (Smart 1971). At some point in development, a basal layer is established, and mitosis becomes restricted to this layer.

Reasner: Furthermore, during embryonic development in the mouse, there are cells proliferating at a high rate that do not look like adult supporting or basal cells. These embryonic stem cells first disappear from the middle of the epithelium, leaving both a basal and a superficial zone of proliferation (Smart 1971, Cuschieri & Bannister 1975). However, at late embryonic ages, cells in the epithelium have the appearance of adult basal, receptor and supporting cells, and proliferation in the superficial zone nearest the lumen is reduced. In adults, the percentage of proliferation occurring superficially is quite small, but still present (which is the point I made earlier). In short, the proliferative compartment for olfactory receptor neurons is located in the basal layer while the little remaining superficial proliferation appears adequate to maintain the supporting cell population.

Ryals: In normal development, is there a layer that is proliferating to form two separate sets of basal and support cells?

Calof: The data aren't very good because they are older anatomical studies which looked at things like mitotic figures in the apical and basal parts of the developing olfactory epithelium. They show a gradual transition, with many intermediate states, from a lot of apical mitoses to, in the mature epithelium, mostly basal mitosis.

It is difficult to specify the cell types that are dividing in those earlier anatomical studies of the embryonic olfactory epithelium. If you use antibody markers that have specific distributions in the adult, these antibodies stain cells in the embryonic epithelium as well, but the location of these cells is not always the same as in the adult epithelium. Our basal cell marker (anti-keratins antibody) is like this. We can see cells stained with anti-keratins in both basal and apical regions of the embryonic olfactory epithelium of the mouse (cf. Calof & Chikaraishi 1989). This is consistent with the observation that the proliferative population has not settled down to its adult, basal distribution.

Raymond: What is the exact embryological origin of the olfactory epithelium? Are all the cells derived from a placode which is ectodermal in origin, as a single layer which thickens in the placodal region?

Farbman: In a detailed study in mouse embryos, the placode is shown to be essentially a piece of neural plate epithelium that has been isolated (bilaterally) before the neural plate folds into the neural tube (Verwoerd & Van Oostrum 1979). The question of where the placode originates and how it is induced has been a matter of concern for many years. The two tissues most plausibly thought to be able to cause the induction were the neural tube and the mesenchyme, both just beneath the placode. Many experiments were done on several species, with no consistent results (reviewed by Farbman 1988). But it is a reasonable hypothesis to consider that the olfactory placode, and perhaps the otic and other placodes, arise as islands of neural plate epithelium that have become detached from the neural plate before it folds over to become the neural tube.

Lewis: For otic placode, there is evidence against that. From our experiments on the induction of the otocyst in the chick, it seems that the old story is probably correct, that you can induce otocyst from epidermis that would not normally become otocyst by putting it adjacent to the right bit of hindbrain (J. Adam, unpublished results).

Farbman: The same was found to be true with the olfactory placode, in some species but not others.

Palay: But the olfactory placode is the first one to appear in early embryonic development of the medullary plate, so the idea that it is really part of neural tube is probably justified.

Cotanche: Julian, how do you define an otocyst—just as an invagination, or do you get hair cell differentiation?

Lewis: All we have shown so far is that you can get an extra otic vesicle forming if you take epidermis from the side of the head of the chick embryo and turn it round, so the otic placode is now anteriorly misplaced; in its new position, the original placode continues its development, while the previously anterior tissue, put in a posterior position, is induced to invaginate to give you a second vesicle.

Cotanche: That is different from a vesicle with hair cells!

Lewis: That's true, but it's a hint; I wouldn't wish to claim more from our results so far.

Oakley: Dr Costanzo, there is some difference between chemically induced degeneration, for example by zinc sulphate treatment, and the way you undercut the olfactory bulb. In your nerve transection procedure, all the neurons which have incompletely extended axons are spared, if they are not transected or injured by the undercutting procedure. You would therefore get somewhat better regeneration with your methodology. I am wondering if those spared axons might form a substrate for entirely new neurons to extend axons.

Costanzo: It could be so. One experiment that we plan to do is a double transection; that is, wait long enough for any of the cells that are immature at the time of the first procedure to grow out, and then transect those. Any cell that is left must be a replacement cell after the initial procedure.

One problem with zinc sulphate is that you have to titrate it carefully or you can burn out the epithelium all the way down to the basal cells (hence our preference for nerve transection methods).

Since you bring up this point about chemically induced degeneration, I wonder what others here think is the effect of the environment in stimulating the olfactory system to undergo continuous neurogenesis. Is that the cause, or is it genetic?

Burd: When Hinds and coworkers (1984) put mice in 'clean' environments, they found that many olfactory receptor cells lived longer than previously reported (up to six months).

Farbman: You can up-regulate or down-regulate the rate of mitosis in the olfactory epithelium by manipulating the animal.

Rubel: Is this the cell cycle time, or the number of cells entering mitosis?

Farbman: The number of olfactory basal cells that will take up [^{3}H]thymidine can be up-regulated by removing the olfactory bulb in rats. Within a week there is a four-fold increase in the number of cells taking up the isotope (Carr & Farbman 1990). On the other hand, if one nostril of a newborn rat is occluded, the rate of [^{3}H]thymidine uptake is down-regulated by 40% (Farbman et al 1988). The degree of stress placed on the olfactory epithelium may determine how rapidly cell division occurs.

The question was raised whether the olfactory epithelium behaves throughout life like an embryonic system that is growing. There is a great deal of overproduction, even in control (untreated) animals. Genesis of new cells occurs in control animals, but in order to maintain the epithelium at constant size an equivalent number of cells must die. We have found that many of the newly formed cells do not live for the average lifespan of 30–40 days; rather, many die precociously, before reaching maturity (Carr & Farbman 1990, see also Breipohl et al 1986). Moreover, in highly stressed (bulbectomized) animals, a greater proportion of newly formed cells die prematurely, as early as one or two days after cell division.

Rubel: Have you, or has anybody else, been successful in dissociating the number of cells dying from the number proliferating? Because usually, as you said, the number of cells that die is correlated with the number of mitotic figures seen. One question that we have been asking is: does there have to be cell death in order to turn on proliferation? Is there any way of dissociating these two events, to answer our question?

Steinberg: Is there a finite lifetime for an olfactory epithelial cell?

Farbman: Some olfactory cells in mice live for a year. Hinds et al (1984) showed that labelled olfactory cells were demonstrable in mice 12 months after a single dose of [^{3}H]thymidine.

Rubel: But is there a manipulation that increases mitosis but doesn't cause more cells to die, or vice versa?

Burd: There is; Mackay-Sim & Beard (1987) showed that basal cell division continues in adult hypothyroid mice, yet the cells don't mature into olfactory receptor cells. This results in a reduction in the thickness of the olfactory epithelium. Therefore, you can have a constant rate of cell genesis without adding new cells to the receptor cell population.

Ryals: So cells continue to be produced, even though cells are not dying; but to keep a constant number, some cells remain immature?

Burd: Yes. They find a reduction in the ability of a cell to mature in a hypothyroid mouse.

Farbman: The olfactory epithelium is less thick, and contains as many mature cells as controls and as many dividing cells. The reduction in epithelial thickness is at the expense of the 'almost mature', or not fully differentiated cells.

Rubel: Do these postmitotic cells have neuronal characteristics?

Burd: They did not really address that question.

Costanzo: We have measured the thickness of the olfactory epithelium and the cell number after nerve transection, and even though the thickness stays low, it looks as if it's approaching some new level of cell density; so the supporting matrix has shrunk, but the number of cells that come back may restore epithelial cell density.

References

Breipohl W, Mackay-Sim A, Grandt D, Rehn B, Darrelmann C 1986 Neurogenesis in the vertebrate main olfactory epithelium. In: Breipohl W (ed) Ontogeny of olfaction. Springer-Verlag, Berlin, p 21–33

Calof AL, Chikaraishi DM 1989 Analysis of neurogenesis in a mammalian neuroepithelium: proliferation and differentiation of an olfactory neuron precursor *in vitro*. Neuron 3:115–127

Cancalon PF 1987 Survival and subequent regeneration of olfactory neurons after a distal axonal lesion. J Neurocytol 16:829–841

Carr VM, Farbman AI 1990 Nongenetic regulation of proliferation and cell death in the olfactory epithelium (OE). Soc Neurosci Abstr 16:837

Cuschieri A, Bannister LH 1975 The development of the olfactory mucosa in the mouse: light microscopy. J Anat 119:277–286

Farbman AI 1988 Cellular interactions in the development of the vertebrate olfactory system. In: Margolis FL, Getchell TV (eds) Molecular neurobiology of the olfactory system. Plenum Press, New York, p 319–332

Farbman AI, Brunjes PC, Rentfro L, Michas J, Ritz S 1988 The effect of unilateral naris occlusion on cell dynamics in the developing rat olfactory epithelium. J Neurosci 8:3290–3295

Hinds JW, Hinds PL, McNelly NA 1984 An autoradiographic study of the mouse olfactory epithelium: evidence for long-lived receptors. Anat Rec 210:375–383

Mackay-Sim A, Beard MD 1987 Hypothyroidism disrupts neural development in the olfactory epithelium of adult mice. Dev Brain Res 36:190–198

Smart IHM 1971 Location and orientation of mitotic figures in the developing mouse olfactory epithelium. J Anat 109:243–251

Verwoerd CDA, Van Oostrum CG 1979 Cephalic neural crest and placodes. Adv Anat Embryol Cell Biol 58:1–75

Regulation of neurogenesis and neuronal differentiation in primary and immortalized cells from mouse olfactory epithelium

Anne L. Calof[1]*, Arthur D. Lander[2] and Dona M. Chikaraishi[1]

[1]*Neuroscience Program, SC-6, Tufts University School of Medicine, Boston, MA 02111 and* [2]*Departments of Biology and Brain & Cognitive Sciences, E25-435, Massachusetts Institute of Technology, Cambridge, MA 02139, USA*

Abstract. We have developed an *in vitro* system for studying molecular events regulating neurogenesis in the mouse olfactory epithelium (OE). Our observations suggest that two types of neuronal precursor may be involved: (1) a transiently existing, immediate neuronal precursor (INP), which generates two postmitotic daughter neurons; and (2) a neuroepithelial stem cell, which may be the basal cell (or some subclass of basal cell) of the OE, and is presumed to be the progenitor of the INP. Using antibody markers that distinguish basal cells and postmitotic receptor neurons *in vitro* and *in vivo*, we have shown that neurogenesis occurs early on in OE cultures, but then ceases because INPs divide only once to generate postmitotic neurons and no new INPs are produced by basal cells. To determine whether the basal cell-to-INP transition, or proliferation and neuronal differentiation of the INP, are regulated by crucial growth factors or cellular interactions, we are testing various polypeptide growth factors and extracellular matrix proteins for their effects on OE neurogenesis *in vitro*. We have also generated immortalized OE cell lines by using retroviruses to transduce oncogenes into cultured OE cells. One such cell line (derived from a primary OE basal cell culture) develops branching processes when transplanted into neonatal mouse brain—a condition in which cells from freshly isolated OE can undergo apparent morphological differentiation into neurons.

1991 Regeneration of vertebrate sensory receptor cells. Wiley, Chichester (Ciba Foundation Symposium 160) p 249–276

In order to understand how neurons in the vertebrate nervous system are generated, whether during embryonic development or during regeneration,

*Present address: Department of Biology, University of Iowa, Iowa City, IA 52242, USA

it is important to identify and to analyse functionally the molecules that regulate neurogenesis. To achieve this it is advantageous to establish an experimental system with which the process of neurogenesis can be directly examined and easily manipulated. Our approach to this problem has been to study neurogenesis in tissue culture. For these studies, we have used the olfactory epithelium (OE) of the mouse.

Several features of the OE make it well suited for *in vitro* studies of neurogenesis. Like the neural tube—the primordium of the vertebrate nervous system—the OE is an epithelial structure in which neurons are generated. However, the OE has a much simpler structure than the neural tube; in its mature form the OE contains only three or four cell types, and only a single type of neuron—the olfactory receptor neuron. In addition, the OE continues to generate neurons throughout life (Graziadei & Monti Graziadei 1978). Finally, because the OE maintains its capacity for neurogenesis, it has been possible for investigators to perform studies that show that neurogenesis in the OE is subject to regulation, presumably by cellular interactions (e.g. Graziadei 1973, Monti Graziadei & Graziadei 1979, Costanzo & Graziadei 1983, Camara & Harding 1984).

We have developed a tissue culture system for the olfactory epithelium and have demonstrated that olfactory receptor neurons are generated in these cultures. By identifying molecular markers for the different cell types present in the OE *in vivo* and *in vitro*, and by combining use of these markers with [^{3}H]thymidine uptake analysis in quantitative studies of neurogenesis, we have begun to characterize the cellular stages in the neurogenic pathway (Calof & Chikaraishi 1988a,b, Calof & Chikaraishi 1989). Our current studies are focused on two goals: (1) to determine the role that cellular interactions play in regulating the proliferation of OE neuronal precursors and the differentiation of these precursors into neurons, both by analysing the effects of identified polypeptide growth factors and extracellular matrix molecules on OE cultures, and by transplanting OE cells back into various regions of the developing nervous system; and (2) to generate immortalized cell lines from OE cultures by retrovirus-mediated gene transfer, in order to facilitate biochemical and molecular genetic experiments on clonal populations of OE cells.

Cellular stages in olfactory neurogenesis

The conditions we use for culturing embryonic mouse OE have been described elsewhere (Calof & Chikaraishi 1989). Briefly, the nasal turbinate regions of Day 14 or 15 embryos are dissected, and the sensory epithelium is purified away from the underlying stroma by enzymic digestion and mechanical trituration. Pieces of purified OE are then explanted into culture in serum-free, defined medium, onto glass or plastic substrata that have been coated with various extracellular matrix molecules. Three types of cells appear in cultures prepared in this way: epithelial cells, which grow in flat sheets and are the *in vitro* equivalents of the basal cells; cells with phase-bright bodies and neurites, which are immature olfactory receptor neurons; and, at early times (<24 hours) in culture, small round cells, which we now know to be the precursors of olfactory receptor neurons (see below).

Our ability to perform quantitative *in vitro* studies of olfactory neurogenesis has been aided by the development of molecular markers for these different cell types. Two markers in particular, antibodies that distinguish the basal cells and the receptor neurons of the OE, have proved particularly useful. The basal cells of the OE are distinguished by a commercially available antiserum to keratins (Calof & Chikaraishi 1989). The olfactory receptor neurons are stained by a monoclonal antibody that recognizes a cytoplasmic epitope common to the 140 and 180 kDa forms of the neural cell adhesion molecule, N-CAM (Calof & Chikaraishi 1989, DiFiglia et al 1989). In OE cultures, the antiserum to keratins exclusively stains the sheets of flat epithelial cells, indicating that these are the *in vitro* equivalents of the basal cells, and the monoclonal anti-N-CAM exclusively stains the neurite-bearing cells, which are the olfactory receptor neurons (Fig. 1; cf. Calof & Chikaraishi 1989).

Previous studies of neuronal regeneration in the OE *in vivo* suggested that the basal cells are self-renewing stem cells that ultimately give rise to the neurons in this system (Harding et al 1977, Monti Graziadei & Graziadei 1979). However, by means of quantitative [^{3}H]thymidine uptake experiments performed in combination with immunocytochemical analysis, we have demonstrated that the small round cells, which appear at early times in OE cultures and are both morphologically and antigenically distinct from either basal cells or olfactory receptor neurons, are actually the immediate precursors of olfactory receptor neurons (Calof & Chikaraishi 1989). We have named these cells the immediate neuronal precursors (INPs) of olfactory receptor neurons. Our current hypothesis is that INPs are the *in vitro* equivalents of the so-called 'globose' basal cells that have been described occasionally in anatomical studies of olfactory epithelium (e.g. Graziadei & Monti Graziadei 1979).

The molecular markers we have developed for OE cultures have also made it possible to characterize the nature of INP divisions. Pulse–chase [^{3}H]thymidine uptake autoradiography, combined with anti-N-CAM immunocytochemical analysis of OE cultures, demonstrated that proliferating INPs do not express N-CAM, but that more than 80% of their progeny are postmitotic, N-CAM-positive cells. This indicates that the majority of INP divisions produce two, N-CAM-positive sister neurons (Calof & Chikaraishi 1989). Thus, most INPs do not appear to behave like stem cells *in vitro*, but appear to be unipotential cells that give rise only to neurons.

Our observations on olfactory neurogenesis *in vitro*, combined with the results of others' investigations of olfactory receptor neuron regeneration *in vivo*, suggest a sequence of events that is schematized in Fig. 2. In this scheme the

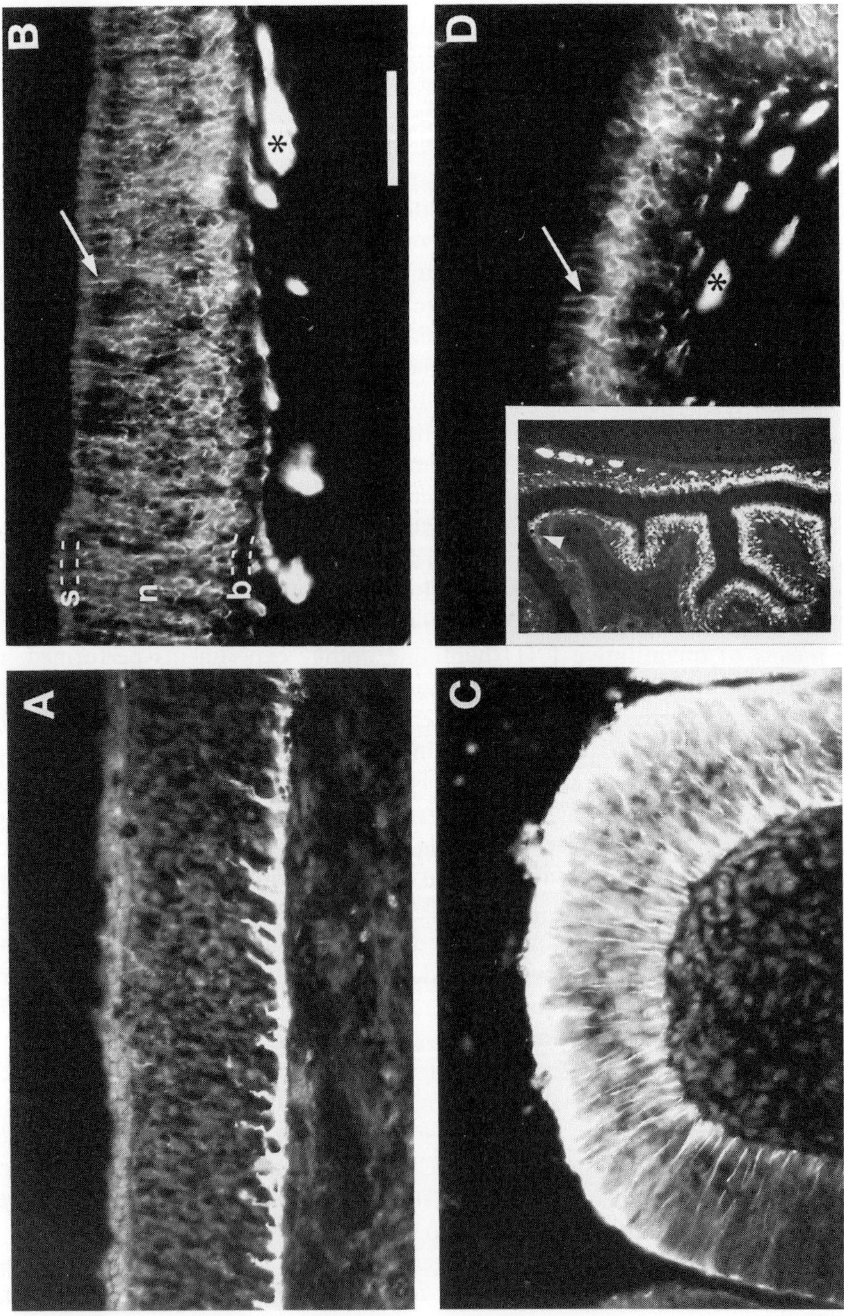
A
B
C
D
s
n
b

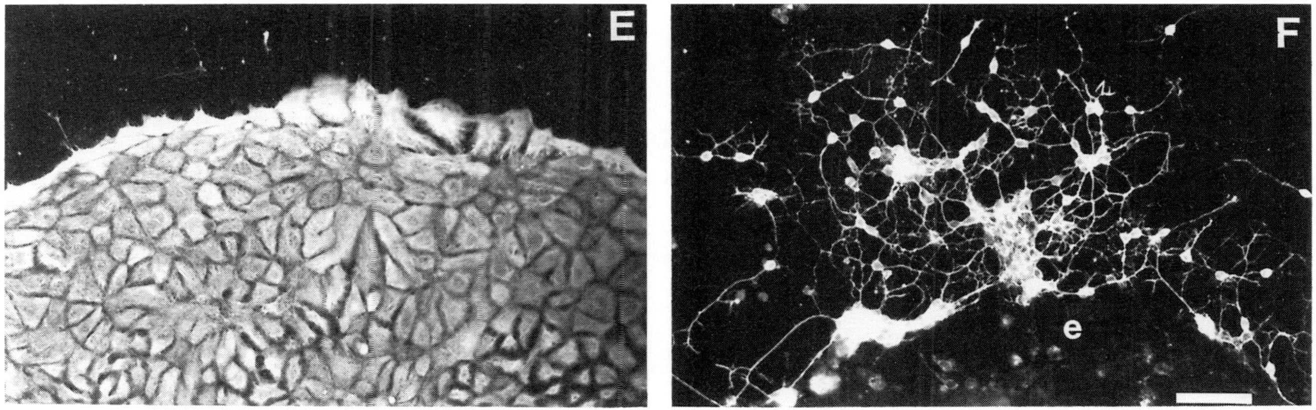

FIG. 1. Immunocytochemical distinction between mouse basal cells and olfactory receptor neurons *in vivo* and *in vitro*. A, C and E show staining with rabbit anti-keratins antiserum in postnatal mouse olfactory epithelium (OE), embryonic (E15) OE, and 36 h OE cultures, respectively. B, D and F show staining with monoclonal anti-N-CAM antibody in the same tissues. *In vivo*, anti-keratins antiserum exclusively stains the layer of basal cells immediately adjacent to the basement membrane in postnatal OE (A); in E15 OE (C), this antiserum stains about half of the cells, many of which extend from the basal to the apical surface. This staining pattern is consistent with the observation that, in OE at this age, the nuclei of proliferating cells are present throughout the basal–apical extent of the epithelium, and only later become restricted to the basal layer (Smart 1971). *In vitro*, the anti-keratins antiserum stains the sheets of cells with epithelial morphology (E). Monoclonal anti-N-CAM (B) exclusively stains neurons in the extensive receptor neuron layer (n), but does not stain cells in the basal cell layer (b) or the luminal sustentacular cell layer (s). Note that both the apical dendrites of olfactory receptor neurons (arrows, B, D) and the bundles of olfactory neuron axons (*, B, D) are stained by the antibody to N-CAM. The anti-N-CAM antibody also clearly demarcates olfactory from nearby respiratory epithelium (arrowhead, D, inset). *In vitro*, monoclonal anti-N-CAM antibody stains the neurite-bearing cells (F). Bars, 50 µm. The bar in B also applies to A, C and D; the bar in F also applies to E. (Taken, with permission, from Calof & Chikaraishi 1989 © 1989 Cell Press.)

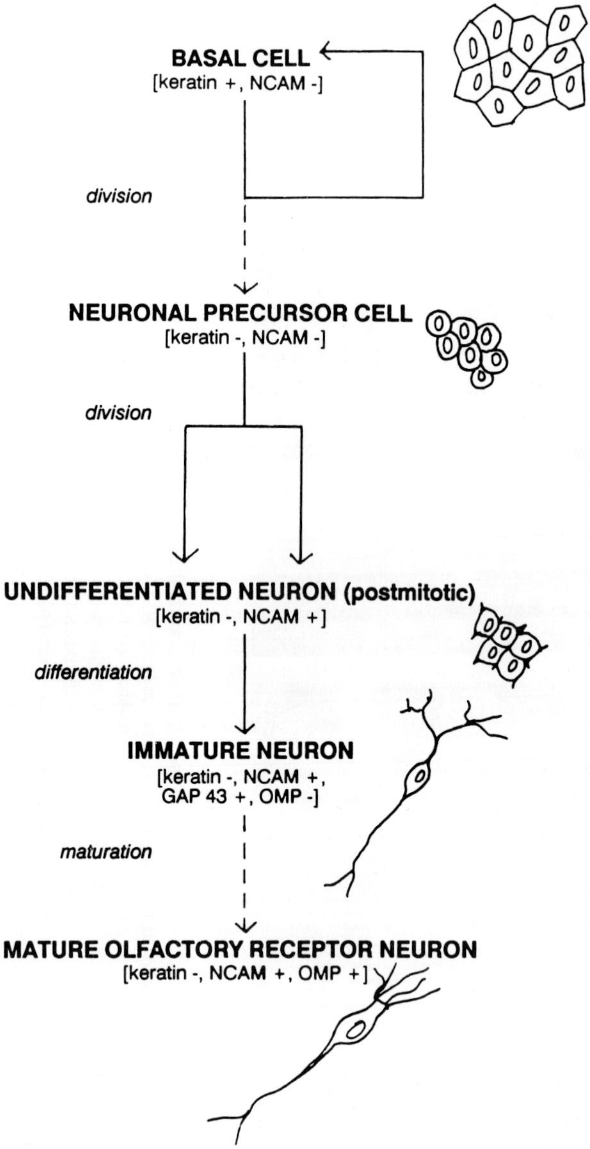

FIG. 2. Neurogenesis in the olfactory epithelium: a proposed sequence of events. The diagram shows a proposed series of steps in proliferation and differentiation leading to the production of olfactory receptor neurons via neuronal precursor cells (immediate neuronal precursors, INPs) (see text). Solid arrows indicate steps whose existence is directly supported by our observations. Dashed arrows indicate steps whose occurrence has been suggested previously by *in vivo* studies. (Taken, with permission, from Calof & Chikaraishi 1989 © 1989 Cell Press.)

solid lines represent events that we can confirm in tissue culture; the dashed lines are hypothesized steps based on suggestions made from *in vivo* studies. Our current experiments are directed towards answering three major questions that this scheme makes apparent:

1. Is the keratin-positive basal cell really a neuroepithelial stem cell? This question subsumes several further questions. Can we obtain evidence from *in vitro* experiments that the keratin-positive cell behaves like a stem cell—i.e. divides asymmetrically to regenerate itself and an INP? What is the stimulus for stem cell division and commitment to neuronal differentiation? Are there specific genes that direct neuronal determination?
2. Is it possible for an INP to undergo several rounds of division, analogous to the amplification divisions of committed progenitor cells in the haemopoietic system, before its progeny differentiate into neurons?
3. What are the conditions necessary for the long-term survival and biochemical maturation of olfactory receptor neurons?

Regulation of olfactory neurogenesis by intercellular interactions

We are using several approaches to gain some understanding of the way in which intercellular interactions might regulate neurogenesis in the olfactory epithelium. The first approach is to test the effects of polypeptide growth factors and conditioned media on the process of neurogenesis in OE cultures. If any of these agents affects OE neurogenesis *in vitro*, it is reasonable to assume that this should be detectable as (a) a change in the mitotic index of basal cells; (b) a change in the mitotic index of INPs; or (c) a change in the number of neurons generated in OE cultures.

Table 1 shows the results of experiments designed to determine whether various agents can sustain neurogenesis in OE cultures. Under normal conditions in OE cultures, the proliferation of INPs is ceasing by 24 hours; this is reflected in the low labelling index of INPs observed at this time in culture in pulse-fix [^{3}H]thymidine uptake analyses (cf. Calof & Chikaraishi 1989, Fig. 3). In the experiments shown in Table 1, mouse OE cultures were grown in the continuous presence of the indicated agent. [^{3}H]Thymidine was added after the first 24 hours of culture. After a further 24 hours in culture the labelling index of INPs + neurons was determined. In this type of experiment, an increase in the labelling index above control levels suggests that an overall increase in neurogenesis is being effected. However, it should be noted that such an increase could reflect any or all of the following biological events: an increase in basal cell-to-INP transitions; an increase in the percentage of INPs that proliferate; or a lengthening of the INP cell cycle.

Two effects are noteworthy (Table 1). The first is an apparent inhibition of neurogenesis by epidermal growth factor. EGF has a dramatic effect in promoting basal cell proliferation, but in long-term OE cultures it does not

TABLE 1 Effects of growth factors and conditioned media on neurogenesis in mouse olfactory epithelium cultures

Growth factor/conditioned medium (Concentration)	*Labelling index (%) (SD)*
No growth factor or conditioned medium	6.5 (1.9)
Basic fibroblast growth factor (5 ng/ml)[a]	11.4 (1.3)
Acidic fibroblast growth factor (10 ng/ml)[a]	6.0 (0.6)
Transforming growth factor-β (5 ng/ml)[b]	4.4 (1.0)
Epidermal growth factor (50 ng/ml)[c]	2.9 (0.4)
Nerve growth factor (2.5S; 50 ng/ml)[d]	4.4 (0.3)
BRL-CM (serum-free conditioned F12; 33%)[e]	6.5 (1.3)
STO-CM (serum-free conditioned F12; 33%)[e]	5.6 (1.2)

This test was designed to determine whether any of the added agents can promote the sustained generation of neurons in olfactory epithelium cultures. E15 mouse OE cultures were grown in serum-free, defined low-calcium medium as described in Calof & Chikaraishi (1989), except that carrier protein was increased to 5 mg/ml crystalline bovine serum albumin (ICN Biochemicals); this is the protein concentration that would be achieved in medium supplemented to approximately 8% calf serum. Cultures were grown on glass coverslips coated with laminin and fibronectin as described (Calof & Chikaraishi 1989) for 24 h in 1 ml of medium with growth factor or conditioned medium added at specified concentrations. After 24 h in culture the medium was removed and cultures were re-fed with 1 ml of fresh medium (again with growth factor or conditioned medium added) containing 0.1 μCi/ml [^{3}H]thymidine; 24 h later (total of 48 h in culture), cultures were fixed and processed for autoradiography as described in Calof & Chikaraishi (1989). Slides were exposed for eight days at −80°C, developed in D-19 developer, and stained with Hoechst 33258 (bisbenzamide, 1 μg/ml). Labelled nuclei of INPs + neurons were counted using a 40× phase objective. The labelling index is the percentage of INPs + neurons with silver grains over their nuclei. For each data point a minimum of two cultures was examined; for each culture, approximately 1000 cells in a minimum of 20 fields were scored.

[a]Basic and acidic FGF were purified from bovine brain by heparin affinity chromatography using published procedures (Lobb & Fett 1984).

[b]TGF-β, purified from human platelets, was obtained from Collaborative Research, Inc. (Bedford, MA, USA).

[c]Recombinant human EGF was obtained from Amgen Biologicals (Thousand Oaks, CA, USA).

[d]2.5S NGF, purified from male mouse submaxillary gland, was obtained from Collaborative Research, Inc.

[e]Ham's F12, supplemented with defined culture additives (cf. Calof & Chikaraishi 1989) and glucose (0.3%), was conditioned for 48 h over confluent cultures of Buffalo Rat Liver cells or STO embryonic mouse fibroblasts (generous gifts of Dr Elizabeth Robertson, Dept of Genetics, Columbia University College of Physicians and Surgeons, New York, NY, USA).

appear to cause basal cells to generate INPs and thereby neurons (unpublished observations). It is possible that EGF promotes squamous differentiation of basal cells *in vitro*, thereby inhibiting basal cell-to-INP transitions and causing a net decrease in neurogenesis, as reflected in this assay. The other effect observed is a modest increase in labelling index caused by basic fibroblast growth factor. Here again, maintaining cultures for long periods of time in bFGF does not seem to trigger basal cells to generate neurons continuously, so the effect of bFGF on neurogenesis in this assay may be exerted directly

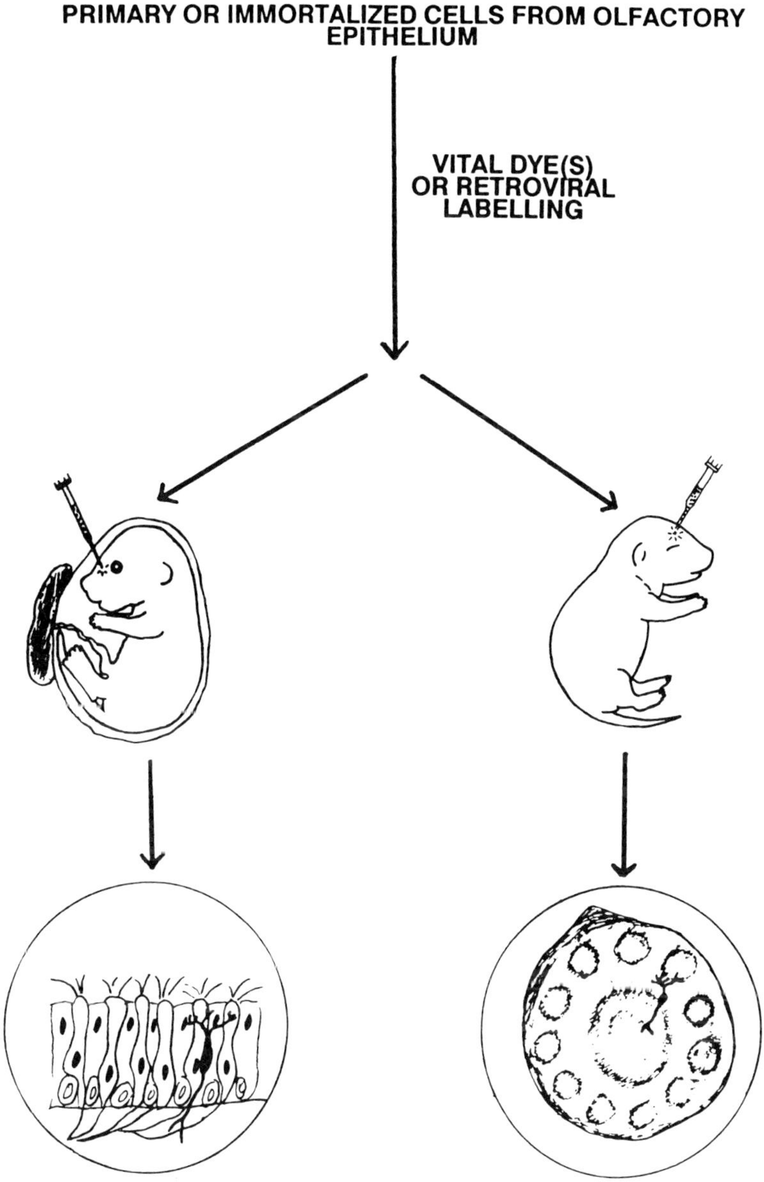

FIG. 3. Transplantation of mouse olfactory epithelium cells. Primary or immortalized cells from olfactory epithelium are first labelled, either by means of a chemical dye, or by transducing a marker gene into them using a recombinant retrovirus. Labelled cells are then transplanted either into the olfactory epithelium of E14 host embryos or into the olfactory bulb of neonatal host mice. After five days the host animal is killed and its tissues are sectioned and examined for the presence of labelled cells with the characteristics of differentiated neurons.

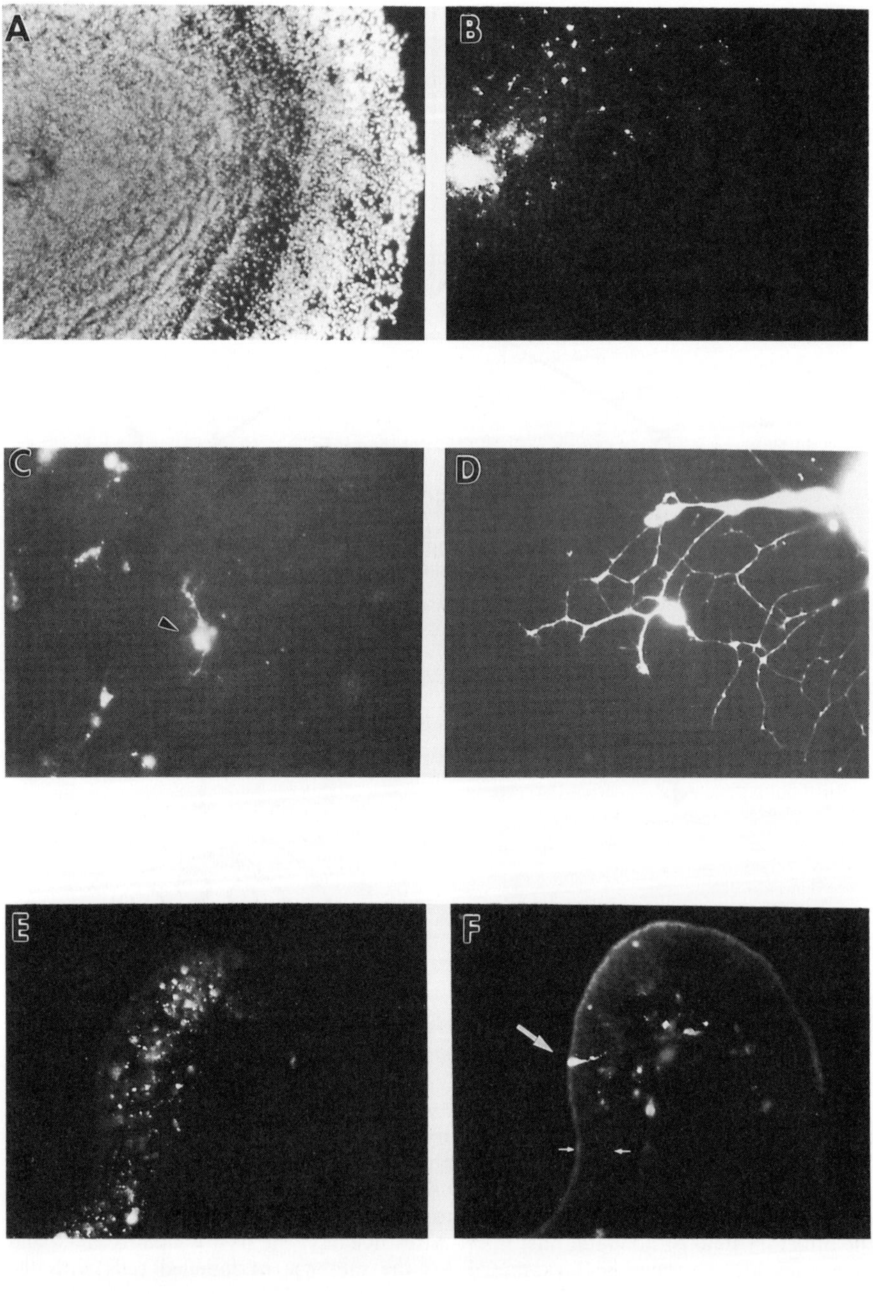
A
B
C
D
E
F

on INPs (by, for example, promoting INP proliferation), rather than on basal cells. A more detailed experimental analysis of these effects is currently in progress.

So far, we have not succeeded in obtaining culture conditions for the olfactory epithelium in which neurogenesis continues for prolonged periods. In fact, in

---

FIG. 4. Primary mouse olfactory epithelium cells can undergo neuronal differentiation after transplantation. Olfactory epithelium was purified from the turbinate regions of mouse embryos as described in Calof & Chikaraishi (1989). Tissue was labelled as described below and then transplanted into either the olfactory bulb of a neonatal host mouse (A–C) or into the olfactory epithelium of an embryonic (E14) host mouse (E,F) using *exo utero* surgical procedures (Muneoka et al 1986). Five days later the host animals were killed and host tissue was fixed in 4% paraformaldehyde in phosphate buffer and then sectioned on a vibratome (A–C) or cryostat (E,F).

A–D: Pieces of purified OE were incubated for 1 h at 37 °C in 40 μg/ml DiI (Honig & Hume 1986) in serum-free, defined medium ('normal-calcium medium'; cf. Calof & Chikaraishi 1989). The tissue was washed 10 times in a large excess of Hank's buffered salt solution (HBSS) containing antibiotics, 5 mM Hepes (pH 7.4), and fetal bovine serum (10%; first eight washes) or crystalline bovine serum albumin (BSA) (0.5 mg/ml; last two washes). Tissue was held in the dark on ice until transplanted by pressure injection through glass micropipettes. At the end of the surgical procedure the remaining tissue was placed into culture in low-calcium medium as described previously (Calof & Chikaraishi 1989).

A. Low power view of the host olfactory bulb stained with Hoechst 33258 (bisbenzamide, 1 μg/ml) to visualize cell nuclei. B. Low power view of DiI-labelled graft, same section. C. High power view of DiI-labelled cells in the transplant, same section. Note cell with bipolar neuronal morphology (arrowhead). D. DiI-labelled olfactory receptor neuron in cultures of the same tissue used for the graft shown in A–C. Note the morphological similarity between the transplanted cell in C (bipolar, with one thicker, more dendritic branching process, and one thinner branching process), and the cultured neuron in D.

E,F: Purified OE was spun to pellet (5 min at 100 × **g**), resuspended in 70 μl of 10 mg/ml lysinated rhodamine dextran (LRD, 10 000 M_r; Molecular Probes) in phosphate-buffered saline (cf. Bronner-Fraser & Fraser 1988), and transferred to the cuvette of a Baekon-2000 Macromolecular Transfer System (Baekon, Inc, Saratoga, CA, USA). After 'Baekonization' (8 kV; 2^{10} pulses; 0.8 s burst; 9 cycles; OPT 160), 20 μl of 1 mg/ml DNase I were added to the cuvette and the tissue was transferred to a sterile centrifuge tube and resuspended in a total volume of 6 ml of Hank's buffered saline solution. The suspension was underlaid with 3 ml of 4% crystalline BSA in HBSS and centrifuged (5 min at 100 × **g**) and the supernatant was removed. Labelled tissue was mixed with 3 μl of fluorescent-green latex microspheres (Fluoresbrite; Polysciences) and pressure-injected through a glass micropipette into the olfactory epithelium of mouse embryos *in situ*.

E. Fluorescent latex microspheres incorporated throughout the basal-to-apical extent of the OE, plus the underlying stroma, in the region of the host olfactory epithelium that received the graft. F. LRD-labelled cells in the same section. There are a number of LRD-labelled cells in the OE and underlying stroma, some out of the plane of focus. Note one labelled cell (large white arrow) integrated into the apical region of the OE, which appears to be extending a process toward the basement membrane of the epithelium. The basal-apical extent of the OE in this section is indicated by the two small white arrows.

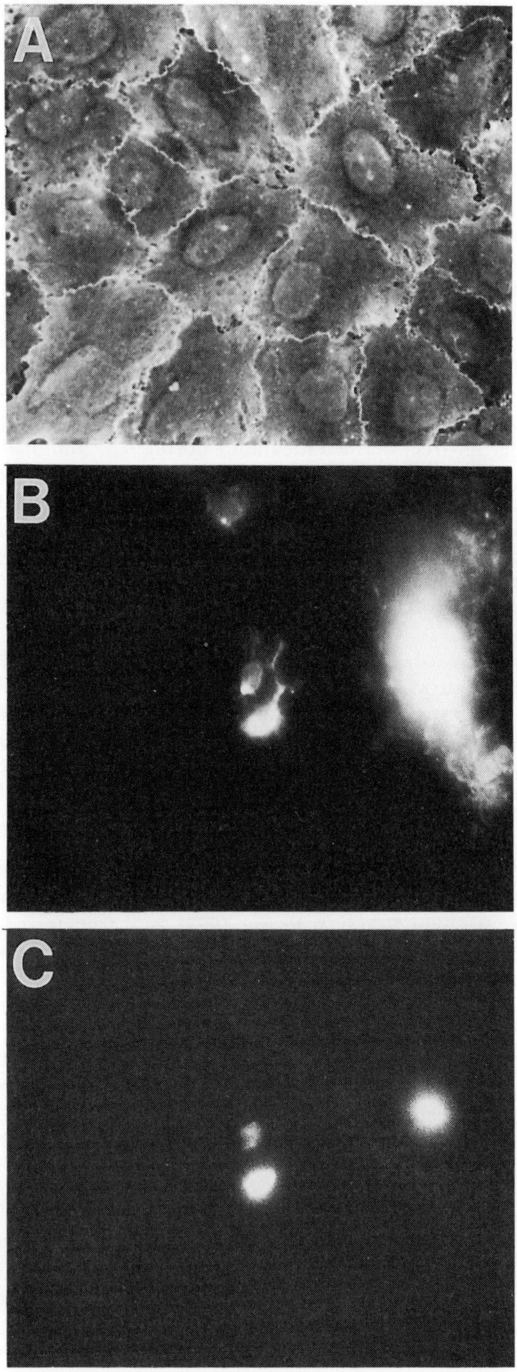

our standard culture conditions using serum-free defined medium, all cells but basal cells die after about a week in OE cultures. However, these basal cells, while continuing to proliferate, do not generate INPs and neurons (cf. Calof et al 1989). There are at least two possible explanations for these results. First, the basal cell-to-INP transition may be a critical step in differentiation, requiring very specific conditions that we cannot yet reproduce *in vitro*; or, second, the keratin-positive basal cells may not be the progenitors of INPs and therefore ultimately of olfactory receptor neurons. It is thus of some importance to determine whether basal cells can, in fact, generate INPs and thereby neurons, and we are continuing to search for different conditions that might permit prolonged OE neurogenesis *in vitro*. However, we are also pursuing an alternative approach for determining whether the keratin-positive basal cells in OE cultures can generate neurons, and what conditions might permit the basal cell-to-INP transition. In this approach we transplant these cells back into an appropriate tissue to see if they show neurogenic potential *in vivo*. The basic strategy for these experiments is depicted in Fig. 3.

Our initial transplantation studies have focused on determining whether purified OE cells, prior to any time in culture, can survive and differentiate when transplanted into either the olfactory bulb or the olfactory epithelium of a host animal. Figure 4 shows examples of transplants into both the olfactory bulb of a neonatal host mouse (A–C) and the olfactory epithelium of an embryonic (E14) host mouse (E,F). OE cells can be labelled either by incubation with the carbocyanine dye, DiI (dioctadecyl-3,3,3′,3′-tetramethylindocarbocyanine perchlorate) (A–D), or by 'Baekonization' (see legend) in the presence of lysinated rhodamine dextran (E,F). After these labelling procedures, OE cells will survive and undergo at

---

FIG. 5. Immortalized basal cells extend processes when transplanted into the olfactory bulb. The experiments used an immortalized OE cell line, OEinfE1A8VB9T8, derived from an OE basal cell culture.

A. Fluorescence micrograph of OEinfE1A8VB9T8 cells *in vitro*, stained with monoclonal antibody OKT8. Note the epithelial morphology of these cells, similar to the morphology of OE basal cells in primary culture (Fig. 1).

B,C. OEinfE1A8VB9T8 cells extend branching processes when transplanted into the olfactory bulb of a neonatal host mouse. Subconfluent cultures of cells were double-labelled with DiI and Fast Blue before transplantation. Cells were incubated for 1 h at 37 °C in 25 μg/ml DiI in complete medium (DME/F12 with 10% fetal bovine serum), then rinsed three times in complete medium and three times in calcium- and magnesium-free HBSS. Cells were then incubated for 15 min at room temperature in 17 μg/ml Fast Blue in CMF-HBSS (cf. McConnell 1988), rinsed four times, dissociated in trypsin-EDTA, spun through a gradient of 4% crystalline BSA in CMF-HBSS, and pressure-injected into the host olfactory bulb through a glass micropipette. Five days later the host tissue was fixed in 4% paraformaldehyde, sectioned on a vibratome and examined for double-labelled cells with neuronal morphologies. B shows two DiI-labelled cells, one of which is clearly extending a branching process. C shows that both of these cells are also labelled with Fast Blue, confirming that both cells are derived from grafted tissue.

least some degree of morphological differentiation into neurons, either *in vitro* (D) or *in vivo* (C,F).

These results now make it possible to contemplate experiments that should answer a critical question about neurogenesis in this system: can purified basal cells, when transplanted back into either of these two locations, generate olfactory receptor neurons? Although we do not yet have the answer to this question, the results of the experiments discussed below suggest that the answer might be yes.

Immortalized cell lines from olfactory epithelium

We have derived immortalized cell lines from primary mouse olfactory epithelium cultures as potential sources of material for biochemical and molecular biological experiments on neurogenesis. Such cell lines are generated by infecting cultures with recombinant murine retroviruses. The viruses contain the coding sequences for various oncogenes, linked to a dominant selectable marker (the *neo* gene, which encodes a phosphotransferase that confers resistance to the neomycin analogue G418). Olfactory epithelium cultures are infected with virus and then grown in the presence of G418. G418-resistant cells are then passaged, cloned and characterized.

Figure 5 shows the results of experiments with a clonal OE cell line, which was derived by infecting an OE basal cell culture with a retrovirus encoding an adenovirus E1A12S oncogene (Cone et al 1988, Calof et al 1989). *In vitro*, these cells have an epithelial morphology, like that of basal cells in primary OE cultures (Fig. 5A). After transplantation into the olfactory bulb of a neonatal mouse, these cells (or their progeny), extend branching processes (Fig. 5B). These results suggest that these immortalized cells may be undergoing some degree of neuronal differentiation after transplantation. At present, however, we have evidence only for this morphological change; in order to determine whether these cells have undergone neuronal differentiation at the biochemical level it will be necessary to demonstrate that they express neuronal markers. To facilitate such studies, we are currently testing various marker genes, which we can use to detect transplanted cells, and, using double-label immunocytochemical analyses, assess the expression of neuronal markers by these cells as well. For example, OEinfE1A8VB9T8 cells have been made to express a foreign marker gene, encoding the human lymphocyte glycoprotein, CD8 (Fig. 5A) (Littman et al 1985). The CD8 cDNA was transduced into these cells with the retroviral vector MV7-T8 (Maddon et al 1986). Expression of human CD8 (=T8) can be detected immunocytochemically using the monoclonal antibody OKT8 (American Type Culture Collection Catalogue) in tissue culture or in tissue sections (Calof & Jessell 1986, Trevino et al 1990). MV7-T8 can infect mouse neuronal precursor cells and human CD8 is expressed by the neuronal progeny of these infected cells, whereas human CD8 is not normally expressed in any mouse tissues (Calof & Jessell 1986, Trevino et al 1990). This marker gene should be useful in transplantation experiments in which double-label immunocytochemical techniques are required to assess the biochemical state of differentiation of transplanted cells.

If these immortalized OE cells do prove to generate neurons *in vivo*, this result would be of significance for two reasons. First, it would suggest that the basal cells that are observed in primary OE cultures are indeed neuroepithelial stem cells; and, second, it would indicate that OE cells can still retain their neurogenic capacity after oncogene-induced immortalization. Cell lines such as these could be useful in several ways. They could be manipulated pharmacologically, to allow the identification of molecules such as growth factors, growth factor receptors, and intracellular signalling molecules that may be important in triggering or sustaining neurogenesis. In addition, such cell lines may be useful for the isolation, molecular cloning, and functional testing of genes that act as switches, controlling the cell-type transitions that occur between neuroepithelial stem cells and neurons (e.g. Davis et al 1987).

Summary

The olfactory epithelium provides a unique system for studying cellular and molecular events in neurogenesis. *In vitro* studies of this tissue have shown that neurogenesis is a multistage process, involving a committed immediate neuronal precursor which divides to generate two daughter neurons. *In vivo* studies suggest that this precursor may be the progeny of a neuroepithelial stem cell that retains its neurogenic potential throughout life *in situ*. Finding the conditions that allow this transition from stem cell to committed precursor to occur *in vitro* will provide information about the kinds of factors that regulate neurogenesis; such factors may prove to have similar roles in all embryonic neuroepithelia, and will be candidates for agents that regulate regeneration of sensory epithelia in adult animals as well. Immortalized neural stem cells, derived from olfactory epithelium, may be useful sources from which to isolate genes that are important in directing the differentiation of neurons from their precursors.

Acknowledgements

The authors thank Ken Muneoka for helpful advice on *exo utero* surgical techniques, Roger Cone and David Williams for the gift of the Psi-2 cell line packaging the DOL-E1A12S recombinant retrovirus, and Dr Tai-Kin Wong for the loan of the Baekon-2000 Macromolecular Transfer System. This work was supported by a grant from the Alzheimer's Disease and Related Disorders Association (PRG-87-167) and a National Research Service Award from the National Institutes of Health to A.L.C., and a David and Lucile Packard Fellowship award to A.D.L.

References

Bronner-Fraser M, Fraser SE 1988 Cell lineage analysis reveals multipotency of some avian neural crest cells. Nature (Lond) 335:161–164

Calof AL, Chikaraishi DM 1988a *In vitro* analysis of neurogenesis in a mammalian neuroepithelium. Soc Neurosci Abstr 14:162

Calof AL, Chikaraishi DM 1988b Identification of a proliferating neuronal precursor cell in cultures of a mammalian neuroepithelium. J Cell Biol 107:510a (abstr)

Calof AL, Chikaraishi DM 1989 Analysis of neurogenesis in a mammalian neuroepithelium: proliferation and differentiation of an olfactory neuron precursor *in vitro*. Neuron 3:115–127

Calof AL, Jessell TM 1986 Use of a retroviral vector to transfer and express a foreign gene in murine neurons and neural ectoderm. Soc Neurosci Abstr 12:183

Calof AL, Lander AD, Chikaraishi DM 1989 Regulation of neurogenesis and neuronal differentiation in primary and immortalized cells from mouse olfactory epithelium. J Cell Biol 109:57a (abstr)

Camara CG, Harding JW 1984 Thymidine incorporation in the olfactory epithelium of mice: early exponential response induced by olfactory neurectomy. Brain Res 308:63–68

Cone RD, Grodzicker T, Jaramillo M 1988 A retrovirus expressing the 12S adenoviral E1A gene product can immortalize epithelial cells from a broad range of rat tissues. Mol Cell Biol 8:1036–1044

Costanzo RM, Graziadei PPC 1983 A quantitative analysis of changes in the olfactory epithelium following bulbectomy in hamster. J Comp Neurol 215:370–381

Davis RL, Weintraub HL, Lassar AB 1987 Expression of a single transfected cDNA converts fibroblasts to myoblasts. Cell 51:987–1000

DiFiglia M, Marshall P, Covault J, Yamamoto M 1989 Ultrastructural localization of molecular subtypes of immunoreactive neural cell adhesion molecule (NCAM) in the adult rodent striatum. J Neurosci 9:4158–4168

Graziadei PPC 1973 Cell dynamics in the olfactory mucosa. Tissue & Cell 5:113–131

Graziadei PPC, Monti Graziadei GA 1978 Continuous nerve cell renewal in the olfactory system. In: Jacobson M (ed) Handbook of sensory physiology, vol IX: Development of sensory systems. Springer-Verlag, New York, p 55–83

Graziadei PPC, Monti Graziadei GA 1979 Neurogenesis and neuron regeneration in the olfactory system of mammals. I. Morphological aspects of differentiation and structural organization of the olfactory sensory neurons. J Neurocytol 8:1–18

Harding J, Graziadei PPC, Monti Graziadei GA, Margolis FL 1977 Denervation in the primary olfactory pathway of mice. IV. Biochemical and morphological evidence for neuronal replacement following nerve section. Brain Res 288:289–295

Honig MG, Hume RI 1986 Fluorescent carbocyanine dyes allow living neurons of identified origin to be studied in long-term cultures. J Cell Biol 103:171–187

Littman DR, Thomas Y, Maddon PJ, Chess L, Axel R 1985 The isolation and sequence of the gene encoding T8: a molecule defining functional classes of T lymphocytes. Cell 40:237–246

Lobb RR, Fett JW 1984 Purification of two distinct growth factors from bovine neural tissue by heparin affinity chromatography. Biochemistry 23:6265–6299

Maddon PJ, Dalgleish AG, McDougal JS, Clapham PR, Weiss RA, Axel R 1986 The T4 gene encodes the AIDS virus receptor and is expressed in the immune system and the brain. Cell 47:333–348

McConnell SK 1988 Fates of visual cortical neurons in the ferret after isochronic and heterochronic transplantation. J Neurosci 8:945–974

Monti Graziadei GA, Graziadei PPC 1979 Neurogenesis and neuron regeneration in the olfactory system of mammals. II. Degeneration and reconstitution of the olfactory sensory neurons after axotomy. J Neurocytol 8:197–213

Muneoka K, Wanek N, Bryant SV 1986 Mouse embryos develop normally *exo utero*. J Exp Zool 239:289–293

Smart IHM 1971 Location and orientation of mitotic figures in the developing mouse olfactory epithelium. J Anat 109:243–251

Trevino C, Calof AL, Muneoka K 1990 Position specific growth of 3T3 cells in the developing mouse limb. J Cell Biochem Suppl 14E:100

DISCUSSION

Presson: I thought there was a possibility that these small round cells, and the cells that they turn into, might be migrating LHRH (luteinizing hormone releasing hormone) cells?

Calof: Two separate groups (at the NIH and also at Rockefeller University) have shown that LHRH neurons of the hypothalamus are derived from the olfactory placode; early in development they migrate out of the placode, via the terminal nerve (which is another derivative of that placode) and migrate into the brain, to form the LHRH neurons (Wray et al 1989, Schwanzel-Fukuda & Pfaff 1989). That's one reason why I am interested in the migratory properties of these cells. The question as to whether or not the small, round cells—which we call the immediate neuronal precursors—or their neuronal progeny express LHRH *in vitro*, or whether all olfactory cells that migrate express LHRH, even *in vivo*, is still open. I have the marker that Dr Schwanzel-Fukuda used and I hope to do that experiment. I suspect that the immediate neuronal precursors (INPs) *in vitro* will not all be LHRH positive; I don't think it's a requisite for migration.

Presson: But isn't there a distinct possibility that these small round cells are not olfactory neuronal precursors at all, but are the LHRH precursors?

Calof: I don't think so; all the evidence we have suggests that they give rise to olfactory receptor neurons. I am more interested in the possibility of something like a transdifferentiation effect, and whether, for example, getting out of the olfactory epithelium and on to a particular pathway induces LHRH expression. There's evidence from a Japanese group that the hypothalamic environment is important for full-blown LHRH expression (Daikoku-Ishido et al 1990).

Farbman: The differentiation of the LHRH-positive cells comes from the medial part of the placode, and you are looking at embryonic cells of the turbinate regions, which derive from cells in the lateral part of the placode, so they are unlikely to be LHRH positive.

Calof: It would be interesting if LHRH was a determined characteristic of cells of that medial part of the placode, or whether any of these neuronal precursors that had a migratory pathway open to them might be capable of expressing LHRH, and might be 'plastic', essentially.

Farbman: It could be!

Reh: Your antiserum to keratins stains basal cells and your monoclonal to N-CAM seems to stain everything else *in vivo* except the support cells; but the round cells in culture are not stained by either. So what is their equivalent *in vivo*?

Calof: These round cells have been referred to by several names, including the 'globose basal cell' by Graziadei & Monti Graziadei (1979), and the 'light basal cell'—the term that Mackay-Sim uses. In the latter study (see Breipohl et al 1986), as has also been confirmed by Schwartz et al (1989), the cells in the regenerating olfactory epithelium that take up [^{3}H]thymidine after a pulse injection are not the keratin-positive 'dark' basal cells; they are the cells immediately above them that are referred to as the 'light' or 'globose' basal cells, or what I call the immediate neuronal precursors (INPs). So the cell type is there, but it is hard to see in the adult olfactory epithelium; whereas, at about E15, this epithelium is growing fast and therefore the numbers of round cells (INPs) are greater. I suspect that in the regenerating epithelium, what I call the 'round cell' or INP is the globose or light basal cell.

Reh: Does it undergo further cytokinesis, after it becomes one of those cells?

Calof: It divides once, to generate two neurons, as far as we can tell.

Reh: And does it make a single division *in vivo*, as *in vitro*?

Calof: Nobody knows. To show this you would need to be able to mark a single globose cell *in vivo*, and then look much later at the olfactory receptor neurons in the same animal, to see if there were more than two olfactory receptor neurons with the label.

Burd: Majorie Schwartz and John Kauer (1990) have evidence in support of a single division. In salamanders, they injected single globose basal cells (which you call immediate neuronal precursor cells) with rhodamine dextrans. If they wait long enough, that single cell becomes two neurons.

Reh: Dr Calof is looking for a multipotent neuroepithelial cell; I wonder if this might be a separate subpopulation that is undergoing only one round of cytokinesis?

Calof: I don't know! We have this cell that I call 'immediate neuronal precursor', which is probably analogous to a committed progenitor cell in the haemopoietic system. So is there really a *stem cell* in the olfactory epithelium that gives rise to the INP, and which cell is it? A subpopulation of the globose basal cells could be that stem cell population; or perhaps we are just not looking at the right conditions to allow the keratin-positive basal cells to act as stem cells *in vitro*. It's also possible that the keratin-positive basal cells are very slowly generating INPs all the time, and that the cellular interactions that up-regulate the division and renewal of the epithelium act only on the globose basal cell (i.e. the INP *in vivo*). Over the long term you would have a basal cell just blebbing out globose cells, which would be susceptible to whatever the stimulus is for regeneration. I think it's an open question. There's no evidence *in vitro* that the keratin-positive basal cell acts as a stem cell.

Corwin: In your explants, have you seen the globose cells producing neurons, *in vitro*? The staining should have shown it, if proliferation is going on in the tissue.

Farbman: You would have to section the explanted tissue to see this.

Calof: All our evidence indicates that the globose cell-equivalent (the INP) gives rise to neurons *in vitro*. To see whether the keratin-positive basal cells can give rise to INPs, I have looked at later cultures, which were essentially pure basal cells and where I hoped to see basal cell-to-INP transitions; but in no experiments of that type have I seen anything convincingly going through the entire proposed pathway. So I am trying to be conservative about it. It may be true; the keratin-positive basal cells may be the neural stem cells of the olfactory epithelium, but I have no convincing *in vitro* evidence of that.

Raymond: Where are the support cells, in your cultures?

Calof: Are there support cells in the olfactory epithelium of the mouse embryo at this stage (E15), and how would one find them? I see no evidence for support cells in these explants; the basal cell marker (anti-keratins) and the neuronal marker (anti-N-CAM) that we use don't recognize support cells in the adult, or, as far as I can tell, in the embryonic mouse olfactory epithelium (cf. Calof & Chikaraishi 1989). *In vivo*, at E15, cytokeratin-positive cells span the thickness of the epithelium (Fig. 1C of my paper). If we dissociate the epithelium into single cells and count the cells that express the basal cell marker and the neuronal marker, we find that about 60% of the cells are keratin positive and, of the remainder, three-quarters of them are N-CAM positive; one-quarter of the cells are negative for both markers. Those numbers fit with the behaviour of the round cells, the neurons, and the keratin-positive epithelial cells *in vitro* (cf. Calof & Chikaraishi 1989). I don't see sustentacular (support) cells as such. There is a marker for such cells, SUS-1, but it doesn't appear until around E20 in the rat. Is that the first time an organized support cell layer forms, Dr Farbman, or is this the time the antigen is first seen?

Farbman: The sustentacular cell layer is well established in the rat fetus on the 15th day of gestation, although they are not mature cells at this stage (Menco & Farbman 1985). At the 20th or 21st day, a few sustentacular cells are immunoreactive with SUS-1, a sustentacular cell-specific monoclonal antibody (Hempstead & Morgan 1983). The number of immunoreactive cells increases postnatally. We have been unable to correlate SUS-1 immunoreactivity with morphological changes seen with the scanning or transmission electron microscope.

Calof: So what is the evidence for an established sustentacular layer? I don't see anything in the embryo that looks like this cell layer, and in the adult our markers don't recognize sustentacular cells. It is possible that the sustentacular cells could form, *in vitro*, part of the keratin-positive population. But in the adult that's certainly not true. So either they are dying in our cultures from E15 mouse embryos, or they are not there to begin with.

Farbman: They may not be dying in culture; they may remain in the body of your explant, and not migrate away from it as neuronal precursor cells do. You would probably be able to visualize the supporting cells in the main body of the explant by sectioning it and examining the sections with a microscope.

Raymond: Are you suggesting that there are two separate lineages, one for sustentacular cells and one for neurons?

Farbman: Yes.

Calof: Most of the data seem to indicate that, both in the regenerating olfactory epithelium and in development.

Farbman: When [^{3}H]thymidine is injected into an animal, one rarely sees uptake in olfactory sustentacular cells. In the relatively rare instances when labelled sustentacular cells are seen they are usually in pairs, suggesting that the cells divide *in situ*.

There is an interesting side issue alluded to by Dr Costanzo (p 237), namely that reconstitution of olfactory epithelium occurs differently depending on the type of injury. If axotomy is done, reconstitution occurs by an increase in the number of cell divisions in the basal cell population and then differentiation of these cells into neurons. However, if the entire epithelium is destroyed with zinc sulphate, something very different occurs, because the basal cell progenitor cells are destroyed by this procedure. Reconstitution of the epithelium presumably occurs by division of epithelial cells from Bowman's glands or their ducts, but the epithelium reconstitutes as a respiratory epithelium, not an olfactory epithelium. Later, the olfactory epithelium gradually reappears (Smith 1951).

Corwin: Does the re-epithelialization nucleate around the Bowman's glands?

Farbman: I don't think this has been followed in strict detail. I was thinking of this when Dr Costanzo mentioned the boundary between olfactory and respiratory patches; the suggestion was that this was a re-epithelializing patch over an injury site. But there is complete loss of the basal layer after zinc sulphate treatment, and the so-called stem cells come from somewhere else—presumably from the Bowman's glands or their duct cells.

Burd: Within a couple of days the whole epithelium is covered by these cuboidal cells.

Calof: I thought that after zinc sulphate lavage in adult mouse, Dr Matulionis saw some sustentacular cell migration out of Bowman's glands. In support of that view, I believe the SUS-1 antigen is also present in Bowman's gland cells?

Farbman: Yes. The SUS-1 mAb stains the cytoplasm of almost all the supporting cells of the olfactory epithelium and Bowman's gland cells.

Watt: Dr Calof, were your E1A-infected basal cells keratin positive? I gather that when you transplant them, they produce neuronal processes?

Calof: They were keratin positive when first immortalized; after many passages and recloning, the cells of this clone are not keratin positive. I don't know whether that's the notorious fickleness of intermediate filament expression in transformed cells, or whether the cell is actually not a basal cell; it has epithelial morphology, but perhaps is more like an immediate neuronal precursor (INP) that has been immortalized.

Watt: But basal cells that are not immortalized divide in culture?

Calof: Yes; they produce more keratin-positive basal cells in culture.

Watt: Have you transplanted them and seen neuronal differentiation?

Calof: Yes; we did this experiment once, and we thought we saw Olfactory Marker Protein expression by the transplanted cells. But this needs to be repeated, and it needs to be followed longitudinally as well.

Watt: So it might be better to stick with the uninfected basal cells?

Calof: If one wants to determine the developmental potential of the keratin-positive basal cell, yes. The reason for making cell lines is to be able to use them as sources of material for biochemical purification. One thing that is interesting about the E1A-immortalized basal cell line is that if it is really undergoing neuronal differentiation after transplantation, this suggests that the basal cell we observe *in vitro* may really be a neuroepithelial stem cell that can be immortalized and still retain neurogenic potential.

Watt: Does the cell line differentiate on a dish, as well?

Calof: No. It acts just like a primary basal cell in that respect, dividing to give rise to more of the same cells, but never generating immediate neuronal precursors (INPs) and then neurons.

Burd: My laboratory is studying the role of thyroid hormone in the development of the olfactory system. We have examined the influence of early administration of thyroid hormone and the effects of blocking thyroid hormone synthesis on the number of cells dividing in the olfactory epithelium and the number of mature olfactory receptor cells at different stages of development in the African clawed frog, *Xenopus laevis*. These results on the developing olfactory epithelium may be relevant to adult plasticity, since the olfactory epithelium can be replaced in adulthood and since thyroid hormone levels influence chemosensory functions in adult vertebrates. For example, McConnell and colleagues (1975) showed that hypothyroid patients have olfactory deficits that are reversed with the administration of thyroid hormones. In addition, Mackay-Sim & Beard (1987) presented evidence suggesting that thyroid hormones may be necessary for normal maturation of the olfactory receptor cells in adult mice. Therefore, while the following experiments are focused on neural development, they are likely to be applicable to adult plasticity as well.

Development in *Xenopus* tadpoles can be divided into three major periods: embryonic development (stages 1 to 35), larval development (stages 35 to 58), and metamorphic climax (stages 58 to 66). In addition, stages 49 to 54 are part of the premetamorphic period and stages 54 to 58 are part of the prometamorphic period. These stages of development were defined by Nieuwkoop & Faber (1956) and are based on external and internal body characteristics.

There are two forms of thyroid hormone, thyroxine (T_4) and 3,5,3′-triiodothyronine (T_3). Thyroxine predominates as the circulating hormone, but is converted to T_3 in peripheral tissues; T_3 is the active form of the hormone. Amphibian metamorphosis is dependent upon thyroid hormone; without it, larvae grow but do not undergo metamorphosis. Plasma levels of thyroid hormones are very low in *Xenopus* prior to stage 54, begin to rise after stage 54, peak at stage 62, and decline slowly in later stages of metamorphosis (Leloup & Buscaglia 1977).

In all vertebrates, many aspects of neural development depend on thyroid hormone. A study of normal development of the olfactory nerve showed that the number of olfactory axons in the olfactory nerve of *Xenopus* tadpoles increased more rapidly after stage 54, the stage when thyroid hormone secretion begins (Burd 1991). We wondered to what extent development of the olfactory epithelium in *Xenopus* required or was influenced by thyroid hormone. To examine this, we implanted small pellets of 10% thyroxine in cholesterol and oil unilaterally, adjacent to the nasal capsule (hyperthyroid treatment), or grew larvae in 0.01% propylthiouracil (PTU), a drug that blocks synthesis of thyroid hormone (hypothyroid treatment). In both experiments, the treatment started at stage 48, well before the normal rise in thyroid hormone at stage 54.

Seven days after thyroxine pellets were implanted, larvae were injected with [^{3}H]thymidine, to label dividing cells; three days later they were killed for autoradiography and electron microscopy. We observed that within ten days, the precocious thyroxine treatment stimulated a dramatic increase (two-fold) in the number of olfactory receptor cell axons, and hence an increase in the number of mature olfactory receptor cells (Burd 1988, 1990). In addition, from the autoradiography results we determined that there were at least 12 times the number of newly generated cells in the olfactory epithelium of hormone-treated larvae (Burd 1990). These results indicate that thyroxine can stimulate a striking increase in cell division and neuronal maturation in the olfactory epithelium of animals, before the normal rise in hormone levels.

We also examined the effect of thyroid hormone on normal development of the olfactory epithelium by studying hypothyroid larvae. Animals were placed in PTU at stage 48 and were killed along with age-matched siblings when the siblings reached stage 52 (before thyroid hormone levels rise) or when the siblings reached stage 58 (when thyroid hormone levels are high and rising). Some PTU-treated animals were also killed long after siblings had gone through metamorphosis. Quantitative analysis of the number of olfactory receptor cell axons revealed that the PTU treatment had no effect on axon number by comparison with stage 50 larvae, when both control and PTU-treated animals have very low levels of thyroid hormone. The number of olfactory axons in

PTU-treated animals, however, was only half the number in stage 58 siblings, suggesting that the hypothyroid condition either prevented receptor cell maturation or blocked neurogenesis by acting on cell division of stem or precursor cells (Burd 1988). Paternostro & Meisami (1989) also find that hypothyroidism affects the development of the rat olfactory epithelium.

To determine whether hypothyroid conditions influenced cell genesis in the olfactory epithelium, we raised animals in PTU beginning at stage 48 and injected them and age-matched siblings with [^{3}H]thymidine when the siblings reached stage 50 or stage 58. After one day of survival, the animals were killed and processed for autoradiography and light microscopy. While there was no difference in the thickness of the epithelium or in the number of labelled cells in the epithelium of PTU-treated animals compared with their age-matched, stage 50, control siblings, these parameters were statistically different between PTU-treated animals and the age-matched, control animals at stage 58. The olfactory epithelium was thinner and there were five times fewer [^{3}H]thymidine-labelled cells in the hypothyroid larvae and significantly fewer labelled cells than in the younger PTU-treated animals (Burd & Thomas 1990). It is possible that at some point in larval development the olfactory epithelium requires thyroid hormones for cell genesis. These findings demonstrate that hypothyroidism during larval development affects cell genesis in the olfactory epithelium. This may account for the reduced number of receptor cells. Thyroid hormones may also be necessary for neuronal maturation, but we have not yet directly addressed this question.

Mackay-Sim & Beard (1987) showed that euthyroid hormone levels are necessary for normal maturation of the olfactory receptor cells in adult mice. We found that thyroid hormone levels may play a role in cell genesis in the developing olfactory epithelium and can accelerate neuronal maturation. I would like to suggest that thyroid hormone levels may also be important for receptor cell turnover in other sensory systems, and recommend that you keep this in mind as you design your future experiments.

Rubel: Your suggestion is very good; your study, and Anne Calof's, remind me however of something that those of us who were studying nervous system development many years ago went through, when most of the models came from regeneration. We discovered that regenerating systems didn't necessarily follow the same rules as developing systems. Here we are interested in the reverse; we are interested in regeneration. In both these cases we need to find out if the same kinds of rules apply during regeneration as during initial development. I don't think we have evidence in either of these experimental paradigms.

Margolis: In my lab we have tried to address this question in the mammalian olfactory system by studying the relative temporal sequence of expression of B50/GAP43 and the Olfactory Marker Protein (OMP), both in normal development and during regeneration in response to olfactory lesions. We showed (Verhaagen et al 1989, Biffo et al 1990) that during normal ontogeny,

B50/GAP43 is expressed in immature but not mature olfactory neurons, while the profile of OMP expression is quite the reciprocal of this and is expressed only in mature olfactory neurons. In the regeneration studies (Verhaagen et al 1990) this sequence of events is fully replicated for B50/GAP43 and OMP proteins and mRNAs, so long as the target olfactory bulb is present. However, if the olfactory bulb is extirpated, then the sequence is arrested and only immature B50/GAP43-positive neurons are seen. This model illustrates that with appropriate molecular probes regeneration can be shown to recapitulate, at least partially, the sequence of events seen during ontogeny. However, that is not to say that one is a complete, exchangeable and unambiguous model of the other.

Costanzo: I would like to ask Gail a hypothetical question. We have learned from experience in the clinics we have set up for patients with taste and smell disorders, resulting from, for example, polyps or nasal sinus disease, that there is in fact very little we can do to improve these senses. Corticosteroids help in some patients, but this is a short-term solution. In the case of head injury, we just have to wait and hope for improvement during recovery. Do you think that giving thyroid hormone might work?

Burd: In fact, as shown by McConnell and colleagues (1975), hypothyroid patients have olfactory deficits that can be reversed within about 14 days by administering thyroid hormone. So if the reason your patients can't smell is because they are hypothyroid, it is probably a good idea to try giving them thyroid hormone. However, if they have undergone some kind of trauma and the loss is due to an inability of the axons to contact the olfactory bulb, then thyroid hormone would probably not be beneficial.

Costanzo: These are not all trauma cases. There are many nasal diseases that involve presumably cellular mechanisms.

Lewis: Cancer patients who receive chemotherapy with cytotoxic drugs have unpleasant disorders of taste and smell. I am unaware of studies of how far this is a consequence of damage to the proliferating cells of the olfactory epithelium. Cytotoxic drug treatment for cancer is extremely common, and a large number of people are seriously affected by this. Distortions of taste and smell may contribute to nausea and loss of appetite.

Oakley: One view of that is conditioned aversion; people who get sick after eating will associate the malaise with food, and therefore are put off eating.

Lewis: That is the explanation given in the textbooks, but they don't seem to consider the possibility of damage to olfactory or taste epithelium and resultant disturbance of the sensory nerve connections.

Oakley: That would seem to be an additional factor. Besides conditioned aversion, there could be damage to the chemical senses.

In relation to hormonal effects on sensory receptors, years ago Allara (1952) and subsequently Zalewski (1969) found that testosterone treatment of rats caused a few taste buds to form on the tongue at sites where they were normally apparent only during development. It seems as though some quiescent stem cells were reactivated in adults. This observation has not been followed up.

Fernald: I would certainly agree that thyroxine would be an interesting hormone to study. The September 1990 issue of the *Journal of Neurobiology* is devoted to metamorphosis. A postdoctoral fellow in my lab, Dr Barbara Evans, has evidence that in flounder (*Pleuronectes americanus*) which undergo metamorphosis, thyroxine triggers rod production (Evans & Fernald 1990). Premetamorphic animals have a retina with one, cone-like kind of photoreceptor. At metamorphosis, concomitant with the appearance of T_4 and probably triggered by it, is the production or rods. A wave of neurogenesis is thus associated with thyroxine.

Rubel: Do you know if that's a direct triggering of the stem cells?

Fernald: Almost! A Japanese group showed that thyroxine, put into the eye of a flounder, causes it to migrate before it would normally have done so (i.e., before metamorphosis) (Inui & Miwa 1985). We are doing similar experiments and it looks as if thyroxine is one of the major triggers for the remodelling of the retina that occurs at metamorphosis. That issue of the *Journal of Neurobiology* contains articles on a range of metamorphic phenomena that occur in the natural life of the animal.

Reh: On the question of functional recovery in the olfactory system, to what extent is the appropriateness of the connections necessary for recovery of function?

Costanzo: There is a good topographical projection of the olfactory epithelium onto the olfactory bulb. One hypothesis suggests that olfaction is encoded by a spatial pattern of activity. An experiment that we hope to do is to selectively block axon projections from specific areas of the epithelium, preventing them from projecting back to the olfactory bulb, after regeneration. We could then test to see if hamsters are capable of discriminating odours. These hamsters would be compared to those where the projections are allowed to go back to their normal targets. My guess is that any unique projection pattern that can be discriminated from other patterns can be used to identify different odours. For example, you may have one pattern for the odour produced by an orange and another for that of an apple. After regeneration the 'orange' pattern may change to an 'apple' pattern in the brain, but if you keep associating it with an orange, you may relearn that that's an orange. In other words, unique patterns can be used to code sensory information but the patterns don't have to be the same every time in every individual.

Ryals: Using HRP retrograde labelling of olfactory neurons, could you follow endings back to their original position? In other words, is there a map of the periphery in the olfactory bulb?

Costanzo: So far, it looks as if axons go to the closest and most convenient spot. In order to see if they go back to selected targets, one would need to block the target area and see if they go around the block and go back to their target, or if they go around it and then get diverted. Those experiments have not been done yet.

Rubel: It's my understanding that there is a beautiful topographical relationship between the epithelium and the bulb. However, the patterning mechanism for olfactory stimuli is quite controversial and whether the unique topography is important for the patterning mechanism is still unresolved.

Costanzo: Some people believe there are selective projections and others believe there are diffuse projections; some believe that both exist. My belief is that it is a mixture, and there are some selective areas that are more focused than others and there is a distribution of information across a population of cells. Either way, you still get a unique pattern; regardless of how the information is projected from the periphery, you can produce a unique cortical pattern that can be associated with a unique sensation.

Oakley: It is also fair to say that there's not uniform agreement that there is any significant topographical representation in the olfactory system. John Kauer (1991) and I feel that anatomical topographical projection is marginal. In rainbow trout, David Riddle and I (unpublished work) have found no evidence for topographical projections. We made focal injections of rhodamine and fluorescein-conjugated beads in the glomerular layer of the olfactory bulb, where they were retrogradely transported to the mucosa. You cannot tell from the mucosal labelling pattern whether the bead injection sites were widely separated or close together in the trout olfactory bulb.

Palay: But it could be different in different species?

Oakley: There are certainly species differences. In the rat, some very careful work shows a degree of regional topographical representation of receptor axons (Astic et al 1987).

Costanzo: In the frog *Rana catesbeiana*, we recorded from second-order cells in the olfactory bulb, showed that by stimulating different positions along the nasal epithelium, some cells will only be excited by a single area, whereas adjacent cells could receive input from all areas of that epithelium (Costanzo & Mozell 1976). Thus, there may be a mixture of topographical and diffuse projection onto cells in the bulb.

References

Allara E 1952 Sull'influenza esercitata degli ormoni sessauli sulla stuttura dell formazioni gustative di mus rattus albinus. Riv Biol (Perugia) 44:209–229

Astic L, Saucier D, Holley A 1987 Topographical relationships between olfactory receptor cells and glomerular foci in the rat olfactory bulb. Brain Res 424:144–152

Biffo S, Verhaagen J, Schrama LH, Schotman P, Danho W, Margolis FL 1990 B-50/GAP43 expression correlates with process outgrowth in the embryonic mouse nervous system. Eur J Neurosci 2:487–499

Breipohl W, Mackay-Sim A, Grandt D, Rehn B, Darrelmann C 1986 Neurogenesis in the vertebrate main olfactory epithelium. In: Breipohl W (ed) Ontogeny of olfaction. Springer-Verlag, Heidelberg, p 21–33

Burd GD 1988 Thyroxine stimulates neural development of the peripheral olfactory system in the clawed frog, *Xenopus laevis*. Soc Neurosci Abstr 14:708

Burd GD 1990 Role of thyroxine in neural development of the olfactory system. In: Doving K (ed) ISOT X: Proceedings of the Tenth International Symposium on Olfaction and Taste. University of Oslo Press, Oslo, p 196–205

Burd GD 1991 Development of the olfactory nerve in the African clawed frog, *Xenopus laevis*. I. Normal development. J Comp Neurol 304:123–134

Burd GD, Thomas L 1990 Development of the olfactory epithelium in normal and hypothyroid *Xenopus* larvae. Chem Sens 15:558

Calof AL, Chikaraishi DM 1989 Analysis of neurogenesis in a mammalian neuroepithelium: proliferation and differentiation of an olfactory neuron precursor *in vitro*. Neuron 3:115–127

Costanzo RM, Mozell MM 1976 Electrophysiological evidence for a topographical projection of the nasal mucosa onto the olfactory bulb of the frog. J Gen Physiol 68:297–312

Daikoku-Ishido H, Okamura Y, Yanaihara N, Daikoku S 1990 Development of the hypothalamic luteinizing hormone-releasing hormone-containing neuronal system in the rat: *in vivo* and in transplantation studies. Dev Biol 140:374–387

Evans, BI, Fernald RD 1990 Metamorphosis and fish vision. J Neurobiol 21: 1037–1052

Graziadei PPC, Monti Graziadei GA 1979 Neurogenesis and neuron regneration in the olfactory system of mammals. I. Morphological aspects of differentiation and structural organization of the olfactory sensory neurons. J Neurocytol 8:1–18

Hempstead JL, Morgan JI 1983 Monoclonal antibodies to the rat olfactory sustentacular cell. Brain Res 288:289–295

Inui Y, Miwa S 1985 Thyroid hormone induces metamorphosis of flounder larvae. Gen Comp Endocrinol 60:450–454

Kauer JS 1991 Contributions of topography and parallel processing to odor coding in the vertebrate olfactory pathway. Trends Neurosci 14:79–85

Leloup J, Buscaglia M 1977 La triiodothyronine, hormone de la metamorphose des Amphibiens. CR Hebd Seances Acad Sci Ser D Sci Nat 284:2261–2263

Mackay-Sim A, Beard MD 1987 Hypothyroidism disrupts neural development in the olfactory epithelium of adult mice. Dev Brain Res 36:190–198

McConnell RJ, Menendez CE, Smith FR, Henkin RI, Rivlin RS 1975 Defects of taste and smell in patients with hypothyroidism. Am J Med 59:354–364

Menco BPM, Farbman AI 1985 Genesis of cilia and microvilli of rat nasal epithelia during pre-natal development. I. Olfactory epithelium, qualitative studies. J Cell Sci 78:283–310

Nieuwkoop PD, Faber J 1956 Normal table of *Xenopus laevis* (Daudin). North-Holland, Amsterdam

Paternostro M, Meisami E 1989 Selective effects of thyroid hormonal deprivation on growth and development of olfactory receptor sheet during the early postnatal period: a morphometric and cell count study in the rat. Int Dev Neurosci 7:243–255

Schwanzel-Fukuda M, Pfaff DW 1989 Origin of luteinizing hormone-releasing hormone neurons. Nature (Lond) 338:161–164

Schwartz MA, Kauer JS 1990 Examination of the lineage of adult olfactory receptors by autoradiography and intracellular injection of vital dyes. Chem Sens 15:635

Schwartz MA, Chikaraishi DM, Kauer JS 1989 Characterization of the precursor population in mouse olfactory epithelium after bulbectomy. Soc Neurosci Abstr 15:749

Smith CG 1951 Regeneration of sensory olfactory epithelium and nerves in adult frogs. Anat Rec 109:661–671

Verhaagen J, Oestreicher AB, Gispen WH, Margolis FL 1989 The expression of the growth associated protein B50/GAP43 in the olfactory system of neonatal and adult rats. J Neurosci 9:683–691

Verhaagen J, Oestreicher AB, Grillo M, Khew-Goodall Y-S, Gispen WH, Margolis FL 1990 Neuroplasticity in the olfactory system: differential effects of central and peripheral lesions of the primary olfactory pathway on the expression of B-50/GAP43 and the olfactory marker protein. J Neurosci Res 26:31–44

Wray S, Grant P, Gainer H 1989 Evidence that cells expressing luteinizing hormone-releasing hormone mRNA in the mouse are derived from progenitor cells in the olfactory placode. Proc Natl Acad Sci USA 86:8132–8136

Zalewski AA 1969 Neurotrophic–hormonal interaction in the regulation of taste buds in the rat's vallate papilla. J Neurobiol 1:123–132

Neuronal-epithelial interactions in mammalian gustatory epithelium

Bruce Oakley

Department of Biology, University of Michigan, Ann Arbor, MI 48109, USA

Abstract. There are two neurally controlled morphogenetic programmes in the fungiform papillae of the rodent tongue. When gustatory axons are present they not only promote the expression of taste receptor cells but also simultaneously suppress the outgrowth of an ectopic filiform spine. Innervation during development is essential for the initial formation of taste buds. Consequently, denervation during development irreversibly prevents the establishment of a permanent line of stem cells for taste buds, in contrast to the reliable re-emergence of taste buds after the reinnervation of denervated adult gustatory epithelium. Monoclonal antibodies to keratins selectively recognize cells in particular differentiated states and may contribute to an understanding of the cell lineages and cell–cell interactions that lead to taste buds and filiform spines. Regeneration may recapitulate the late steps in taste development but not the early steps, which are slower and subject to irreversible alteration with denervation. Arguments are set forth that the nerve dependence of taste buds and some other cutaneous receptor cells provides an important selective advantage. Nerve dependence ensures that a more protective integumental covering will replace any superficial, secondary sensory cells rendered useless by denervation.

1991 Regeneration of vertebrate sensory receptor cells. Wiley, Chichester (Ciba Foundation Symposium 160) p 277–293

The study of taste buds offers several topics of interest to cell biologists who study receptor cell regeneration. These include the neural induction of receptor cells, cell commitment, cell turnover, synaptic plasticity, tonic lateral inhibition, and sensory codes and their stability during receptor cell turnover. Taste receptor cells are neurally induced during perinatal development (Hosley et al 1987a). Early denervation during a sensitive period will permanently impair the capacity of the gustatory epithelium to express taste buds. While the turnover of adult receptor cells helps to assure functional taste buds, turnover necessarily challenges the stability of sensory messages by disassembling synaptic connections as cells are replaced (Oakley 1985, 1990). Recent evidence indicates that signal processing in mammalian taste buds can be subject to profound lateral inhibitory interactions that may last for several minutes (Riddle et al 1990).

Positive and negative regulation of morphogenetic programmes, neural induction, neurotrophic dependence, and distinctions between receptor cell development and regeneration are processes to be discussed here that must have important counterparts in other sensory systems.

The factors that control the differentiation of sensory receptor cells are poorly understood. In the gustatory system of mammals we know that taste axons are the agents responsible for the differentiation of the taste buds. In a series of studies we have shown that the structure and function of taste buds depend upon the transport of materials in taste nerves. Shortly after the transport of radiotagged molecules is blocked, taste responses fail and taste buds begin to degenerate, even though impulse traffic can pass through the blocked nerve site (Oakley et al 1981, Sloan et al 1983). The present paper attempts to identify more fully the scope of the morphogenetic processes whose molecular interactions must ultimately be described. Four objectives of the present treatment are: (i) to describe the multiple roles of taste axons in fungiform papillae; (ii) to examine markers for cells in the gustatory epithelium; (iii) to compare taste bud development with regeneration; and (iv) to reflect on the adaptive significance of nerve dependency in secondary sensory cells.

The multiple roles of taste axons

At the apex of each filiform papilla in the mammalian tongue is a conical spine. This protruding filiform spine arises from several upwardly migrating columns of keratinocytes that meld and compress together into tapered layers (Farbman 1970, Hume 1983, Hume & Potten 1976). Analogous 'epidermal proliferative units' generate an upwardly mobile stack of flattened basal cell progeny that comprise the hexagonal keratinized squames of the skin surface (Potten & Allen 1975). When the rat or gerbil tongue is viewed from above, the nearly 90 fungiform papillae per side resemble scattered, circular clearings in a dense forest of caudally curving filiform spines. At its apex, which is flat and thinly keratinized, the fungiform papilla is breached by the tip of a solitary taste bud.

The nerve dependency of taste buds is well established. The nerve dependence of additional morphological features of gustatory papillae is an issue that warrants further scrutiny. A vallate papilla that has been bilaterally denervated from three to 90 days post partum lacks taste buds, but matures to about three-quarters of its normal size with patent ducts and trenches ($n = 3$ rats, unpublished observations). The transformation of adult-denervated fungiform papillae has recently been described (Oakley et al 1990). We examined fungiform papillae in serial sections of adult rat and gerbil tongues, after unilateral transection of the chorda-lingual nerve had caused many fungiform taste buds to degenerate. Most taste buds disappeared; residual taste buds were either atrophic, averaging about half of the volume of a normal taste bud (Fig. 1B), or misshapen remnants of taste buds, averaging about one-quarter of the normal volume, and made

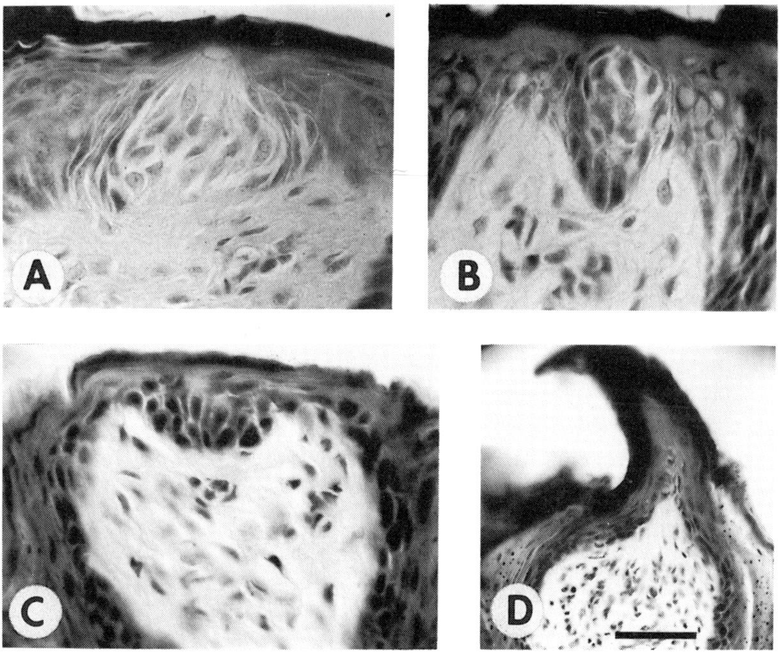

FIG. 1. Normal fungiform taste bud of rat (A). After interruption of the rat chorda-lingual nerve most normal taste buds degenerate and completely disappear, or are reduced to still ovoid, but atrophic taste buds (B), or to even smaller remnants of taste buds (C). Three-fifths of fungiform papillae that lacked a taste bud developed an ectopic filiform spine (D). All photomicrographs are of 10 μm paraffin sections of rat tongue, fixed in buffered 4% paraformaldehyde and stained with iron haematoxylin. The scale bar is 25 μm for A–C and 60 μm for D.

up of small rounded cells and, on occasion, an elongated cell (Fig. 1C). Of rat fungiform papillae that lost all trace of their taste bud after denervation, three-fifths grew a solitary filiform-like spine from the fungiform papilla's apical surface (Fig. 1D). An ectopic spine never appeared in the presence of normal (Fig. 1A) or atrophic taste buds (Fig. 1B). The two-fold greater width of fungiform papillae, and the occasional association of an ectopic spine with a taste bud remnant, provided clear evidence that the spines observed were ectopic outgrowths from fungiform papilla, rather than from large filiform papillae. These results are consonant with observations of tongue explants in which pointed keratinized outgrowths were noted on fungiform papillae (Farbman 1971) and residual taste buds were occasionally associated with a filiform-like spine (Zalewski 1972).

Trigeminal innervation of the tongue reduced the incidence of ectopic filiform spines in proportion to the cross-sectional area of the trigeminal nerve branches that were tested (the mylohyoid nerve, the lingual nerve, lingual and mylohyoid or lingual and auriculotemporal nerves). However, the chorda tympani nerve was 60 times more effective than were trigeminal branches in preventing ectopic filiform spines. Evidently, gustatory axons are much more potent suppressors of the outgrowth of filiform spines from fungiform papillae than are motor axons or non-gustatory sensory axons. By preventing the outgrowth of ectopic spines on fungiform papillae, taste axons maintain a thinly keratinized apical surface that can be breached by the taste receptor cells. Consequently, each fungiform papilla must contain at least two morphogenetic programmes involving the interplay of dozens of cells. One morphogenetic programme leads to a taste bud, the other to a filiform spine. Taste axons act to invoke the programme for taste buds and to suppress the programme for the filiform spine that would form by default on a denervated fungiform papilla.

Figure 2 depicts the critical role of the taste nerve both in controlling the development and regeneration of fungiform taste buds and in preventing the expression of ectopic filiform spines.

Markers for the gustatory epithelium

Keratins are a reasonable choice of markers with which to identify the cells constituting the gustatory epithelium. Panels of antibodies can distinguish among the diverse keratin peptides and allow one to identify specific types of epithelial

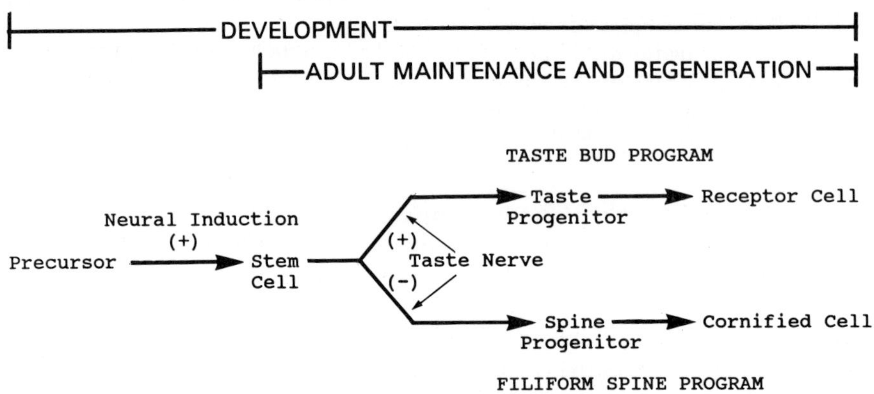

FIG. 2. This diagram outlines possible steps in the development of mammalian taste buds. Taste axons control two morphogenetic programmes: they promote the development of taste buds and prevent expression of a filiform spine. This could involve separate stem cell populations or, as depicted here, a single population of stem cells with a bifurcated differentiation path gated toward receptor cells by taste axons. Some of the suprabasal progenitor cells are likely to engage in amplifying division. As indicated, adult maintenance and regeneration probably do not recapitulate the initial developmental steps (see text, p 281–284).

cells by their keratin profile. At least 19 soft keratins are present in various combinations in human tissues. Anti-keratin monoclonal antibodies differentially stain gustatory epithelia (K19 stained by mAb K4.62, K10 et al. by mAb KL-1). In the rat, gerbil and rabbit lingual epidermis, keratin 19 is found exclusively in the elongated cells of the taste buds (Fig. 3A). We have used three different monoclonal antibodies to demonstrate keratin 19 in fungiform, foliate, vallate, naso-incisor, palatine and epiglottal taste buds. Keratin 19 is also present in the salivary ducts of von Ebner's glands and at the luminal surface of other ducts and visceral organs, including the ureters, bladder, lung, mammary ducts and bronchioles (Bartek et al 1986, Stasiak et al 1989).

In the rabbit tongue, KL-1 staining is restricted to the basal cells, except in the gustatory epithelium where the suprabasal cells adjacent to, but not within, the taste buds are also stained by mAb KL-1 (Fig. 3B). Thus, specific keratin expression is found both in the taste receptor cells and also in those suprabasal cells that surround the taste bud. Differentiation markers such as antibodies to keratins may help us to reach the goal of elucidating the cell lineages and interactions that lead to the development and maintenance of taste buds.

Differences between taste bud development and regeneration

Newborn rats have no mature vallate or foliate taste buds and have few mature fungiform taste buds (Hosley & Oakley 1987, Oakley 1988). A mature bud is defined as one with a taste pore. It is neural induction that establishes a population of stem cells and suprabasal cells which respond to adult taste axons by generating replacement taste cells. Induction by axons cannot be proved from the chronic absence of taste buds after denervation in neonates, because all existing taste buds would rapidly disappear simply for lack of neurotrophic support. Consequently, it required more than a century to provide the experimental evidence to verify the conjecture that nerves induced the formation of taste buds (Hosley et al 1987a), although the proof is uncomplicated. In the rat's bilaterally innervated vallate papilla, when one IXth nerve is present, 228 taste buds develop. Yet when both IXth nerves are present, 154 more taste buds develop than twice the 228 buds associated with one nerve. Clearly, the development of the additional 154 taste buds *required* the presence of *both nerves*. But this is what we set out to prove—that the presence of nerves is necessary for the development of taste buds (Fig. 4). Similarly, at postnatal Day 90, counts of taste buds associated with nerves crushed at postnatal Day 3 revealed substantial synergism from bilateral innervation (30 taste buds with one IXth nerve intact and 144 with two) (Hosley et al 1987b).

Denervation shortly after birth permanently prevents most rat vallate taste buds from ever forming, even if the vallate papilla is subsequently reinnervated by the IXth nerve. We wished to determine whether these early diminished neuronal–epithelial interactions impaired the development of the gustatory

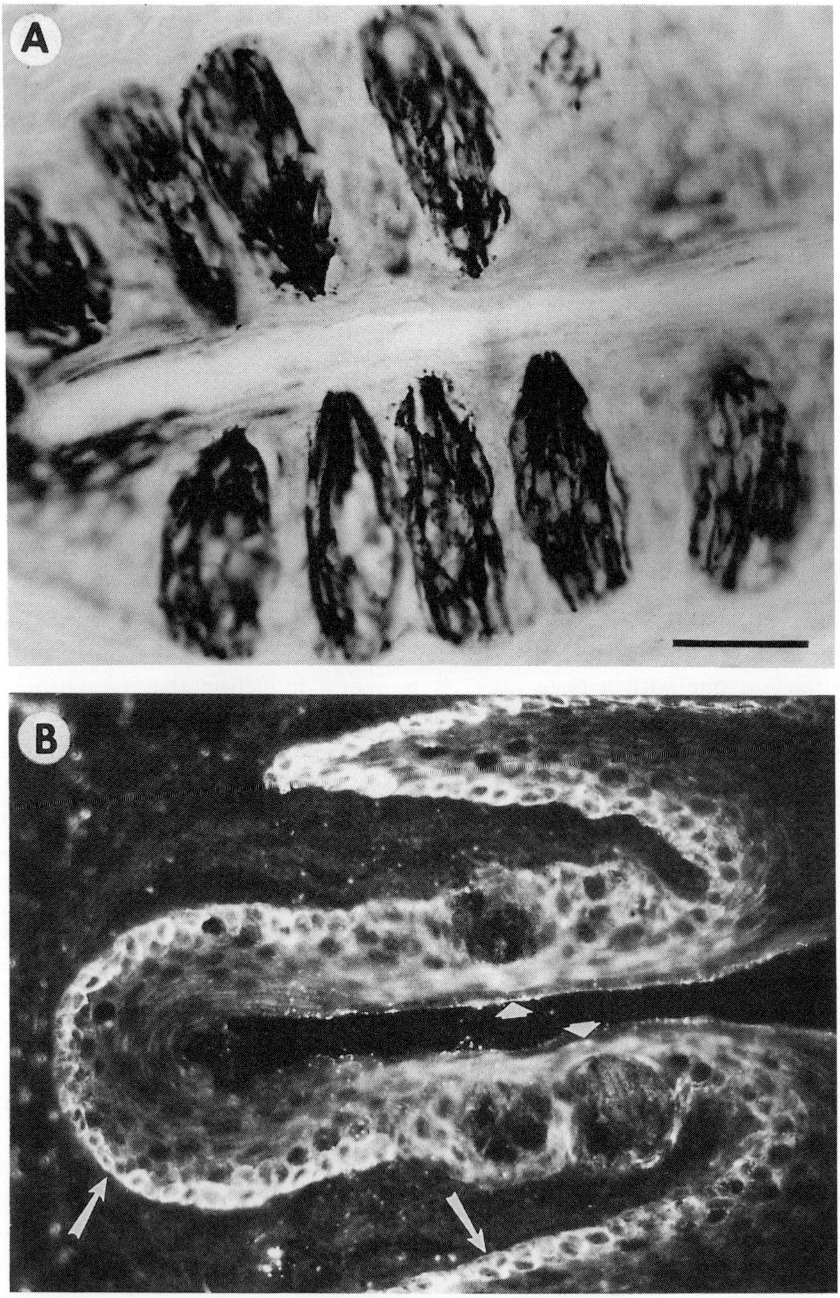
A
B

epithelium or the survival of the taste axons. The vallate papilla was denervated from postnatal Day 3 to 13 by avulsing one IXth nerve and crushing the other. Nerve-crossing experiments, subsequently carried out when the rats were at least 75 days old, revealed that innervation of the vallate papilla by the chorda tympani nerve was four times more effective in those vallate papillae spared early denervation (214 vs 51 taste buds). We concluded that there is a developmental window in the induction of taste buds when innervation endows precursor cells with the potential to contribute to the gustatory system. Apparently, in the absence of interactions with taste axons, the gustatory precursor cells either die or adopt alternative non-gustatory fates.

Three lines of evidence suggest that as taste buds regenerate they probably do not fully recapitulate their developmental history. First, there is no parallel in regeneration to the irreversible failure of taste bud development resulting from denervation during the neonatal sensitive period (Hosley et al 1987a,b, Oakley 1988). Second, the development of vallate taste buds requires about 10.5 days (Oakley et al 1991), whereas regeneration will occur in the three-day interval between the arrival of regenerating axons in the epithelium and the reappearance of mature vallate taste buds (Hosley et al 1987b). Similarly, it requires about three days for taste buds to regenerate and function after chorda tympani axons reinvade the fungiform papillae in adult gerbils (Cheal & Oakley 1977, Cheal et al 1977). The moment of fungiform taste bud maturation in the gerbil was determined by recording the initial taste responses, which appeared coincidentally with the formation of the taste pore. Evidently, taste bud regeneration is several times faster than taste bud development. Third, development and regeneration (maintenance) differ in that one IXth nerve can maintain or cause the regeneration of about twice as many taste buds in adults as it can induce in neonatal development (Hosley et al 1987b). For example, one IXth nerve can maintain or cause the regeneration of more than 400 taste buds, but in development it can induce only 228 taste buds. Similarly, a IXth nerve crushed at three days of age induces by itself about 30 taste buds, although

FIG. 3. A. Cells in nine taste buds of the rat vallate papilla show intense keratin 19-like immunoreactivity. The monoclonal antibody K4.62 (Sigma, 1 : 200 dilution) was recognized by biotinylated secondary antibody (goat anti-mouse IgG, Sigma, 1 : 200 dilution) using the Vectastain ABC reaction with diaminobenzadine enhanced with nickel chloride (0.5 mg/ml DAB, 0.04% $NiCl_2$, 0.001% H_2O_2). B. The monoclonal antibody KL-1 intensely stains the row of basal cells throughout the rabbit lingual epithelium (arrows). Suprabasal cells, including cornified cells, are reactive around the foliate taste buds (arrowheads). The tissue section was reacted with the monoclonal antibody KL-1 (Biodesign, Inc.; 1 : 50 dilution) followed by biotinylated secondary antibody (goat anti-mouse IgG, Sigma, 1 : 200 dilution) and streptavidin Texas Red (Bethesda Research Laboratories, 1 : 200 dilution). Rats and rabbits were perfused with Locke's Ringer's solution, followed by a fixative of 10% acetic acid and 70% ethanol. The tongue tissue was sectioned in a cryostat at 10 μm and stained on gelatin-coated glass slides. The scale bar is 40 μm in A and 70 μm in B.

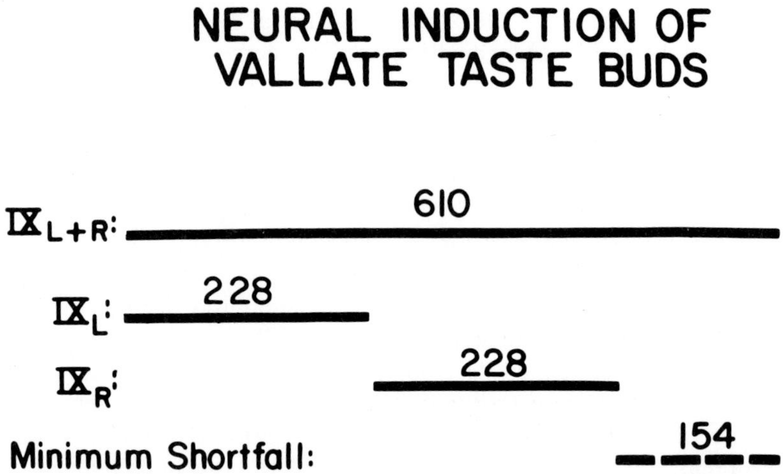

FIG. 4. Evidence for the neural induction of taste buds. The sum of the developmental contributions of the right and left IXth nerves in the rat leaves a shortfall of 154 vallate taste buds in comparison with normal development when both IXth nerves are present. Since axons from both nerves must be present for the additional 154 taste buds to develop, innervation must be necessary for the induction of these and probably all vallate taste buds. The 154 taste buds represent the minimal extent of bilateral synergism, for the right and left IXth nerves might innervate some overlapping sites where either nerve induces the same taste buds among the 228.

it can maintain 100 adult taste buds. Hence, it is easier to regenerate or maintain taste buds than to induce them developmentally. That regenerating taste buds reappear more rapidly and in greater numbers in comparison with development suggests either that the developmental steps are more demanding, or that the adult axons have a greater trophic impact.

Notwithstanding these important differences between development and regeneration, there is no compelling evidence that regeneration fails to recapitulate the latter part of development. That is, while the process that regenerates taste buds may not revert to the initial developmental events, it may use the final steps leading to taste bud formation (see Fig. 2).

On the adaptive significance of the nerve dependence of secondary sensory receptor cells

The aim of this section is to provide a plausible explanation for the nerve dependency of some adult receptor cells.

Cutaneous receptor organs, like Meissner and Pacinian corpuscles (Zelena 1964) and taste buds (Hosley et al 1987a), develop only when prompted by arriving sensory axons. Such neural induction of cutaneous receptors ensures that the CNS receives the correct kind of sensory information, since receptor cells presumably develop if, and only if, they encounter a sensory axon of the appropriate kind. The pronounced developmental nerve dependence of deep mechanoreceptors may subside with maturation. When denervated in adults, muscle spindles or deep cutaneous receptors such as Pacinian corpuscles or Meissner corpuscles undergo only a slow partial atrophy. However, more superficial adult secondary sensory cells may fully degenerate after loss of their sensory innervation. Why should denervation cause the more superficial secondary receptor cells to disappear promptly and completely?

Nerve-dependent adult receptor cells are associated with renewing epithelia. Mitotically active basal cells endow renewing epithelia with the capacity to restore damaged or degenerated tissue. A selective advantage in eliminating denervated receptor organs may be evident in a reduced risk of cellular infection or damage. Perhaps the restorative capabilities of skin include programmes to replace denervated receptor cells with cells that provide more protection.

It is primarily the superficial cutaneous receptors, such as taste buds, some encapsulated/corpuscular endings, and a subset of Merkel cells, that are profoundly nerve dependent in adults. Hence, we propose that nerve dependence may reflect the importance of protecting the skin at sites where denervation has eliminated sensory function. Superficial cutaneous receptor cells are a potential liability. Because they were assembled to confer exquisite sensitivity to environmental stimuli, such sensory organs are necessarily less effective at shielding the organism from mechanical, chemical or microbial damage. Merkel cell mechanoreceptors are an example of cutaneous cells whose pattern of neurotrophic dependence seems closely related to protection. Those deeply situated Merkel cells that envelop the base of a vibrissa or lie in the heavily calloused footpad are not trophically dependent; they survive denervation (Nurse & Farraway 1988). In contrast, nerve dependence is the rule for the superficial Merkel cells that ring the tops of vibrissae and pelage hairs, or are found in the superficial touch domes of the skin. To evaluate the hypothesis that nerve dependence allows exposed, at-risk epidermal regions to seal themselves whenever sensory function has been eliminated, we can consider epidermal structures after denervation. Denervated touch domes lose most of their Merkel cells and develop markedly thickened layers of keratinized cells (Nurse et al 1984). If evolutionary pressure has favoured fortifying epithelia at sites where mechanosensory axons and their function have been eliminated, then similar rules may hold for taste buds. Taste innervation maintains a thin layer of keratin that can be breached by receptor cells. Interruption of the chorda tympani nerve causes many fungiform papillae to develop an ectopic filiform spine (Oakley et al 1990). When taste axons disappear, the main sensory role of the fungiform papillae also vanishes, and the tissue is armoured with a keratinized spine. It seems likely that other kinds of denervated gustatory papillae will also revert to the structure that characterizes the adjoining non-gustatory tissue.

Summary

Future investigations of cell lineage and cell–cell interactions in gustatory epithelia may be aided by the recognition that multiple morphogenetic programmes are under neural control. We suggest that positive and negative regulation of epithelial tissue are normal traits of taste axons. In fungiform papillae, taste axons simultaneously promote the formation of taste buds and prevent the expression of ectopic filiform spines. Cellular differentiation and cell lineages within these morphogenetic programmes may be usefully explored using antibodies to keratins. The irreversible neurally inductive steps that precede the establishment of a gustatory stem cell line appear to be restricted to development; regeneration may recapitulate the terminal series of developmental steps. Selection for nerve dependence may have occurred for those superficial secondary sensory cells whose requirements for transduction open unavoidable chinks in the epidermal armour.

Acknowledgements

I am indebted to A. Lawton and L. Wong for their valuable help and expertise. Supported in part by NIH grant DC-00083.

References

Bartek J, Bartkova J, Taylor-Papadimitriou J, Reijthar A, Kovarik J, Lukas Z, Vojtesek B 1986 Differential expression of keratin 19 in normal human epithelial tissues revealed by monospecific monoclonal antibodies. Histochem J 18:565–575

Cheal ML, Oakley B 1977 Regeneration of fungiform taste buds: temporal and spatial characteristics. J Comp Neurol 172:609–626

Cheal ML, Dickey WP, Jones LB, Oakley B 1977 Taste fiber responses during reinnervation of fungiform papillae. J Comp Neurol 172:627–646

Farbman AI 1970 The dual pattern of keratinization in filiform papillae on rat tongue. J Anat 106:627–646

Farbman AI 1971 Differentiation of foetal rat tongue homografts in the anterior chamber of the eye. Arch Oral Biol 16:51–57

Hosley MA, Oakley B 1987 Postnatal development of the vallate papilla and taste buds in rats. Anat Rec 218:216–222

Hosley MA, Hughes SE, Oakley B 1987a Neural induction of taste buds. J Comp Neurol 260:224–232

Hosley MA, Hughes SE, Morton LL, Oakley B 1987b A sensitive period for the neural induction of taste buds. J Neurosci 7:2075–2080

Hume WS 1983 Stem cells in oral epithelia. In: Potten CS (ed) Stem cells: their identification and characterisation. Churchill Livingstone, New York, p 232–270

Hume WS, Potten CS 1976 The ordered columnar structure of mouse filiform papillae. J Cell Sci 22:149–160

Nurse CA, Farraway L 1988 Development of Merkel cell populations with contrasting sensitivities to neonatal deafferentation in the rat whisker pad. Somatosen Motor Res 6:141–162

Nurse CA, Macintyre L, Diamond J 1984 A quantitative study of the time course of the reduction in Merkel cell number with denervated rat touch domes. Neuroscience 11:521–533
Oakley B 1985 Trophic competence in mammalian gustation. In: Pfaff DW (ed) Olfactory and gustatory influences on the central nervous system. Rockefeller University Press, New York, p 92–103
Oakley B 1988 Taste bud development in rat vallate and foliate papillae. In: Hnik P, Soukup T, Vejsada R, Zelena J (eds) Mechanoreceptors. Plenum Publishing Corporation, New York, p 17–22
Oakley B 1990 On the neurotrophic support of sensory receptor cells. In: Doving KB (ed) Olfaction and taste X. GCS Press, Oslo, p 186–195
Oakley B, Chu JS, Jones LB 1981 Axonal transport maintains taste responses. Brain Res 221:289–298
Oakley B, Wu LH, Lawton A, DeSibour C 1990 Neural control of ectopic filiform spines in adult tongue. Neuroscience 36:831–838
Oakley B, LaBelle DE, Riley RA, Wilson K, Wu LH 1991 The rate and locus of development of rat vallate taste buds. Dev Brain Res 58:215–221
Potten CS, Allen TD 1975 Control of epidermal proliferative units (EPUs). An hypothesis based on the arrangement of neighboring differentiated cells. Differentiation 3:161–165
Riddle DR, Hughes SE, Belczynski CR, DeSibour CL, Oakley B 1990 Inhibitory interactions among rodent taste axons. Brain Res 533:113–124
Sloan HE, Hughes SE, Oakley B 1983 Chronic impairment of axonal transport eliminates taste responses and taste buds. J Neurosci 3:117–123
Stasiak PC, Purkis PE, Leigh IM, Lane EB 1989 Keratin 19: predicted amino acid sequence and broad tissue distribution suggest it evolved from keratinocyte keratins. J Invest Dermatol 92:707–716
Zalewski AA 1972 Regeneration of taste buds after transplantation of tongue and ganglia grafts to the anterior chamber of the eye. Exp Neurol 35:519–528
Zelena J 1964 Development, degeneration and regeneration of receptor organs. In: Singer M, Schade JP (eds) Mechanisms of neural regeneration. Elsevier, Amsterdam (Prog Brain Res 13) p 175–213

DISCUSSION

Margolis: Is keratin 19 intracellular or extracellular?

Oakley: I assume it's intracellular, but we have done no electron microscopy to determine the exact location.

Pujol: So you cannot say which kinds of cell in the taste bud are stained by keratin 19. Are they support cells, or sensory cells, or basal cells?

Farbman: I thought that all cell types were stained?

Oakley: One view is that the cells in taste buds differ as a function of age, and this accounts for functional differences among cells. Or there may be several lineages. It's not clear if there are supporting cells in taste buds. In any event, the first question to ask is whether most or all of the elongated cells have keratin 19. From our initial observations, I suspect that all do; but we are doing further experiments to evaluate this.

Pujol: Do you mean that the cell which, at the time you do the experiment, is playing a sensory role, is labelled for keratin 19?

Oakley: Yes. It's quite improbable that the receptor cells are keratin 19 negative, because so many cells of the taste bud are labelled.

Farbman: I gather that the epidermis isn't labelled with keratin 19. Where does this keratin come from, if it isn't labelling epidermis?

Oakley: Keratin 19 is generally found in simple epithelia, such as the lining of ducts. It is found at surfaces that are exposed to chemical and osmotic conditions unlike those of extracellular fluid.

Farbman: So it is found invariably in mucous membranes, and not epidermis?

Oakley: With some exceptions.

Cotanche: With staining by monoclonal KL-1 you showed that basal cells in the taste buds stain, and also suprabasal cells around the taste bud. What happens when you denervate those papillae? Do you lose the staining by KL-1 in those cells?

Oakley: We have examined only two rabbits, but the result is that the suprabasal cells seem to lose KL-1 binding ability. So, according to this principle, the default condition with no innervation is to return the gustatory epithelial tissue to the status of the neighbouring tissue.

Cotanche: This suggests that the KL-1 keratins keep the basal and suprabasal cells from becoming the filiform spine.

Oakley: That's an interesting idea.

Calof: Have you looked at keratin 19 immunoreactivity in other sensory epithelia, such as the olfactory epithelium?

Oakley: We haven't looked at olfactory epithelium.

Rubel: I am confused about the progenitor cells. I gather that when the filiform spine develops, the same cells are turning into this spine that would normally turn into taste receptor cells. But do we know that there are not *two* progenitor populations? Is proliferation going on at the same rate but the differentiation programme is different, or are they two different lineages? I am confused!

Oakley: You are right, and we know very little about this. There are two alternatives. Either there is one stem cell population and two differentiation pathways, depending on whether the nerves are present or not. Or, alternatively, there are two stem cell populations and the one for taste buds is quiescent unless nerve fibres are present. Nerve fibres would activate the taste stem cell population and also repress the stem cells that would otherwise give rise to the spine. We don't have the information on which to decide between these two possibilities.

Corwin: Can you tell us more about the early embryonic formation of these receptors? We shall be hearing from Harold Zakon about another system that develops from placodes. I don't know the early embryogenesis of these taste buds and whether we are dealing with a population of potential progenitor cells that are laid down through a placodal migration, or something of that kind.

Oakley: We have no markers for these hypothetical precursor cells that await the arrival and influence of nerve fibres. Some would argue that because keratin is present in taste bud cells, as we demonstrated, this indicates that the cells are of epithelial origin. Beyond that I would ask Dr Farbman to comment on the early development and origin of taste buds.

Farbman: Taste bud cells are presumably all epithelial. Although their survival in adults seems to be dependent on their innervation, during development changes can be detected in cells destined to be taste bud cells before innervation (see discussion of Dr Palay's paper, p 17). All our techniques show that the presumptive taste cells in early development are different from what Dr Oakley is calling the default condition of the tongue surface. We do not understand what influences the early changes that makes them different, but we are convinced it is not a neural influence.

Corwin: I would distinguish between epithelium and epidermis as categories here. We wouldn't think that the taste bud cells are not epithelial, but they could become a part of the epidermal epithelium by originating in some place other than the epidermis and migrating into it. The lateral line cells migrate into the epidermis, so that they become part of the epidermal epithelium, but the migratory progenitor cells are not epidermal in origin; they migrate from a cephalic placode, moving beneath the epidermis and then moving up to the surface between epidermal cells when they become mature sense organs. A similar thing might occur here, and perhaps you could demonstrate it, in the way that Harrison (1903) and Stone (1933) demonstrated the placodal, migratory origin of the lateral line neuromast cells by labelling a region of the early embryo and then transplanting into an unlabelled host. If the progenitor cells were to migrate down the tongue, they would be apparent because of the label.

Farbman: We placed fragments of the dorsal tongue tip from rat fetuses (on the 13th and 14th days of gestation) in organotypic cultures. The placodal bumps referred to earlier (see discussion of Dr Palay's paper, p 20) are first seen on the 15th day. Our results showed that 14-day fetal tongues expressed the placodal bumps in culture, whereas the 13-day tongues did not. It is possible that something happens to the tongue epithelium between the 13th and 14th day of gestation that permits expression of the small placodal bumps, and that the origin of the permissive factor may be extrinsic.

Raymond: When does the nerve come into the tongue?

Farbman: The nerve enters the epithelium on Days 16–17, so the bumps develop earlier.

Raymond: Can that observation be reconciled with what Bruce Oakley is saying about the neural induction of progenitor cells?

Farbman: Probably not, because he is talking about the actual taste bud itself, but the bump that we are seeing will become the papilla, and on top of that papilla is the taste bud, in the fungiform region of the tongue.

Oakley: Our premise is that there are specialized sites in the tongue that are gustatory epidermis. They probably develop in the absence of nerve fibres. Taste nerve fibres can affect that tissue; if you deflect gustatory axons elsewhere, they won't generate taste buds.

Farbman: The presumptive fungiform papillae, in the absence of nerve, would probably develop into what you are calling the default condition—that is, filiform papillae—as we have shown in our cultures.

Raymond: It would be interesting, in light of the keratin staining patterns that Bruce Oakley demonstrated, to see if staining by KL-1 appears in this 'competent' gustatory epithelium, before the nerve comes in. Would that keratin be a potential marker for competence to form a taste bud?

Farbman: It could be!

Calof: Your comparison with Merkel cells was very interesting, Dr Oakley. What is known about the development of Merkel cells? Are they epidermal in origin? You said that denervated touch domes lose most of their Merkel cells; in the absence of sensory innervation, will Merkel cells form? Do they form before sensory innervation? I thought it was an interesting comparison.

Rubel: Nerve dependency was shown in neonates; so it may be that these same Merkel cells would be nerve dependent during development (Nurse & Farraway 1988).

Farbman: In the electron microscope, Merkel cells don't have an epithelial character, because they don't have demosomes, and cells lacking desmosomes are probably not true epithelial cells.

Calof: I didn't expect that Merkel cells would be epithelial, but are they derived from the epidermis?

Farbman: I don't really know. There are other cells in in epidermis that are not epithelial cells, such as the Langerhans cell, which is an immune cell.

Calof: I am just curious about the origin of Merkel cells and whether or not they might direct innervation in some sort of chemotropic manner. Maybe that's not known.

Corwin: I would not think that a cell must have desmosomes to be considered a 'good' epithelial cell. Hair cells lack desmosomes; they have zonula adherens and tight junctions, and these are characteristic of epithelial cells, rather than desmosomes, which are associated with intermediate filaments. A cell wouldn't need to have desmosomes or intermediate filaments to be considered an epithelial cell, but if it had tight junctions (zonula occludens) and zonula adherens, then it would be an epithelial cell.

Farbman: I don't think Merkel cells have these either!

Potten: If I understood you correctly, Dr Oakley, taste buds grow back with time, after damage and reinnervation. I am having difficulty in visualizing how that happens; if the structure has grown a pseudo-filiform papilla, where are the stem cells for the taste bud located in relation to that three-dimensional structure? Are they still in the stratified epithelium, or are they sitting in the connective tissue in the middle of the structure? In which case, how do the stem cells break through again?

Oakley: We don't know where the stem cells might be; we don't have a specific marker for them.

Potten: I suppose it would be like looking for a needle in a haystack, to look for some sort of isolated keratin 19-labelled cells?

Oakley: We have no evidence that any but the differentiated, elongated, receptor cells express keratin 19, so I don't think this keratin is a marker for the stem cells that produce the taste bud.

Potten: Where would you locate those cells beneath the taste bud?

Oakley: I can't say yet. However, we find that the labelling by mAb KL-1 in the basal cell layer is continuous except for an oval area directly beneath the taste bud, where there's no reactivity. I don't yet know whether that's because there are no basal or suprabasal cells there, or because those cells lack KL-1 reactivity. We shall see!

Farbman: Will a papilla that has sprouted a filiform spine go back and produce a taste bud?

Oakley: I presume that in time, it will, on the quantitative argument that there are enough denervated papillae with spines that if you count the number of taste buds that regenerate, some nascent spines have to be cast off to account for the number of taste buds found. The process of spine formation can occur rapidly; we have seen ectopic spines as early as eight days after denervation.

Rubel: Do these spines develop if you do a colchicine block of mitosis?

Oakley: I haven't looked at that.

Rubel: What happens to proliferation, and mitotic activity, when you denervate?

Oakley: We looked at this in a preliminary way. We tried to avoid the transitional period immediately after denervation, where one might expect a mitotic burst as the taste bud collapses and degenerates, which probably occurs fairly rapidly. This would be a healing period as the taste bud degenerates and is phagocytized. We looked 2–3 weeks after denervation. We had no evidence that the mitotic rate differed in denervated and innervated gustatory papilla tissue.

Rubel: How about during reinnervation?

Oakley: We haven't looked at that.

Palay: I am interested in the critical 0–10-day period after birth when denervation produces irreversible failure of taste bud development. Is that period dependent on the glossopharyngeal (IXth) nerve, or can it be modified by having, say, trigeminal nerve input during that time?

Oakley: We haven't tested that because when we take the chorda-lingual nerve out in the newborn or three-day-old rat, the animal is so small that we have to remove both the lingual and the chorda tympani nerve. I don't know what would happen if one removed predominantly the gustatory fibres—whether the lingual nerve would sustain normal development.

Cotanche: Can you cut the chorda tympani in the middle ear? It runs right through the ear drum.

Pujol: It is possible to destroy the middle ear in the newborn rat.

Oakley: It is impossible to visualize the area well, so you are doing a blind operation.

Zakon: In the experiments where you were claiming that the stem cell population is influenced by innervation, I gather that you destroyed innervation to the vallate papillae, bilaterally, and then months later you brought the chorda tympani in and got fewer taste buds developing than you expected. Do you know how many taste fibres were present? Is it possible that you just didn't get as many nerve fibres as might have been there, or do you know whether the chorda tympani nerve has the capacity to induce taste buds as effectively as the original nerves?

Oakley: We did strictly comparable operations in which the chorda tympani was forced to grow into a vallate papilla which had developed normally, compared to a papilla with a history of denervation during the first two weeks after birth. The route that the chorda tympani took, and the pathway it had to track, were the same in the two cases. Cross-reinnervation was carried out in the adult, so the chorda tympani was never influenced by the early operations. So the only difference is the early history of treatment of IXth nerve innervation of the vallate papilla. If it was not denervated early on, we got four times as many taste buds (214 as against 51).

Presson: Can you manipulate the time it takes a nerve to come back in, and, if you make it longer, does it then take longer to induce differentiation? Can you step back, in other words, by denervating for a longer period of time, to the earlier population and therefore have a handle on what's different between the progenitors and the stem cells?

Oakley: We haven't attempted to look at the variable with the time of re-entry of the nerve fibres; typically, because of where we crush the nerve, the fibres come back in 10–13 days. We could try to do some of those experiments. If the return of chorda tympani fibres is delayed, it might be more difficult to slough off the ectopic spine and re-establish a taste bud.

Rubel: In that respect, your earlier studies on the relationship between the nerve and the number of taste buds were based on the size of the nerve and various manipulations of the size. Obviously, taste buds are not being controlled by the nerve, *per se*; they are being controlled by the axons. So is there any control for the number of axons with those different manipulations? It's known that when you cut nerves, the axons often sprout multiple processes.

Oakley: It's hard to treat that experimentally. The gustatory fibres are also a subpopulation, which we can't specifically identify. The suppression of ectopic filiform spines by trigeminal branches is proportional to the cross-sectional area, which should reflect the volume of axoplasm.

Farbman: How do you explain the results of those who cut the chorda tympani nerve to the fungiform papilla and observed taste buds surviving as 'aneural' taste buds (Whitehead et al 1987, Hård af Segerstad et al 1989)?

Oakley: I described the atropic or remnant taste buds that we see after transection of the chorda-lingual nerve in the gerbil. All the fungiform papillae that had a taste bud were innervated. Whenever we see a papilla that has no axons, it has no taste bud, remnant or atrophic. So in spite of Mark Whitehead's group's considerable efforts to determine whether nerve fibres were present, I suspect that there were a few taste nerve fibres that were not identified. We see taste buds surviving denervation if, and only if, there are nerve fibres. The moment the nerve fibres grow back into the papilla the taste buds reappear, in the gerbil.

Ryals: Is it true that if you have reduced innervation, the taste bud looks atrophic? In other words, is there a gradient of reinnervation?

Oakley: Yes; that's the way to produce these small residual taste buds, by reducing the level of innervation.

References

Hård af Segerstad C, Hellekant G, Farbman AI 1989 Changes in number and morphology of fungiform taste buds in rat after transection of the chorda tympani or chorda-lingual nerve. Chem Sens 14:335–348

Harrison RG 1903 Experimentelle Untersuchungen über die Entwicklung der Sinneorgane der Seitenlinie bei den Amphibien. Arch Mikrosk Anat Entwicklungsgesch 63:35–149

Nurse CA, Farraway L 1988 Development of Merkel cell populations with contrasting sensitivities to neonatal deafferentation in the rat whisker pad. Somatosen Motor Res 6:141–162

Stone LG 1933 The development of lateral-line sense organs in amphibians observed in living and vital-stained preparations. J Comp Neurol 57:507–540

Whitehead MC, Frank ME, Hettinger TP, Hou LT, Nah H-D 1987 Persistence of taste buds in denervated fungiform papillae. Brain Res 405:192–195

Regeneration of electroreceptors in weakly electric fish

Harold H. Zakon

*Department of Zoology, University of Texas, Austin, TX 78712, USA

Abstract. Weakly electric teleost fish possess two classes of electroreceptors: *tuberous* and *ampullary* organs. Ampullary organs are used for detecting prey, while tuberous organs detect the fish's own electric organ discharges (EODs) and those of conspecifics. EOD frequency varies among individuals within a species and a fish's tuberous receptors are sharply tuned to its own EOD frequency. In young, small fish both tuberous and ampullary afferents innervate only single organs. As fish grow new receptor cells are added to each organ and it divides into two daughter organs. This process continues resulting in numerous organs in a cluster; the afferent nerve innervates all the organs in a cluster. When a patch of skin is removed new skin grows back complete with new receptor organs of both classes. From our studies we have shown that: (1) new organs are found only in the presence of nerve fibres; (2) their morphological development during regeneration is similar to their normal development; (3) organs divide rapidly giving rise to daughter organs until each afferent fibre innervates the correct number of organs for a fish of its size; (4) receptor cells are broadly tuned below the EOD frequency of a given fish and they gradually increase their tuned frequency and sharpness of tuning until they become correctly tuned to that EOD frequency; (5) the correct matching of receptor tuning to EOD frequency occurs in fish in which the spinal cord has been severed or with lesions of the medullary pacemaker nucleus, thereby eliminating the EOD and any possible 'calibration' signal; and (6) basal and capsule cells of receptor organs in the intact skin around the wound divide after skin damage and are a possible source of precursor cells for new receptor organs.

1991 Regeneration of vertebrate sensory receptor cells. Wiley, Chichester (Ciba Foundation Symposium 160) p 294–313

Weakly electric teleost fish (orders Gymnotiformes and Mormyriformes) generate electric fields from a specialized electric organ in the tail. These fields are used for two functions. The first, electrolocation, so called by analogy with echolocation in bats, is the detection and identification of surrounding objects by the distortions that they produce in these self-generated fields (Bullock & Heiligenberg 1986). The second function is social communication—the detection

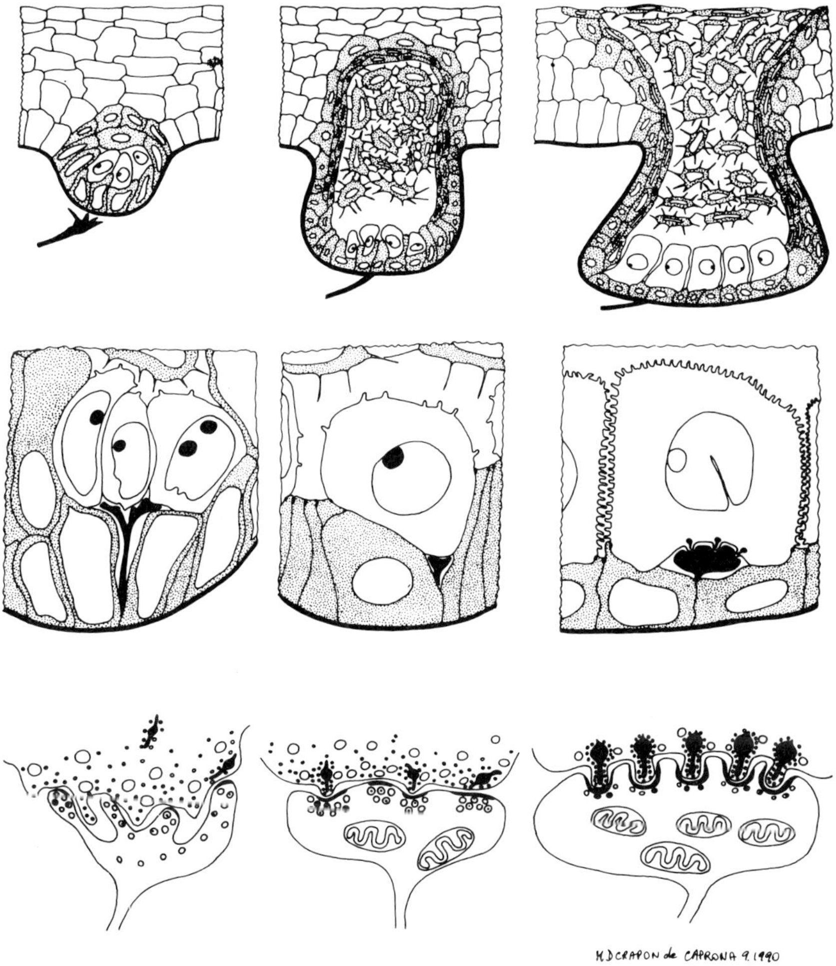

FIG. 1. Schematic illustration of a regenerating tuberous electroreceptor. The top row depicts the whole organ and surrounding epidermis, the middle row depicts the sensory and basal cells, and the bottom row shows the development of the synapse. Basal and plug cells are stippled, sensory cells are clear.

and analysis of the waveform of the electric organ discharges (EODs) of conspecifics (Bullock & Heiligenberg 1986). Each fish detects these electric fields with specialized receptor cells, called *tuberous electroreceptors*. In addition, both groups of weakly electric fish possess a second type of electroreceptor called an *ampullary electroreceptor*. These function to detect the low frequency and virtual DC fields that may be produced by prey (Szabo 1965, Zakon 1986a).

Electroreceptors of the ampullary type are also found in a few other teleosts, notably catfish, a number of non-teleost fish, some larval amphibia and even in monotreme mammals (*Platypus*) (Zakon 1986a). Since the bulk of the work on electroreceptor development and regeneration has been done in the South American gymnotiformes and their close relatives the catfish, I shall limit detailed descriptions of results to those groups, parenthetically remarking upon others where appropriate.

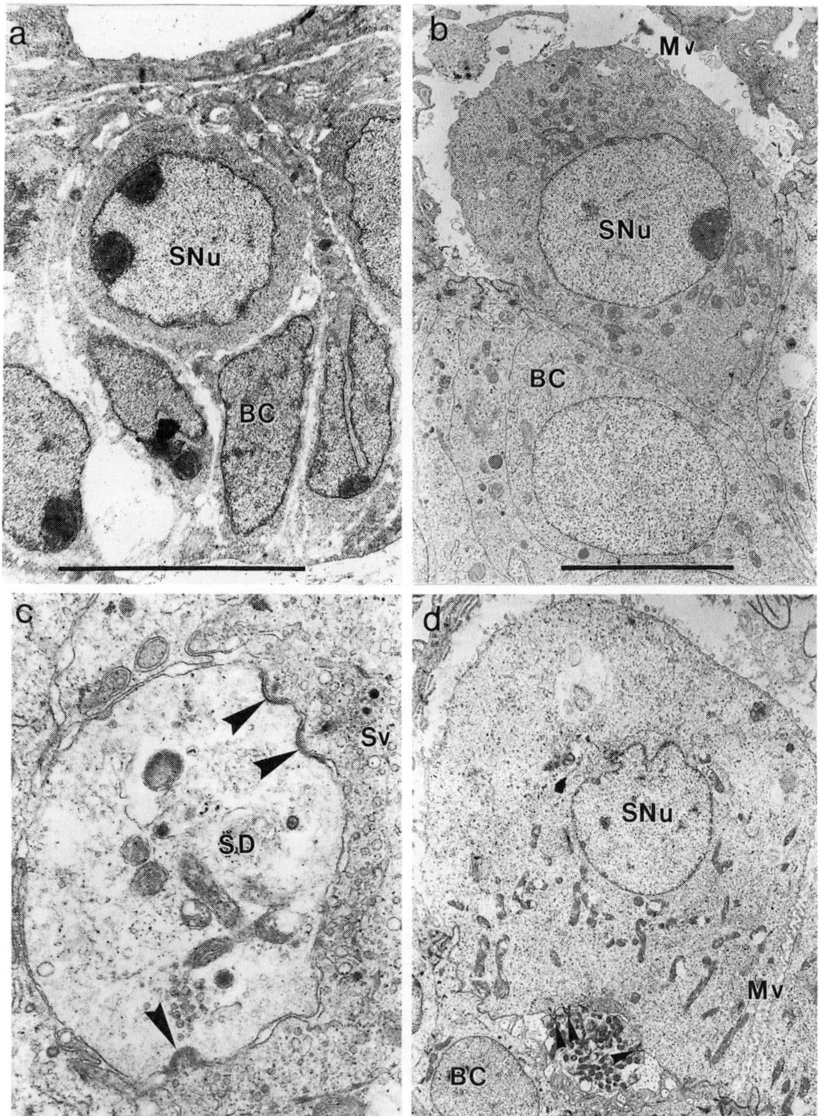

Morphology and development

Both tuberous and ampullary electroreceptor organs are found in the fish's skin, more or less generally over the body. Both types of sensory organ consist of receptor cells encased within a capsule, the interior of which communicates with the water surrounding the fish via a canal. The capsule wall in both types of organs is made of flattened cells connected by tight junctions; these cells are continuous with an epithelium at the base of the organ which is associated with the sensory receptor cells. The cells of the epithelium are called basal or support cells. Ampullary organs are distinguished by the interior of the canal, which appears hollow in histological sections, but is filled with a conductive jelly, and by 8–10 round sensory receptor cells which are embedded into the single layer of basal cells that line the base of the organ. The tuberous organs, on the other hand, have a cellular plug in the canal and 25–30 ovoid sensory cells which project almost entirely into the lumen of the capsule, only their basal faces contacting the basal epithelial layer (Zakon 1986a) (Fig. 1, top row, right). The sensory cells are covered by microvilli and make tight junctions at their basal faces with the basal cells in the epithelium (Fig. 1, top row, right and middle row, right; Fig. 2d).

Each organ is innervated by a single afferent fibre from the lateral line nerve which forms a bouton, or synaptic disc, at its base. The ampullary afferents form large (4 μm) diameter discs, while those terminating on tuberous sensory cells are more modest in size (1–2 μm) (Zakon 1984). Receptor cells in both organs possess an active zone comprising a synaptic bar with a halo of vesicles and, facing this on the postsynaptic side, a typical membrane thickening (Wachtel & Szamier 1966, Szamier & Wachtel 1969, Fritzsch et al 1990). A few scattered medium-sized vesicles (80 mm) may be seen near the active zones in the afferent terminal (Bennett et al 1989, Fritzsch et al 1990) (Fig. 1, bottom row, right; Fig. 2c).

Experiments involving the removal or ectopic implantation of the preotic placode in the axolotl salamander *Ambystoma mexicanum* have demonstrated that the ampullary receptor organs, as well as the lateral line neuromasts on

---

FIG. 2. Regenerating tuberous electroreceptors. (a) At two weeks of regeneration, round sensory cell precursors with round nuclei (SNu, sensory cell nucleus) may be observed above the lower layer of basal cells (BC). (b) At three weeks of regeneration, sensory cells begin to migrate into the lumen. Microvilli (Mv) are apparent on their surface. (c) At six weeks of regeneration the synaptic terminal, or synaptic disc (SD), of the afferent nerve fibre is normal, with synaptic vesicles (Sv) near the active zones (arrowheads). There are still clusters of postsynaptic vesicles close to some of the active zones at six weeks (e.g. to right of lower arrowhead). (d) A normal sensory cell and its underlying afferent terminal. Arrowheads indicate sites of synaptic contact. Bars, 10 μm in (a), 5 μm in (b) and (d) and 1 μm in (c). (From Fritzsch et al 1990, by permission of the *Journal of Comparative Neurology*.)

the head, derive from this placode (Northcutt et al 1990). Unfortunately, whether the electroreceptors derive from placodes in teleosts or, indeed, whether teleosts possess placodes at all, is controversial (R. G. Northcutt, personal communication).

Although their early embryonic origin is unclear, the later development of these organs has been well documented (Srivastava & Seal 1981, Kirschbaum & Denizot 1975) especially by Vischer (1989). The earliest indication of organ formation is a cluster of cells forming above the basement membrane between the epidermis and the dermis. These cells sort themselves into a lower layer of basal cells, which are assumed to be a proliferative epithelium, and a small upper layer of presumptive receptor cells in the tuberous organs, while the receptor cells remain within the epithelium in the ampullary organs. The existing receptor cells grow in size and new receptors are added, presumably as a result of mitosis in the basal cell layer (H. A. Vischer, personal communication). A lumen forms above the sensory cells and the cells of the capsule wall become thinner. A canal forms and becomes filled with cells in the tuberous organs and hollow in the ampullary organs.

As is typical of other sensory cells and neurons in teleost fish (Johns & Easter 1977, Corwin 1983, Popper & Hoxter 1984), there is a continuous addition of new sensory cells in both classes of electroreceptors throughout life (Zakon 1984, 1987). The ampullary organs of all gymnotiformes and catfish (which have only ampullary organs) so far studied accommodate the new electroreceptor sensory cells by dividing as the fish grow (Zakon 1984, 1987, Vischer 1989). A single organ divides into two daughter organs, each of which then divides at a later time. The nerve bifurcates at each division, a branch accompanying each daughter organ. Large specimens of some species may have over 30 ampullary organs innervated by a single highly elaborated afferent.

Tuberous organs divide in some species of gymnotiformes, like *Sternopygus* and *Eigenmannia* (family Sternopygidae), but not others (Zakon 1987). In those species where organs divide, the process is essentially identical to that observed in the ampullary system, except for one difference. Receptor organs are added more rapidly to one group of afferents than another so that large fish come to have some afferents with one or two organs, and another group with many (5–10) (Zakon 1984) (Fig. 3). These two groups become physiologically distinct subtypes of tuberous receptors (Sanchez & Zakon 1990). In those species where the tuberous organs do not divide, the diameter of the organ increases to accommodate the increased number of receptor cells (Zakon 1987).

Regeneration of electroreceptors

Both ampullary and tuberous electroreceptors can regenerate in gymnotiform and mormyriform fish, as can the ampullary receptors of catfish (Roth 1986,

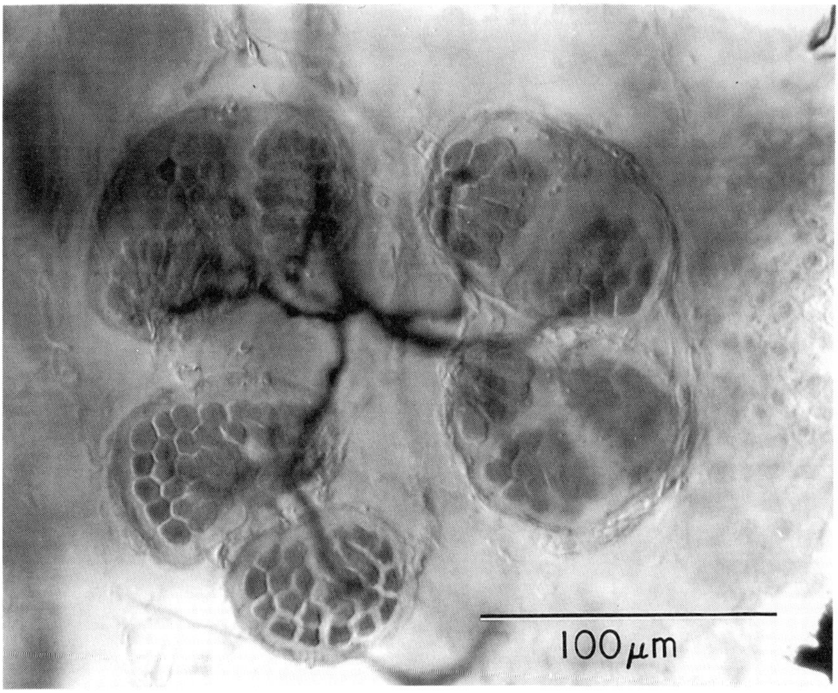

FIG. 3. Whole-mounted tuberous electroreceptor organ showing how single organs divide into daughter organs. Silver-stained afferent fibre is out of the plane of focus.

1987, Roth & Szabo 1969, Denizot & Baillet-Derbin 1969, Zakon 1986b, Fritzsch et al 1990). A recent ultrastructural study of receptor regeneration in *Sternopygus* shows that the differentiative events that occur during regeneration are very close to those that occur during development (Fritzsch et al 1990). In these studies a patch of cheek skin is removed and allowed to regrow. Cells migrate in from the wound margin and replace the missing skin patch and, ultimately, a full complement of electroreceptors appears.

The first indication of the formation of new receptor organs in regenerating skin is the appearance of small clusters of cells at the base of the epidermis that protrude into the dermis. The cells at the base become arranged in two crude layers: the lower layer is a future basal cell layer and the upper layer consists of presumptive sensory receptor cells (Fig. 1, top and middle rows, left). At this time, and earlier, there is no obvious distinction between ampullary and tuberous organs.

A small hollow forms in this precursor organ (Fig. 1, top row, centre and middle row, left) and, eventually, a canal begins to form in the epidermis above the incipient organ, first noticeable as a loosening of the epidermal cells in a column above the organ. Soon, the types of organs become distinguishable as an entirely hollow canal forms above the ampullary organs, while the tuberous canals still have an obvious epithelial plug.

At this time (two weeks after skin removal), new sensory cells begin to form (Fig. 2a). The cells of the inner layer become more rounded and, in the tuberous organ, move up from the epithelial layer into the lumen, leaving a single thick layer of epithelial cells forming the outer capsule. These cells continue to divide and generate new electroreceptor sensory cells which move out into the lumen (Fig. 1, top row, centre and middle row centre; Fig. 2b). The cells of the organ walls become flattened and take on the appearance of capsule cells, while the cells at the base of the organ remain as more or less columnar basal cells; these cell layers remain continuous, however. The organs now have an adult-like appearance except that the tuberous sensory cells are smaller and fewer in number and with fewer microvilli on their surfaces (Fig. 1, middle row, centre). Sensory cells continue to be produced in the basal epithelial layer and move into the lumen in the tuberous organs. Sensory cell addition is probably occurring in the ampullary organ as well, although we have not studied them as extensively.

The tuberous receptor cells are first contacted by the tips of the afferents while they are first forming and moving into the lumen (Fig. 1, bottom row, left). The afferents form numerous thin branches upon each sensory cell. There are no indications of synaptic organelles at this time. Gradually, a small terminal disc forms at the base of the sensory cell, and a synaptic bar and synaptic vesicles can be visualized in the sensory receptor cell (Fig. 1, bottom row, centre). Numerous other medium- and large-sized vesicles can also be seen at these times, as in an adult sensory cell. Meanwhile, a postsynaptic thickening appears in the afferent. Interestingly, a number of medium-sized (80 nm) vesicles also appear in the afferent ending, typically adjacent to the active zones. They are most numerous during weeks 2–4 of regeneration, while the sensory cells are forming and maturing, and decrease to low, adult levels thereafter (Fig. 1, bottom row, right; Fig. 2c). We have hypothesized that they might be the possible carriers of a trophic substance (Fritzsch et al 1990) although this remains to be substantiated.

An interesting question is whether the nerve is necessary for the differentiation and maintenance of the receptor organs. Electroreceptors degenerate if their nerve supply is removed (Roth & Szabo 1969, Szamier & Bennett 1973, Denizot and Libouban 1985); most dramatically, the ampullary sensory cells in the glass-catfish, *Kryptopterus*, die within 24 hours and the accessory structures of the organs within a few more days upon denervation (Szamier & Bennett 1973).

Similarly, the nerve supply is needed for the formation of receptor organs. During regeneration, receptor organs do not appear in patches of skin where nerves have not yet entered or which are permanently denervated (when observed at three weeks of regeneration) (Fritzsch et al 1990, H. H. Zakon, unpublished work 1990). Nevertheless, apparently healthy organs without innervation have been observed early in regeneration (1–2 weeks) in serially reconstructed, thick-sectioned (1 μm) material (Fritzsch et al 1990). Thus, we have proposed that, as occurs in limb regeneration in the newt (Singer 1974), a nerve supply is necessary for the early differentiation of the organ, although the nerve need only be in the vicinity of the differentiating organs, not in actual physical contact. This idea is in agreement with evidence from silver-stained material from juvenile or adult fish, in which apparently newly differentiated organs have been observed with thin nerve fibres advancing towards them (Zakon 1984, Vischer 1989).

As mentioned above, the organization of receptor units in a large fish is complex, in that a single afferent innervates a large number of ampullary organs, whereas the tuberous afferents are divided into one class that innervates one or two organs and another that innervates a number (Zakon 1984). How well does this organization return as the electroreceptors regenerate? Almost immediately after new organs are formed in regenerating skin, these begin to divide and produce new daughter organs. They continue to divide until the correct number of organs per axon is reached, as indicated by the distribution in the contralateral control cheek, and then stop dividing (Fritzsch et al 1990). The receptor organs of a small fish will continue to divide until the correct number of organs is reached by three months, and no further division occurs. At 3–5 months, however, the organs of a larger fish are still dividing and attempting to regain the correct number of organs per axon.

The signal that instructs a receptor unit that the correct number of organs has been regenerated at an afferent is unknown. There is a suggestion that the signal is not the 'knowledge' within each organ that it divide a particular number of times. The evidence for this is that during early regeneration afferents may innervate two or three organs over a stretch of tissue in a linear fashion. Each of these begins to divide. If each organ underwent a predetermined number of divisions, then those afferents that innervated two organs would have twice as many organs in total as the largest unit in the control cheek. The fact that this does not happen suggests that the afferent 'knows' how many organs it innervates, even if it initially innervates more than a single one, and each of these divides fewer times than if that afferent initially innervated only a single organ.

Physiology of regenerating receptor organs

One intriguing attribute of tuberous electroreceptors is that they are tuned—that is, most sensitive to a particular frequency (Hopkins 1976, Zakon & Meyer 1983). A 'tuning curve' can be constructed for each afferent during electrophysiological recordings. A tuning curve depicts the threshold response of the afferent over a range of stimulation frequencies. The important information to be extracted from such a curve is the best frequency (BF)—the particular frequency to which the cell is most sensitive—and the width of the tuning curve at 10 dB above

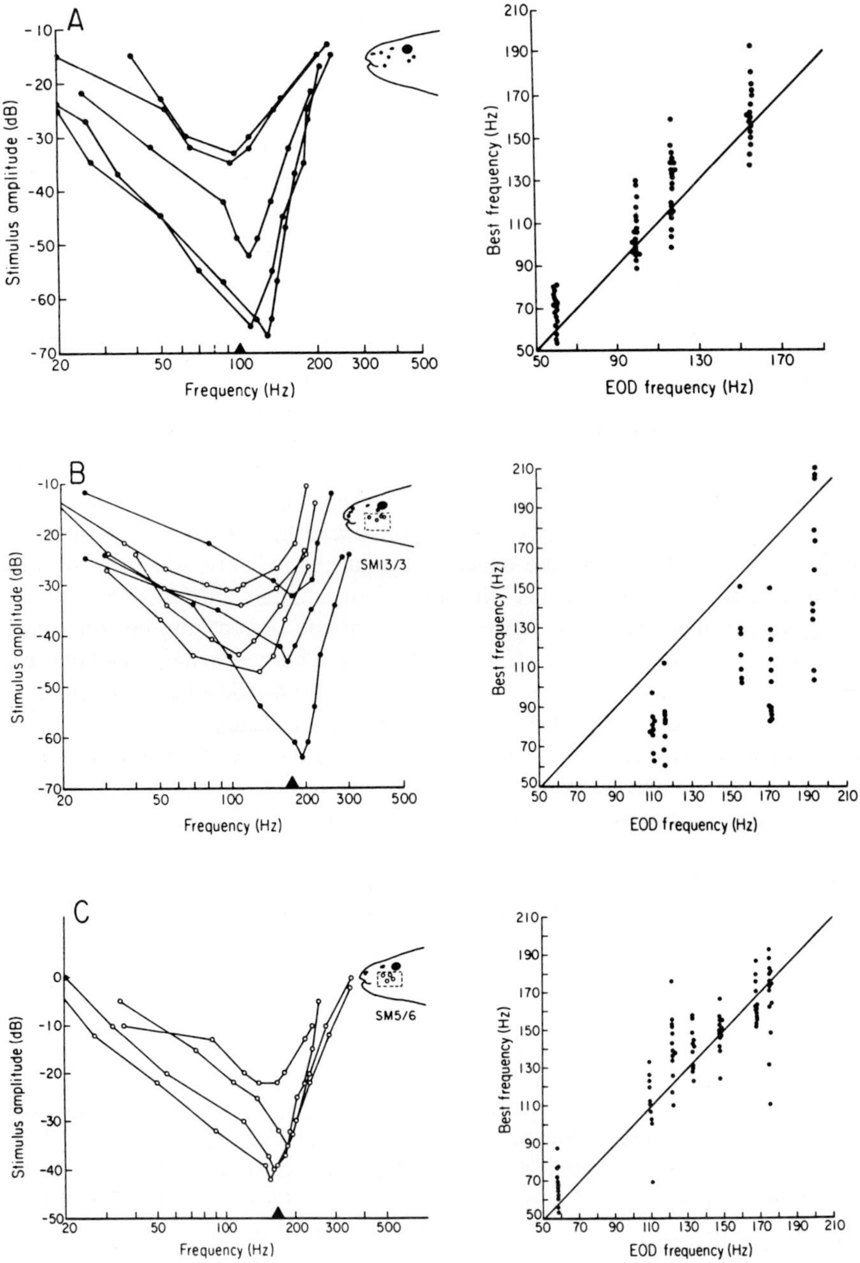
A
Stimulus amplitude (dB)
Frequency (Hz)
Best frequency (Hz)
EOD frequency (Hz)
B
SM13/3
Stimulus amplitude (dB)
Frequency (Hz)
Best frequency (Hz)
EOD frequency (Hz)
C
SM5/6
Stimulus amplitude (dB)
Frequency (Hz)
Best frequency (Hz)
EOD frequency (Hz)

its BF threshold, which indicates how sharply or poorly tuned each afferent is. Such recordings indicate that the tuberous electroreceptors of each individual are generally sharply tuned and are tuned best to its own EOD (electric organ discharge) frequency.

Electrophysiological recordings made from the afferents at various times during regeneration show that, when they can be first recorded at three weeks of regeneration, the tuning curves are broad and the BFs are poorly defined (Fig. 4). Furthermore, the BFs are at much lower frequencies than the fish's individual EOD frequency. By six weeks of regeneration, the tuning curves of most of the afferents are normally sharp and tuned best to the fish's EOD frequency. Thus, the afferent recordings suggest that receptor cells start out poorly tuned and tuned low, and gradually increase their BF until they reach the correct BF for that particular fish (Zakon 1986b). Direct recordings of stimulus-induced receptor potentials from the sensory cells bear this out.

The fact that the new sensory cells become correctly tuned to a particular frequency, the fish's own EOD frequency, suggests that perhaps the sensory cells are 'calibrated' either by the electric fields that they are sensing, or by some signal from the pacemaker nucleus (PMN)—the region of the brain that controls the firing frequency of the electric organ. A series of experiments have indicated that neither of these possibilities is true. If receptors are allowed to regenerate after a high spinal transection which blocks the descending output from the medulla and removes the EOD, or after a lesion of the PMN, they still become correctly tuned to the fish's previous EOD frequency (Zakon 1986b, Ferrari & Zakon 1989). Thus, we suppose that the instructions for correct tuning are intrinsic to the precursor cells, or derive from another source.

It is well known that the tuning of tuberous sensory receptors as well as the firing frequency of the PMN is modified by sex hormones (Zakon & Meyer 1983, Keller et al 1986). Treatment of male *Sternopygus dariensis* and *S. macrurus* with androgens lowers EOD frequency and receptor tuning in tandem, whereas treatment of females with human chorionic gonadotropin raises them (Zakon et al 1990). All the tuberous receptors might be tuned the same in all fish as a 'starting condition', and the levels of various hormones might dictate how receptor tuning changes. Thus, if skin is removed in a fish with

---

FIG. 4. Tuning curves and the distribution of best frequencies from normal and regenerating tuberous receptors. The left-hand column shows representative tuning curves from an individual that is (A) intact, (B) with a regeneration patch three weeks after skin removal, or (C) with a regenerating cheek patch six weeks after skin removal. The right-hand column shows the distribution of best frequencies for a number of fish in each group, graphed according to each fish's EOD frequency. In the left-hand column the open circles indicate receptors that have been localized to the regenerating patch of skin (see inset of fish head); filled circles represent tuning curves of receptors that are from intact skin surrounding the regenerating patch. The arrowhead on the abscissa indicates the EOD frequency of each fish.

a PMN lesion, it might not be surprising that the new receptors become tuned to the pre-lesion EOD frequency even in the absence of a calibration signal; the circulating hormones might provide that information.

Where do the regenerated organs come from?

It is still unclear from where the new electroreceptor organs originate. Previous investigators have suggested that the lateral line nerve could transform differentiated epidermal cells into lateral line hair cells and accessory cells (Brocklebank 1925, Bailey 1937), or electroreceptors (Roth 1987). This conclusion was based on studies in which new receptor organs were induced when the lateral line nerve was deflected under the skin in a portion of the body in which few or no receptor organs exist. However, it is also possible in these experiments that the new receptor organs derive from cells of existing organs that migrate into the epidermis toward the transplanted nerve.

Other studies of lateral line organ regeneration indicate that when organs are destroyed, usually after tail amputation, support cells from existing organs migrate out into the epidermis and found new organs (Corwin et al 1989, Balak et al 1990). In fact, most of the work on regeneration of sensory receptors indicates that new cells come from a particular precursor cell rather than the transformation of an already differentiated cell type (Raymond et al 1988, Corwin et al 1989).

In our studies of regeneration of electroreceptors we have noticed that as many as 20% of the tuberous organs are doubly innervated, a condition which arises in less than 5% of organs in normal skin (Fritzsch et al 1990). This observation seems incompatible with the idea of each afferent being able to contact any epidermal cell and induce it to form into an electroreceptor organ. Instead, this result suggested that the afferents are 'competing' for a cell type that is much less abundant than epidermal cells. Such a candidate could be the basal or capsule cells of intact organs that have migrated into the new skin along with epidermal cells.

To test this idea we have labelled fish with the thymidine analogue 5-bromo-2′-deoxyuridine (BrdU) after removing a piece of cheek skin. We found in the weakly electric fish, *Apteronotus*, that basal, capsule and epidermal cells in skin surrounding the wound label with BrdU at levels higher than control skin from the other intact cheek (H. H. Zakon & S. M. Gustavson, unpublished work 1990). By 18–24 hours after skin removal, capsule and basal cells divide significantly more than in control patches of skin (Fig. 5). Both cell types show elevated levels of division, along with epidermal cells, for almost a week while new skin is forming in the wound. During this time the capsule cells show greater numbers of labelled cells than basal cells. These results support the contention that basal or capsule cells from adjacent organs could give rise to new organs during regeneration. Nevertheless, we have not yet determined whether these cells migrate into the new skin and actually form new organs.

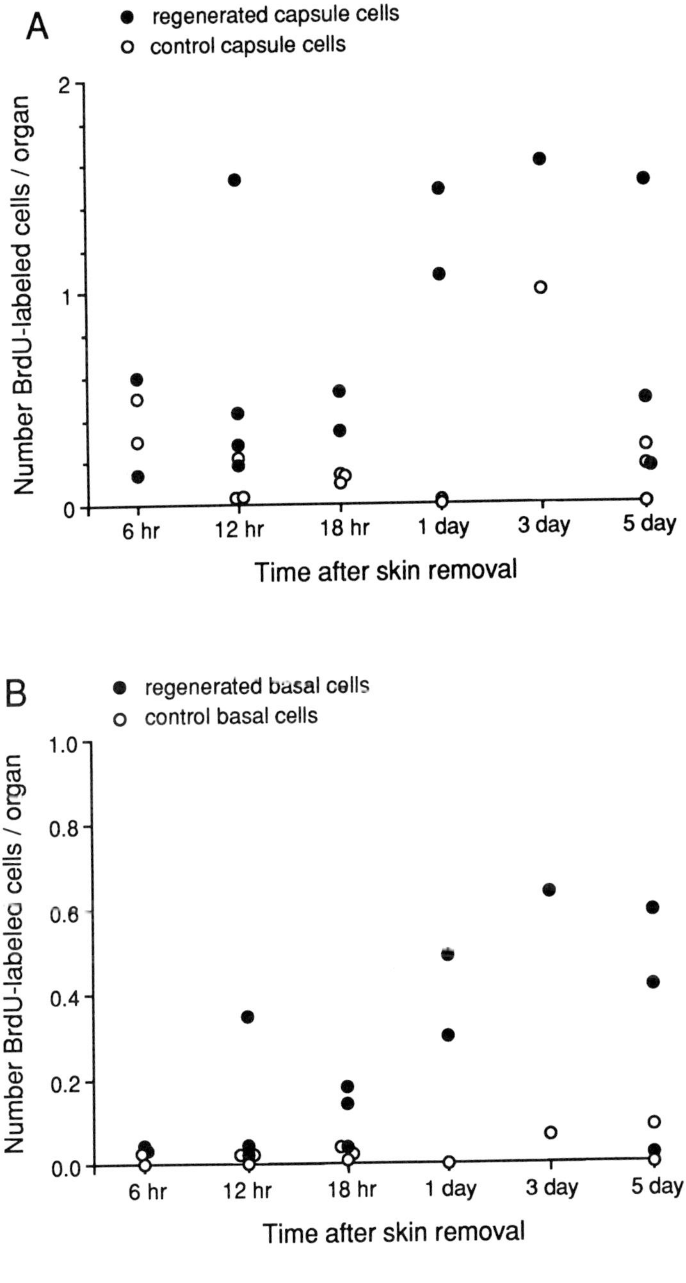

FIG. 5. Rate of uptake of 5-bromo-2′-deoxyuridine (BrdU) in capsule (A) and basal (B) cells of tuberous organs of *Apteronotus*, around the wound margin in regenerating skin (filled circles) and in control skin in the opposite cheek (open circles).

Our results are similar to those of Balak et al (1990) who studied the regeneration of the lateral line neuromasts in the axolotl salamander, *Ambystoma mexicanum*. They also distinguish between two epithelial cell types: inner support cells, which are below the sensory cells, and outer mantle-like support cells, which lie between the epidermis and sensory cells. They found that when the hair cells were destroyed by laser irradiation, mitoses occurred more frequently than normal in both classes of support cells, with inner support cells showing a greater rate of division than the outer mantle-like cells. However, in a normal organ, it seemed to be the outer mantle-like cells (which divided in advance of an organ division) that would throw out a small group of precursor cells to form a new neuromast. Thus, they suggest that cell turnover within an organ is due to the inner support cells while the production of migrating precursors is the job of the outer support cells. We do not yet know whether the basal and capsule cells perform analogous functions in the electroreceptor organs.

Conclusions, speculations and future directions

It seems likely that either basal or capsule cells migrate into the skin to provide precursor cells for new electroreceptor organs in weakly electric fish. These cells, we presume, begin to form small organs which have a critical need for innervation, probably after the organ begins to differentiate. We surmise that the nerve is necessary for continued development of the electroreceptor organs and that the postsynaptic vesicles may provide a trophic factor for their growth.

A number of questions remain to be answered. For example, what is the signal that initiates the division of basal and capsule cells after injury? Is the signal a general mitogen for all cell types in the epidermis or are there specific mitogens for epidermal, basal and capsule cells, of tuberous and ampullary organs? If the new organs do come from basal or support cells of intact organs, how do they migrate into the skin? Do the contents of the postsynaptic vesicles encourage survival of the whole organ, division of the basal cells, or differentiation of the precursor cells into various cell types? These will be challenging questions for the future.

Acknowledgements

I would like to thank Dr Bernd Fritzsch for his collaboration and for stimulating discussions on electroreceptor regeneration; Dr Dominique Crapon de Caprona for drawing Fig. 1; and the National Institutes of Health and the National Science Foundation for support.

References

Bailey WE 1937 An experimental study of the origin of the lateral-line structures in embryonic and adult teleosts. J Exp Zool 76:187–233

Balak KJ, Corwin JT, Jones JE 1990 Regenerated hair cells can originate from supporting cell progeny: evidence from phototoxicity and laser ablation experiments in the lateral line system. J Neurosci 10:2502–2512

Bennet MVL, Sandri C, Akert K 1989 Fine structure of the tuberous electroreceptor of the high-frequency electric fish *Sternarchus albifrons* (Gymnotiformes). J Neurocytol 18:265–283

Brocklebank MC 1925 Degeneration and regeneration of lateral line organs in *Ameiurus nebulosus*. J Exp Zool 42:293–305

Bullock TH, Heiligenberg W 1986 Electroreception. Wiley, New York

Corwin JT 1983 Postembryonic growth of the macula neglecta auditory detector in the ray. *Raja clavata*: continual increases in hair cell number, neural convergence, and physiological sensitivity. J Comp Neurol 217:345–356

Corwin JT, Balak KJ, Borden PC 1989 Cellular events underlying the regenerative replacement of lateral line sensory epithelia in amphibians. In: Coombs S, Görner P, Münz H (eds) The mechanosensory lateral line: neurobiology and evolution. Springer-Verlag, New York, p 161–183

Denizot JP, Baillet-Derbin C 1969 Sur la régénération des organes récepteurs spécifiques cutanés de la ligne latérale du poisson électrique *Gymnotus carapo*. Arch Anat Microsc Morphol Exp 58:249–256

Denizot JP, Libouban S 1985 New formation of sensory cells in the tuberous organ (electroreceptor) of *Brienomyrus niger* (Mormyridae) induced by transection of the afferent nerve. Int J Dev Neurosc 3:323–330

Ferrari MB, Zakon HH 1989 The medullary pacemaker nucleus is unnecessary for electroreceptor tuning plasticity in *Sternopygus*. J Neurosci 9:1354–1361

Fritzsch B, Zakon HH, Sanchez DY 1990 Time course of structural changes in regenerating electroreceptors of a weakly electric fish. J Comp Neurol 300:386–404

Hopkins CD 1976 Stimulus filtering and electroreception: tuberous electroreceptors in three species of Gymnotoid fish. J Comp Physiol A Sens Neural Behav Physiol 111:171–207

Johns PR, Easter SS 1977 Growth of the adult goldfish eye. II. Increase in retinal cell number. J Comp Neurol 176:331–342

Keller CH, Zakon HH, Sanchez DY 1986 Evidence for a direct effect of androgens upon electroreceptor tuning. J Comp Physiol A Sens Neural Behav Physiol 158:301–310

Kirschbaum F, Denizot JP 1975 Sur la différenciation des électrorécepteurs chez *Marcusenius* sp. (Mormyrides) et *Eigenmannia virescens* (Gymnotides), poissons électriques à faible décharge. C R Acad Sci Ser D Sci Nat 281:419–421

Northcutt RG, Fritzsch B, Brändle K 1990 Experimental evidence that ampullary organs of salamanders derive from placodal material. Soc Neurosci Abstr 16:129

Popper A, Hoxter B 1984 Growth of a fish ear. I. Quantitative analysis of hair cell and ganglion cell proliferation. Hear Res 15:133–142

Raymond PA, Reifler MJ, Rivlin PK 1988 Regeneration of goldfish retina: rod precursors are a likely source of regenerated cells. J Neurobiol 19:431–463

Roth A 1986 Afferent nerve fibers induce electroreceptors in the skin of fish. Naturwissenschaften 73:264–266

Roth A 1987 Lateral line sensory organs: specificity of the organ-controlling substances in the afferent nerve fiber. Naturwissenschaften 74:495–497

Roth A, Szabo T 1969 The effect of sensory nerve transection on the sensory cells and on the receptor potential of the tuberous (*knollen*) organ in mormyrid fish (*Gnathonemus* sp.). Z Vgl Physiol 62:395–410

Sanchez DY, Zakon HH 1990 The effects of postembryonic receptor cell addition on the response properties of electroreceptive afferents. J Neurosci 10:361–369

Singer M 1974 Neurotrophic control of limb regeneration in the newt. Ann N Y Acad Sci 228:308–322
Srivastava CBL, Seal M 1981 Electroreceptors in Indian catfish teleosts. In: Szabo T, Czeh G (eds) Advances in Physiological Science, vol 31: Sensory physiology of aquatic lower vertebrates. Pergamon Press, New York, p 1–11
Szabo T 1965 Sense organs of the lateral line system in some electric fish of the Gymnotidae, Mormyridae and Gymnarchidae. J Morphol 117:229–250
Szamier RB, Bennett MVL 1973 Rapid degeneration of ampullary receptor organs after denervation. J Cell Biol 56:46–477
Szamier RB, Wachtel AW 1969 Special cutaneous receptor organs of fish. III. The ampullary organs of *Eigenmannia*. J Morphol 128:264–290
Vischer HA 1989 The development of lateral-line receptors in *Eigenmannia* (Teleostei, Gymnotiformes). II. The electroreceptive lateral-line system. Brain Behav Evol 33:205–222
Wachtel AW, Szamier RB 1966 Special cutaneous receptor organs of fish: the tuberous organs of *Eigenmannia*. J Morphol 119:51–80
Zakon HH 1984 Postembryonic changes in the peripheral electrosensory system of a weakly electric fish: addition of receptor organs with age. J Comp Neurol 228:557–570
Zakon HH 1986a The electroreceptive periphery. In: Bullock TH, Heiligenberg W (eds) Electroreception. Wiley, New York, p 103–156
Zakon HH 1986b The emergence of tuning in newly-generated tuberous electroreceptors. J Neurosci 6:3297–3308
Zakon HH 1987 Variation in the mode of receptor cell addition in the electrosensory system of Gymnotiform fishes. J Comp Neurol 262:195–214
Zakon HH, Meyer JH 1983 Plasticity of electroreceptor tuning in the weakly electric fish, *Sternopygus dariensis*. J Comp Physiol A Sens Neural Behav Physiol 153:477–487
Zakon HH, Yan H-Y, Thomas P 1990 Human chorionic gonadotropin-induced shifts in the electrosensory system of the weakly electric fish, *Sternopygus*. J Neurobiol 21:826–833

DISCUSSION

Steinberg: Is the tuning a function of each cell, or of each group of cells, and does it originate in the time constant of the basal membrane, or in the synaptic release mechanism?

Zakon: I suspect that tuning is at the level of the individual receptor cell, and is probably due to the ion conductances of the basal membrane; it's probably the same kind of mechanism as seen with electrical tuning in other hair cells, such as in turtles or frogs. Tetrodotoxin (a sodium channel blocker) does not block the impulse-induced oscillations. We can block oscillations with either Ca^{2+} channel blockers or various K^+ channel blockers. We suspect that tuning is due to the interplay between calcium and calcium-activated potassium conductances, as in frog hair cells.

Steinberg: And are the cells coupled by gap junctions?

Zakon: No. One possibility is that neighbouring cells entrain each other. Since they are all clustered within a cavity and each cell produces a large potential when it is stimulated, they may entrain to each other's receptor oscillations. We don't know, yet, whether this really happens.

Calof: At the time early in regeneration, three weeks after skin removal, when the regenerating receptors are more broadly tuned, is there any obvious difference in synaptic structure from later on, at six weeks or so, when tuning has returned to normal?

Zakon: We haven't looked closely at that yet. Generally speaking, at about three weeks after skin removal, when the tuning curves are broad, there are of course fewer receptors organs, but they look fairly normal at the light microscope level. We need to examine this on an organ-by-organ basis, using electron microscopy and matching the anatomy with each organ's physiology.

Calof: Presumably you could record from one receptor cell, mark that particular receptor (by dye injection, for example), and then examine it?

Zakon: That would be the obvious experiment to do, I agree.

Rubel: You might not see anything very striking in terms of a correlation between cell morphology and frequency tuning. Paul Fuchs has done comparable studies in the development of the basilar papilla in the chick and finds a very constant picture over time in the electron microscope. So, probably, a change in channel distribution in hair cells is responsible for the changes in tuning.

Calof: So there's no obvious alteration in synaptic structure, with regeneration?

Zakon: Probably not. We hope to study the receptor cells with patch-clamp recordings to find out how they are tuned, and, as they regenerate, how the tuning changes. One could record from an afferent fibre, localize the regenerated organ, then cut that organ out and try to record from the sensory receptor cells from that particular organ with patch-clamp techniques.

Ryals: During development, do you also see these broad, low frequency tuning curves, followed by a more narrow tuning pattern? Does the pattern that you see after regeneration recapitulate normal development?

Zakon: To a first approximation I can say yes, but this hasn't been well studied. Heinrich Vischer, who is working on the development and morphology of electroreceptors, is coming to my lab and we will try to answer that question, in *Eigenmannia* (a related species of weakly electric fish). Where oscillations have been recorded from the surface of the animal, when they are young they are of lower frequency than in the adult. The wave species studied to date all have lower frequency electric discharges when they are larvae. This gradually gets higher, and reaches adult values. The receptor oscillations seem to keep pace with the EOD frequency and match it as fish grow. So there seems to be that same process, during ontogeny.

Pujol: The sensory cell synapse appears to be glutamatergic, or at least excitotoxic, because the afferent dendrite shows the morphology which characterizes excitotoxicity in other preparations such as the cochlea (Pujol et al 1985, 1991).

Zakon: And we find glutamate staining over the sensory cells.

Pujol: The postsynaptic vesicles you showed are not too surprising, because during the development of auditory dendrites, also glutamatergic, within the inner ear of mammals, we also find a lot of postsynaptic vesicles. So your idea about some kind of trophic factor acting during development is something which may be more widely found.

Zakon: We are by no means the first to show this. There are postsynaptic vesicles in other adult mature sensory terminals, in the inner ear of frog, and in taste receptors.

Pujol: And many of them are found during development?

Zakon: Yes. People have often said that these are probably reciprocal synapses (e.g. vestibular hair cells in frog), and maybe that's true; but we are leaning toward the possibility that the vesicles are conveying something important to the sensory cell. One line of research is to try to isolate the 80 nm vesicles. I don't know yet if they are clathrin-coated vesicles. If they are, one could make antibodies to them, run homogenates of the afferent nerve over a column with clathrin antibody-coated beads, and see if one can enrich the fraction of vesicles. Eventually, one could try to see if treatment with the vesicular content prevents denervated receptor cells from dying, or stimulates mitoses in basal progenitor cells.

Pujol: They could be reciprocal synapses, but not in a real physiological sense. They could work in the direction of the hair cell during development, enabling it to finish its development, and then act as the normal afferent synapses in the adult.

Corwin: You mentioned the idea that these electroreceptors came from a placodal origin, and you referred to Glenn Northcutt's finding that removal of the preotic placode in the axolotl led to the disappearance of the electroreceptors and the lateral line mechanoreceptors. Northcutt and coworkers (1990) also transplanted the placode from a pigmented individual into an unpigmented axolotl. They found pigment both in the neuromast mechanoreceptors and in the electroreceptors of that particular line of organs.

Rubel: Were their techniques good enough for them to say that it was a single placode which was transplanted?

Corwin: They did what Leon Stone did, certainly.

Zakon: This work was done in *Ambystoma*, an amphibian, but we have been studying teleosts, and there is controversy about whether teleost fish have placodes. They probably do, but Glenn Northcutt thinks that they appear and disappear so quickly that it's very hard to see. Bernd Fritzsch, looking at some of Heinrich Vischer's sections of embryo *Eigenmannia* material, thought he saw placodes, but it's not definitive yet.

Watt: To get at the question of the progenitor cells, have you tried wounding skin to different depths? If you made a deep enough incision, could you prevent the recovery of the electroreceptor organs? How much skin do you remove?

Zakon: We are just starting studies to try to localize the progenitor cells. For these studies we work on another teleost species, namely *Apteronotus*, the brown ghost, because the skin is not attached to the muscle, as it is in *Sternopygus*. Thus we can take just dermis and epidermis off and leave the muscle layer intact. It is also deep enough to avoid having any elements remaining of the organs that we have removed. We can also get needles underneath the skin to inject, say, growth factors cleanly, and see if we can increase rates of mitosis of putative precursor cells.

Watt: Does the epidermis heal before the new organs appear?

Zakon: Yes. There are a number of different hypotheses. If, for example, you cut an axon in the glass catfish (*Kryptopterus*), within 24 hours the receptor cells die and by a few days after that, all signs of the organ disappear. If you allow the axon to regenerate, you see new organs within a few weeks. So people have suggested that there is transdifferentiation—that is, the axon goes in, picks some epidermal cells, and transforms them into electroreceptor organs. Bernd Fritzsch and I don't really buy that; we have no definitive proof, but our feeling is that, as Jeff Corwin has shown so well in the lateral line system, cells from intact organs migrate into the newly regenerating epidermis and give rise to the new electroreceptor organs.

Cotanche: When the receptor organs divide, by splitting, the nerve appears to be bifurcated already. Is that bifurcation of the nerve one of the first signs that the organ is going to split?

Zakon: I can't say that with certainty, because you only recognize the fission once the organs start moving apart, and by then the axon already seems to have taken on a more distinctive shape, like a pair of pincers. Whether this means that the dividing organ has told the nerve to bifurcate, or the bifurcating nerve tells the organ to divide, I don't know. I suspect that control of this decision is made locally in the organ.

Corwin: That seems to indicate a kind of regulation that we also see in the mechanoreceptive neuromast, where something within the structure decides how big it should be, and then the neuromast organs bud, at least in some amphibian species. In your electroreceptors, it appears even more complicated; a new septum is arising, to make a new canal along with it. It's hard to see how these tissues would be regulating that.

Zakon: We know that there is also a role for the nerve, because the axon appears to 'know' how many organs it 'should' innervate.

Rubel: Is there any ongoing turnover of the receptor cells?

Zakon: We don't know; but we have never seen any degenerating cells in our light or electron micrographs.

Rubel: I noticed some really dark cells in your pictures of normal organs, that looked as if they were degenerating.

Zakon: There is some variation of staining in intact receptor cells, but one never sees the darkly staining degenerating cells that we see in regeneration when, say, uninnervated organs die.

Palay: Do you know anything about the distribution of the vascular supply to these organs, and whether it changes before, after, or during the development of these organ clusters? Maybe the clue to the pattern lies there, rather than in the nerve or in the organ.

Zakon: We haven't looked at that at all, but perhaps we should! We can see the blood vessels easily enough.

Powers: When you cut the spinal cord, this has no effect on the tuning of the receptor cells in adults. What happens if you cut it very early? What is the developmental process of the tuning then?

Zakon: When the receptors are developing the fish are so small that any procedure like that would probably kill them. Our regeneration experiments were an indirect way of asking that question, in the adult, with the inevitable caveat that development and regeneration are not necessarily the same.

Powers: The issue is where the tuning lies; is it somehow determined genetically within the receptor, or is it elsewhere?

Pujol: Most probably during development the electrical emission is changing, so the reception should also change, to be adjusted to the emission of the individual fish.

Fernald: In the spinal transection experiments there appeared to be a higher frequency EOD than in the ordinary regeneration studies; the peak curve was to the right of the fish's EOD, for the tuning curve.

Zakon: That is probably just because of the two curves that I selected. There are systematic changes in receptor tuning with age. We didn't realize until later that the bigger the fish is, the higher is the best frequency. This is likely to play a role. We usually use bigger fish to cut the spinal cord. The main point is that when you compare the best frequency of those regenerated cells with their surrounding control cells, there's no difference in best frequency.

What may be involved in part in determining the tuning is hormones. There's a sex difference in the tuning curves; males have lower frequency EODs and their tuberous electroreceptors are tuned to lower frequencies; and, of course, they make testosterone. If you inject a female fish with androgens you lower her discharge frequency into the male range and the tuning of the receptors shifts down. We have done experiments showing a few different types of lesion—cutting the spinal cord to destroy the electric field, or lesioning the pacemaker so that the receptors have no electric field or calibration signal from a firing pacemaker—and they still shift down in best frequency after androgen treatment. We suspect that sex hormones have a direct effect on the electroreceptor. One way that there can be a correlation between the discharge frequency and the receptor is via hormone levels. But we know that this can't be the whole story.

Pujol: On transmission electron microscopy, during development, you said that you could distinguish a cell which is going to become a sensory cell from the other surrounding cells. At this point in development, is there innervation around this future sensory cell?

Zakon: If we stain the skin with silver stains, we only find organs where there are already axons. So there is some relationship between innervation and growth of nascent receptor organs. Bernd Fritzsch has taken organs from early on in the regeneration process and both by light microscopy (1 μm sections) and by EM (but not serial EM) there are some organs that are not seen to be innervated; but, in the skin, nearby, there will be nerves. So we don't know whether innervation is necessary to induce the organs initially, or merely to sustain them once they appear of their own accord.

Pujol: In a 1 μm section you cannot be certain, because the innervation consists initially of very thin processes, which you cannot see by light microscopy.

Zakon: That's why we should do serial reconstruction at the EM level.

References

Northcutt RG, Fritzsch B, Brändle K 1990 Experimental evidence that ampullary organs of salamanders derive from placodal material. Soc Neurosci Abstr 16:129

Pujol R, Lenoir M, Robertson D, Eybalin M, Johnstone BM 1985 Kainic acid selectively alters auditory dendrites connected with cochlear inner hair cells. Hear Res 18:145–151

Pujol R, Rebillard G, Puel JL, Lenoir M, Eybalin M, Recasens M 1991 Glutamate neurotoxicity in the cochlea: a possible consequence of ischaemic or anoxic conditions occurring in ageing. Acta Oto-largygol Suppl 476:32–36

Final general discussion

Future directions

Rubel: What I want to do first in this final discussion is to ask the cell biologists here to suggest where this field should go.

Watt: I am working on epidermal stem cells and I am therefore particularly interested in the progenitor cell question. If I were working in this field I would be going in the direction that Anne Calof has chosen, trying to develop cell culture models. It's not realistic to attempt to produce a whole sensory organ in culture, but it might be possible to grow cells that retain some useful characteristics, such as the ability to generate differentiated progeny. So firstly I would like to see some more cell culture work.

Secondly, the question of cell type-specific markers has come up frequently in the symposium. Many antibodies to potential markers are available and can be obtained either as gifts from the labs that developed then, or from commercial suppliers. There are monospecific antibodies to individual keratins, for example, and the tissue distributions of most keratins are quite well understood.

I am struck by the importance of cellular interactions with basement membrane in these sensory organs. The last three years have seen a huge increase in our knowledge of integrin receptors for extracellular matrix molecules (reviewed by Hynes 1987, Ruoslahti & Pierschbacher 1987, Hemler 1990) and their functions. Again, monoclonal antibodies specific for individual subunits are available.

In my field, people were using lectins to look for different cell subpopulations about 10 years ago (Watt 1983). This approach is not so popular these days, partly because identifying the molecular markers detected by lectins can be a problem: one lectin may bind to several different glycoproteins on the cell surface. Nevertheless, lectins provide a well-established way of picking out different subpopulations of cells.

So there are two approaches, cell culture models and well-defined cell markers, that one would be looking for in sensory epithelium research during the next 2–3 years.

Potten: I can't add very much to that, other than to emphasize the value of the use of various cell markers, particularly at an individual cell level within the various sensory organs. Papers presented here have shown that cells in the various structures studied generally show keratin 19; but does every single cell within the structure show it? And, if there are some cells that don't, what other markers might they possess? I would also reiterate the value of the use of cell culture techniques. I would also suggest that it will be advantageous to develop and use a much broader range of markers.

I have been trying to think how to tackle the question of the progenitor cell and the labelling problem, and it seems clear that more work is needed here, particularly with *pulse*-labelling and double-labelling techniques, to track down the lineages. It may involve some detailed grain counting, and a quantitative analysis of the stereological relationships between pairs of cells, and pairs of labelled cells. Those sort of approaches might help with the tracking of lineages and their progenitors and progeny in sensory structures.

Farbman: This has been a symposium on regeneration, after all, and something that seems to be lacking here is any information on the molecular signalling that triggers regeneration. Regeneration is a response to an injury; whether it's noise in the ear, or bright light in the eye, or ouabain, or denervation in the taste bud, some signal has to be sent to a responding tissue—the progenitor or stem cell, whatever you might call it. What I don't see is work on what this signal might be, that turns a switch, resulting in the regeneration of the sensory system.

Rubel: I totally agree with you. This is something that Jeff Corwin and I have talked a lot about.

Presson: One of the most striking things about the symposium is the number of different kinds of paradigms that can produce regeneration. This poses a problem for identifying the signal to regenerate, because we have all kinds of different cells that can respond to the signal; in the same systems we appear to have various cells responding to a signal, and we have a myriad different ways of inducing the signal. It is extremely hard, considering, say, damage due to noise and to gentamicin, to try to find out what the commonalities are, to determine the nature of the signal.

Rubel: We have to be careful. Just as with developmental interactions, however complicated the interactions can be, there is no guarantee that the 'signal' is necessarily going to act directly on the missing progenitor cells or stem cells, if they exist. It may be a cascade of events that is important.

Lewis: I agree with Fiona Watt that the key to identifying the signal and figuring out the molecular mechanisms must come from culture systems like Anne Calof's. But there are pitfalls in using those system as models for normal development: for instance, Anne was looking at the differentiation of cells that have migrated out of the epithelium, and are living in a very unnatural environment. Therefore any such tissue culture study has to be accompanied by parallel studies in the more or less intact organism, so that you can distinguish artifacts from normal physiology.

Watt: You do have the option of assaying the cultured cells by transplantation into animals. One problem with the cell culture approach is that many of the species that are being worked on in this field are non-mammalian: the range of culture media available is smaller than for mammalian cells.

Calof: In many of these techniques where you induce the regenerative response, there must be a tremendous amount of inflammatory reaction in the tissue that has been assaulted, with a zinc sulphate lavage, say, or even a bulbectomy. There is a lot of precedent for the potential role of cytokines secreted by cells of the immune system that are involved in inflammatory responses, and potentially affecting the responses of all the cells in the system—not just those in the sensory epithelia or their progenitors, but the underlying stromal cells. These things could be important in the regenerative response. This is another area that we have to look at. I think it can be approached *in vitro*. It's a wide-scale 'factorology' approach—many of these cytokines are available commercially. We might also want to take a look *in vivo* at the extent of these kinds of protective inflammatory responses after the various assaults, and whether those responses might affect sensory neuron regeneration. One reason I tend to look in developing systems is that for some of these tissues it's easier to do a lot of surgical manipulations without causing tremendous tissue damage and bleeding, which occur in adult tissues. But I am sure that inflammatory responses must be important in these regeneration models. I don't know if anybody has looked, in regeneration in the ear or in olfactory epithelium, at cytokine production and the extent of inflammation after, say, zinc sulphate lavage.

Margolis: Intranasal lavage with zinc sulphate is tricky because the extent of the damage is a function of the concentration and mode of application, as well as the species and age of the recipient. Our own experience with intranasal lavage using zinc sulphate or Triton X-100 indicates that each of these parameters needs to be customized for the specific situation. With these caveats in mind, intranasal lavages have been used to generate a number of model lesions useful for the study of olfactory regeneration. The lab of Michal Schwartz at the Weizmann Institute in Israel has been studying optic nerve regeneration in fish and suspects the involvement of interleukin, but I don't know of any reports in the olfactory system that indicate the involvement of any cytokines or related factors in the process of degeneration–regeneration.

An alternative strategy for identifying such molecules is to use molecular biological approaches and compare, for example, a completely degenerated versus a regenerating system, or versus a normal system, to identify what genes are being turned on (or off) and what are the endogenous agents regulating those genes. That is another way of studying regeneration in sensory systems from a molecular orientation. For example, I am impressed by the potential of the gentamicin treatment, where you seem to get complete destruction of hair cells. This could be used to create subtractive cDNA libraries, to look for some of the genes that are intimately involved in the regenerative process.

Reh: One problem that will get in the way of doing those kinds of experiments, ultimately, in many of these systems, is the isolation of the progenitor cells in, say, the lateral line or hair cell system. That may be a potential pitfall. Perhaps the best way is to generate cell lines from these progenitor cells, once they are identified.

Margolis: The technologies have become highly microminiaturized now.

Reh: But how are you going to dissect out a single support cell from a mechanosensory organ? That will be the real challenge.

Margolis: In a recent paper, Van Gelder et al (1990) have demonstrated that one can use individual Purkinje neurons in the whole-cell patch configuration to generate cDNA. Thus, the cell itself becomes the 'test tube' for the initial incubation steps. If this approach can be generalized, many of the problems associated with small numbers of cells and limited availability of material may be surmountable.

Rubel: One complication of that approach is that many changes occur in these tissues after damage in addition to the specific events that trigger mitotic activity. The number of irrelevant cDNA clones you will get, even by subtractive techniques, will therefore be large. But it is certainly an important approach.

Raymond: In regard to looking at the genes involved in regenerative processes, I would reiterate what Julian Lewis spoke about in his paper, namely the advantage of looking to *Drosophila*, where there has been much more rapid progress in identifying genes and gene products that are involved in these kinds of 'fate' decisions and cell–cell interactions. I agree with his argument that it might appear bizarre to expect homologies between the regulatory developmental genes in insects and in vertebrates, which have very different developmental origins, but nevertheless it is apparent that there are such homologies and these could be very useful. So, in looking for 'candidate genes' to target the genetic search for relevant molecules, it is wise to pay attention to the *Drosophila* field.

Corwin: Another problem (like the one Tom Reh raised) that is specific to the organs and cells that we are interested in is that often sensory cells are highly specialized. In almost all the cases we have been talking about, at least two cell types are involved—the sustentacular (supporting) cell and the sensory cell. The extreme example of that is the electroreceptive hair cells that Harold Zakon described. It may be that these two types of cells have some sort of interdependence. The biological basis for the older concept of trophic support by the supporting cells may predict problems that will be encountered when we try to establish these sensory cells in culture. We may have to use an organ culture approach rather than a dissociated cell approach because of that.

Raymond: Many clever experimentalists in several fields have begun to study preparations that are somewhere between the intact system and isolated cells in a dish. I am referring to slice preparations and partially dissected explants that are placed into culture and observed for durations ranging from hours to days. Those kinds of preparations could span the chasm between the cell culture system with all its problems, and the whole animal with its different set of problems.

Reh: Let me reinforce that comment, particularly with respect to our results in retinal regeneration in the chick. You get a very different response of the cells in a dissociated cell culture as opposed to an aggregated culture. We should explore some of these other sensory systems in an organ culture approach, to see if we can reiterate some of these regenerative processes, to learn what other cells are necessary for us to recapitulate the process of regeneration in the sense organ.

Principles of regeneration

Rubel: Russell Fernald is now going to suggest a structured way of putting together some of the information we have heard.

Fernald: I propose that we consider the regeneration of vertebrate sensory cells as lying on a continuum which extends from cellular repair to replacement to system regeneration. By organizing the data in this way, we may be able to understand *whether* and *how* typical cellular maintenance relates to the wholesale repair of sense organs.

If we assess the available experimental results as a function of sensory system type and phylum, two general rules emerge. First, the likelihood that regeneration of a sensory system will occur increases with the proximity of the receptor surface to the environment. Second, the likelihood that regeneration will occur decreases along a continuum from cold-blooded to warm-blooded vertebrates (Table 1). One caveat to these considerations is that, so far, the methods used to produce damage in sensory systems are primitive. Sensory systems have evolved sophisticated mechanisms for identifying subtle changes in the animal's world—systems capable of single photon or single molecule levels of detection. Yet, damage has been produced using massive noise exposure, antibiotic drug overdose, or metabolic poisons—techniques which might generously be called crude. I believe we shall come to understand how sensory system repair is related to cellular renewal when subtle damage is caused and the response to this can be adequately measured. Less abusive insults may more readily reveal the potential of these systems to repair themselves, making normal mechanisms experimentally accessible.

To illustrate these principles, let us first consider vertebrate rod photoreceptors. As in all cells, cellular and molecular repair occurs in photoreceptors continuously. However, in photoreceptors, it is particularly evident because of the striking structural polarity required for detecting photons. Young (1967) first demonstrated that photoreceptor outer segments are renewed in an organized fashion. Disk membranes in the outer segment of rod photoreceptors are continuously renewed, being assembled at the outer segment base, displaced outward by new disks, and eventually shed at the tip. This renewal process is regulated at the molecular level by light and a circadian rhythm (Korenbrot & Fernald 1989).

In their recent study, Faktorovich et al (1990) discovered two important features about photoreceptor renewal in cases where photoreceptor damage is induced or cell death occurs (reviewed by Roy Steinberg, p 219–223). The first

TABLE 1 (Fernald) The continuum of vertebrate sensory cell regeneration

Phyla	*Process* Cellular repair	Cell replacement	System replacement	
Mammals	————			
	------?			
	....	....	........?	
Birds	————	———		*In vitro* only
	--------	--------	--------	Function?
	....	....	........?	
Amphibians	————	————		Function?

	....	....	........?	
Fish (teleosts)	————	————	————	Function?
	--------	------		
	....	....	........?	

Key: ————, vision; -------, hearing; ...., olfaction.

feature was that basic fibroblast growth factor (bFGF) served to mitigate damage to the photoreceptors and the second that the effects of bFGF were maximal when the factor was administered before inflicting the damage. There is no evidence yet that photoreceptors, much less retinas, can be replaced in mammals by intrinsic processes, but this may not be an unreasonable goal for molecular medicine. Since photoreceptors are related to epithelial cells (Land & Fernald 1991), using molecular tricks, we may ultimately be able to make them replace themselves.

In contrast to photoreceptors, in the olfactory system the receptors turn over regularly during life. This means that not just cellular repair but cellular replacement is common. Whether the whole olfactory system can be replaced after damage remains an open question. And for the present, we have to conclude that in mammals, there is no auditory cell or system replacement, from the evidence summarized earlier in this volume.

As we move to the other phyla in Table 1, we see some significant differences in the amount and likelihood of sensory cell regeneration. In birds, the amount of cellular repair seen in visual system cells is limited to work in dishes (*in vitro*) by dish jockeys. And in this case, it is evident that cellular fate can be altered, rather than being true cellular replacement. In the auditory system in birds, cellular replacement clearly occurs, but there is little compelling evidence that this represents functional recovery since adequate behavioural tests are not yet available.

In amphibians, regeneration of the visual system is well documented, although functional analyses of that recovery are not adequate for us to understand how much recovery has occurred. Auditory replacement is unknown, but we might expect that this occurs on evolutionary grounds. I do not know if total system replacement has been demonstrated in amphibian olfaction.

In teleost fish, the same general questions arise: is there behavioural evidence that regenerated sensory systems can function, and how good are the 'new' systems? In olfaction, there is some evidence that the recovery is complete, but compelling behavioural evidence is lacking. Similarly in vision, histological views of the retina suggest that there must be a reduced visual capacity.

In the context of this overview of vertebrate sensory cell regeneration, four general questions seem to characterize discussions at all levels of the sensory systems we have heard about at this meeting. They are:

(1) Does the response to damage resemble a recapitulation of ontogenetic events, and to what extent does that help in understanding repair?

(2) How many types of stem cell are there and how can we identify them?

(3) What rules do stem cells follow when deciding when and where to divide? This is the basic question that we need to answer.

(4) What extracellular and intracellular factors regulate stem cell fate?

How can we answer these questions? Finding answers depends in part on the system used to ask them (Table 2). We have seen here the range from *in vivo* to *in vitro* systems, but organotypic cultures, which lie in the middle of the range, may be an important adjunct to either of these more conventional approaches, for several reasons. First, in a number of cases there is now evidence that organized groups of cells provide a more realistic representation of the *in vivo* reality than do conventional *in vitro* culture systems (e.g. Reh et al 1991: this volume). Second, in organotypic cultures it is possible to have some structural integrity, allowing easier analysis of experimental results (e.g. Mack & Fernald 1991). Finally, the application of extracellular factors can be compared within a single culture, providing within-system controls that are not possible in typical *in vitro* systems.

Rubel: We are really talking about two different things under the heading of regulation. One is the events that regulate the *differentiation* of the postmitotic

TABLE 2 (Fernald) Systems in which four fundamental questions might be asked

	Questions[a]			
System	*1. Recapitulation?*	*2. Stem cells?*	*3. Rules?*	*4. Regulation?*
In vivo	+ +	+ +	+	
Organotypic culture	+	+ +	+ +	+ +
In vitro			+	+ +

[a]See text.

cells; the other is the events that regulate the *mitotic activity* of stem cells or progenitor cells. From what we now know about cell biology, there are likely to be different kinds of genes involved in these two types of regulation.

Presson: And where do we see differentiation coming in? One could include it with stem cell regulation, but what turns on and regulates stem cells might be slightly different from what actually makes a differentiated neuron. But we don't know anything here yet.

Pujol: We need to discuss question (1), of whether the response to damage is really a recapitulation of ontogeny.

Fernald: My feeling is that we should see the study of ontogeny as one way of understanding the response to damage, because we know a lot about ontogeny. It would be nice if the actors that did the work during development were also involved in repair. But this may not be the case, because it may be a superficial comparison.

The question really is the usefulness of making a distinction between the recapitulation of ontogeny and a new mechanism that responds to damage. If it were shown to be a recapitulation of ontogeny, you would then have a range of putative molecules and cells that you can turn to and say that they are probably all still active, or reactivated. Whereas if a whole new mechanism is involved it is a much tougher question and we might be better off going to different systems where there really is a recapitulation.

Pujol: There is something in the chicken hair cell regeneration studies which does not completely recapitulate ontogeny. In a damaged cochlear preparation in the chick, you have regeneration of the sensory cell but, as Brenda Ryals is showing (personal communication), a decrease in neuron number. so there is something completely different here. You can produce new sensory cells in the chick repair model but no new neurons, and this is very different from ontogeny.

Corwin: I don't think there is any doubt that there are very significant differences between ontogeny and regeneration.

Rubel: Would anybody take a strong stand, in any of the sensory systems studied, that the repair we see is a recapitulation of ontogeny?

Reh: In vision, Pamela Raymond and I have both found, in the frog and in the goldfish, that it's not a recapitulation in the sense that it looks the same; however, it seems as if stem cells, or progenitor cells, have to reproduce the entire programme of development and to revert back to some common neuroepithelial ancester and make all of the same types of neurons again, probably in the appropriate sequence. The amphibian retina differentiates during regeneration in the same sequence as in its original development. So I would argue that once you have regenerated a neuroepithelial cell, it will recapitulate the normal sequence of retinal development.

Raymond: I would agree with that, for both frog and fish. If you look at the level of a single cell and its progeny, you are probably looking at a recapitulation of an embryonic pattern, whereas if you look at the system as a whole, there are clearly differences, of both spatial and temporal order. The process as a whole does not happen in the same way during regeneration as during development. For example, the size is different: in the embryo you start with a tiny structure, but in regeneration the structure to be replaced is much larger. However, at the level of the individual progenitor cell and in terms of the cascade of events that take place in a developing clone of cells, one could make a good case that regenerative events are probably a recapitulation of development.

Steinberg: There is one highly significant proviso with the retina; we have to regenerate both the retina and its connections to the brain. Whether we can recapitulate that in any of these experimental systems is unclear and is very difficult to determine. At the cellular level, of course, the regrowth of the outer segment does appear to recapitulate development.

Corwin: We can make a generalization that the tissue differentiation that occurs during regeneration recapitulates the differentiation seen in the embryonic development of the same structures. I would even extend that to predict that if we can stimulate the production of undifferentiated 'progenitor type' cells, in for instance a structure such as the mammalian ear, which currently is not known to regenerate, there will be a high likelihood that regeneration will occur there with differentiation following that normal developmental path. If we consider the course of regeneration in planaria, or in regeneration of various appendages in invertebrates or vertebrates (see Goss 1969 for reviews), the limiting factor appears to be the production of a group of proliferating, undifferentiated cells. If they exist, or if they form at the site of a wound, then the repair process will go on and follow the embryonic pattern.

Reh: We could add that neuroepithelial cells not only differentiate in their normal pattern; they also retain their identity to a large extent. For example, a neuroepithelial cell of the retina, whether derived from a pigment cell, or derived from a rod precursor cell, is going to make retinal neurons; it's still a dividing cell, but it's not going to make olfactory neurons. This is probably also a generalizable feature of stem or progenitor cells, that they will retain their identity.

Fernald: That is untested and can only be tested by explantation to another environment.

Cotanche: People have taken the visual system of the animal, destroyed the visual cortex, and allowed it to grow into the auditory system; they get an electrically functional visual cortex in the auditory area.

Rubel: That's a separate type of situation, because the receptor system is left intact.

Cotanche: Tom Reh is saying there's no evidence that these things could convert to other systems.

Reh: There is considerable evidence from embryological transplant experiments that neuroepithelial cells retain their regional identity. When eye primordial cells (amphibian or chick) are transplanted to ectopic locations in the embryo, they retain their ability to form eyes (Adelmann 1930).

Corwin: In more general terms, all of this starts with a single cell, the zygote, which has the capacity to produce all of the specialized cells. Gradually, as determination occurs, there is restriction in the normal outcome of differentiation for the various types of cells that are produced and what they can produce.

There is an interesting story with regard to the positional memory of cells and retinoic acid's ability to modify and re-set what a cell knows it is, or has the capacity to form, during limb regeneration. In the salamander, *Ambystoma mexicanum*, if you amputate a forelimb below the elbow, blastema cells that aggregate at the stump do not normally produce a second new elbow and upper (proximal) limb; instead they form the appropriate bones of the distal forelimb, the wrist, and the hand. If they are treated with retinoic acid, their self-identity or positional 'memory' can be re-set, so that they are 'proximalized' and will form structures that duplicate those that are proximal to the level of amputation (Thoms & Stocum 1984, Crawford & Stocum 1988).

There seems to be a dose dependency: at lower doses you can get a re-setting to above the elbow, to produce a new upper limb; at higher doses you can even re-set the blastema cells so that they first produce a new shoulder girdle (Thoms & Stocum 1984). This fits with what we have talked about here, that perhaps these undifferentiated cells may have to go through a number of divisions in order to re-set their state of specialization and the restriction of potential fates.

As I mentioned earlier (see p 107 and 124), I have been impressed with the implications of Jay Jones's finding that when he traced back in time-lapse recordings from the point when two regenerated hair cells had been produced in a laser-treated neuromast there was a sequence of two cell divisions in the lineage of cells that gave rise to each of those regenerated sensory cells. The original progenitor that was present at the time of the laser treatment was the 'grandmother' (as we call it) and appeared to be a supporting cell that gave rise to two daughters, and one of those two cells gave rise to two granddaughters, one of which became a hair cell. We don't know whether that will prove to be a constant feature of this sensory regeneration, even in the lateral line system. If that were the case—if the original cells that were present at the time of the trauma had to go through a certain number of cell divisions to re-set their state of specialization—then that would be consistent with the existence of processes during cell division that have the effect of reversing the restriction of the capacity to produce new types of cells.

Rubel: Going on to the stem cell, are we convinced in any of these receptor systems that we really know the identity of stem cells, or the identity of the progenitors of the receptor cells?

Farbman: In the olfactory system, we know what we think the stem cells are! There are the basal cells which fall into two groups: one of them is a progenitor and one is a stem cell, according to the current view.

Calof: There are what people call these two kinds of basal cells, although I'd like to throw out 'basal cell' as a term. The two types are the lighter globose kind, which I call the immediate neuronal progenitor, and another kind, called by various people the horizontal cell and the dark basal cell. I would say that the light (globose) basal cell is the immediate neuronal precursor, but it is still not clear whether the horizontal basal cell, or some subset of these cells, or a subset of the lighter cells, is the true stem cell in the olfactory epithelium. I really think that the stem cell, if there is one—the true ultimate progenitor of the olfactory receptor neurons—has yet to be clearly identified.

Fernald: Part of the problem is that we need to know what the signal for cell division is. The idealized way to ask this question would be to place 'the signal' for a stem cell to divide onto cells in sequence. You could ask directly how many are competent to become stem cells. However, there may be, under different circumstances, different populations that are competent. In the cases of extreme damage, we may bring into action all sorts of normally extraneous cells which are driven because of the extremity of the situation; whereas 'real' stem cells are far fewer. It may be that there are different populations of potential stem cells depending on the extent of the destruction we impose.

Reasner: The severity of the perturbation is especially relevant in systems where there are ongoing replacement and growth mechanisms, such as the olfactory system. For example, there is continual proliferation at the margins of the respiratory epithelium in adult mice (unpublished observations). In both mice and rats, there is continued growth of the olfactory epithelium after the age of sexual maturity (Hinds et al 1984, Sichlau et al 1990). So, if you could use a paradigm of intense sensory stimulation which does not massively destroy the sensory epithelium, you might realistically expect to see different regenerative mechanisms from those following more invasive paradigms.

Potten: I agree that the problem with identifying stem cells in various systems is context dependent (Potten & Loeffler 1990). The answer to the question 'what are the stem cells?' will depend on the system that you are looking at; you get a different answer if you are studying an amphibian system rather than a mammal. We heard planaria mentioned. These flatworms have some extremely primitive (stem) cells in the adult that can regenerate the entire worm. That isn't the case in mammals! Amphibia can regenerate a limb; that is also not the case with mammals. So the answer to the question depends on the system being examined and the tests being applied. Because of this context dependence of the stem cell identification and characterization, we have to be as careful in transferring information from *in vitro* to *in vivo* studies, from mouse to man, from birds to mammals, and particularly from amphibia, reptiles or fish to mammals, as from invertebrates to vertebrates and mammals.

This also involves the question of the severity of damage in experiments. In many of these sensory systems, if you destroyed a single cell within a structure, and asked which cells replace it, you might get a very different answer from the experiments where you destroy the whole structure and ask which cells replace that entire structure. The electroreceptors in the teleost fish clearly indicate this. There must be different types of stem cells, firstly those that are capable of going through many divisions and differentiating in a variety of ways when there is the fission of the ampullary organ that Dr Zakon described. However, these cells are different from those seen when a piece of skin is cut out and this then regenerates the organs, because Dr Zakon told us that these stem cells come from somewhere else: they are epidermal cells that have had some interaction with the nerve fibre. So the answer to the question of which are the stem cells in this electroreceptor system is dependent on the experiment you do. That is probably true for all the sensory systems that we have been looking at.

Raymond: The obvious question, that has been asked many times, is whether the capacity to regenerate is dependent on continued growth. Does regeneration occur only in those systems that are continually growing?

Lewis: The answer is clearly 'no'!

Raymond: That would seem to be the case, if we believe that hair cell proliferation does not normally occur in the avian ear. However, in all the other cases of regeneration that have been described at this meeting, there is evidence for proliferation in the intact sensory system.

Margolis: To what extent is the definition of a stem cell a function of who the neighbouring cells are? We seem to be talking of an isolated stem cell, but we know that during regeneration and development the so-called stem cells, and in fact the whole cellular environment, will be different from the normal quiescent tissue. Is a stem cell a function of where it is?

Rubel: We have to run the process backward in time to determine the answer to that question.

Watt: In adult mammalian tissues we tend to think of stem cells as being present in tissues where there is a high rate of turnover of the terminally differentiated cells under normal conditions, for example in the epidermis and the haemopoietic system (Hall & Watt 1989). In at least some of the sensory organs you would expect that in the absence of environmental damage, there is not much cell replacement going on. So the sensory systems might be more analogous to a tissue such as liver, where you don't normally see much cell division, but if you remove half the organ, many of the cells are stimulated to divide until the damage is repaired.

Rubel: What happens to these stem cells? Are they quiescent cells that are suddenly kicked into function in the liver, or are they cells that have some function and differentiated property and then revert to being stem cells when the right stimulus is presented?

Watt: They are probably not a discrete subpopulation.

Potten: I don't think that's entirely clear! In a mouse liver, if you cut out three-quarters of it, there is a fairly synchronous burst of cells entering DNA synthesis, but not every cell in the liver enters S phase—only 20% or so. But you don't know what regenerative function the cells that appear in that first wave really have. Are they the cells that go on to divide several times and regenerate the total mass, or are they just cells that respond most dramatically to the stimulus 'divide once and stop', and other cells either come along later or emerge from the cells in the first wave, and these go through several divisions, and eventually are responsible for the regeneration of missing liver cells?

It comes back to asking the question the other way round: when the adult liver mass is reached, how does the system switch itself off? Does an entire cell lineage freeze, with all the cells in G0, or do the cells work through their predetermined lineage and just leave some stem cells in G0? This would be a small subpopulation of the liver—a part, or maybe even all, of the 20% that respond. We don't know the answers to these questions, even for the liver, which has been fairly well studied.

Oakley: There are comparable problems in the regeneration of minced mammalian striated muscle, where it's presumed that there are small quiescent satellite cells that are activated by the destructive process, and may be responsible for the substantial regeneration of the muscle (Carlson 1968).

Lewis: I want to turn the attack against the stem cell mystics! If you see a population of cells which all look similar, by all the criteria available to you, and some of these divide and do something, the first hypothesis should be (because it's the simplest) that they are indeed all alike and that it is by some random process that some of them go on to divide and do something while others, at random, don't. It's true that there *may* be hidden differences within that population of cells; some members may be determined to behave differently from others, and we should bear that possibility in mind; but until you have demonstrated such a difference, the first hypothesis should be that they are all the same.

Potten: In the mouse liver and other tissues (such as bone marrow and gut), we know that they are not; even though they look all the same, only some are capable of a regenerative response. It depends on how you are looking at them.

Lewis: Of course, I wouldn't dispute that there are cases where there are indeed subtle functional differences that people can demonstrate by appropriate techniques. But in the auditory epithelium, for instance, I see no reason to believe that there's a subpopulation of stem cells that are alone capable of regenerating hair cells and are distinct from the other supporting cells.

Presson: In teleost fish, I think there are. I don't know, again, whether they are a different part of the cell cycle of the same cell, but there are clearly, to me, subtle morphological differences between a cell that takes up [^{3}H]thymidine and a cell that doesn't.

Corwin: There is one thing we didn't mention when we discussed your data on that, namely the type of cell that you call the basally located S phase cell. That cell type, and supporting cells, are both distinguished in your definitions by having basally located nuclei.

Presson: They have very different biochemical and cytochemical characteristics. The basally localized S phase cells do not have endogenous peroxidases, for example, whereas supporting cells do.

Reh: Pamela Raymond raised the question of whether continued neurogenesis is always associated with the ability to regenerate. This has been brought up with respect to the nervous system in terms of axonal regeneration. Are we certain that continued neurogenesis is *not* needed? Surely the auditory epithelium is the exception in the bird, in that there is regeneration, yet people say that there is just an occasional labelled cell. Are those who study the bird certain that there is no continued neurogenesis under normal conditions? Slight damage occurs all the time, and perhaps cells are replaced; is that the same as saying that there is a continuous level of slight damage in the olfactory epithelium?

Rubel: You would have to look at a bird in the wild to answer that question.

Fernald: That gets back to my Table 2, where the conditions of damage have been so extreme that we can't yet answer that question; we need to have subtle, intermediate levels of destruction.

Corwin: I don't think so. Brenda Ryals can probably answer that best. In earlier discussion of Dr Jørgensen's paper we may have left the wrong impression, that there is more labelling and proliferation in undamaged avian cochleas than we were willing to admit. To me, it is striking how little labelling there is. I can't recall seeing a labelled cell in an undamaged region of the cochlea after injection of [^{3}H]thymidine for seven days.

Rubel: It is *very* rare, but does occur.

Ryals: Yes, it is rare. We saw only a total of four labelled cells out of thousands of cells in two control animals (Ryals & Westbrook 1990). Dr Jørgensen saw no labelled cells in the auditory papilla after 19 days of [^{3}H]thymidine injections, and he looked at a much larger cell population than I did (Jørgensen & Mathiesen 1988). So, if there is normal turnover in the auditory papilla, it is clearly not at the level of that seen in the vestibular end organ.

Presson: How do you interpret these cells? Are they new cells, or artifacts?

Ryals: I think they are new cells. What I don't know is what the trigger may have been for their production. My birds came to me in a carrier in a plane—a situation which is potentially traumatic. However, they were in my lab for two months before we studied them, so one would expect that any response to that trauma would have been completed by then. Nevertheless, we cannot rule out the possibility of some local cell death, for unknown reasons, as a stimulus for this new cell production.

Rubel: That is the way we interpret these rare instances of labelled cells in control tissue, but it is only an interpretation.

Fernald: The other caveat is that we should remember the vast number of vertebrate species. We are looking at a miniscule fraction of that number, and these are domesticated! If we look at the wilder species, we may see repair as a common feature.

Ryals: In fact, I saw [^{3}H]thymidine label in hair cells of normal birds in quails, which are somewhat more wild than domestic chickens!

Lewis: I wonder whether territorial birds, which gain territory by singing as loud as they can, may damage their own auditory systems?

Rubel: I think you should do that experiment!

What I would like to say, in conclusion, is that I think we have all learned something here, or at least learned how little we know. I would hope that through this symposium we have gained the ability to better define the questions that we shall be asking in our research over the next few years. Many of us may also have learned a broader range of approaches to these questions. If we can define the questions better, one of the big hurdles will have been overcome. I hope some of us will go back to our laboratories and do slightly different and somewhat better experiments than we would have done before this meeting. If that happens, I think we have achieved our goal.

References

Adelmann HB 1930 Experiment studies on the development of the eye. J Exp Zool 54:219–317

Carlson BM 1968 Regeneration of the completely excised gastrocnemius muscle in the frog and rat from minced muscle fragments. J Morphol 125:447–472

Crawford K, Stocum DL 1988 Retinoic acid coordinately proximalizes regenerate pattern and blastema differential affinity in axolotl limbs. Development 102:687–698

Faktorovich EG, Steinberg RH, Yasumura D, Matthes MT, LaVail MM 1990 Photoreceptor degeneration in inherited retinal dystrophy delayed by basic fibroblast growth factor. Nature (Lond) 347:83–86

Goss R 1969 Principles of regeneration. Academic Press, New York

Hall PA, Watt FM 1989 Stem cells: the generation and maintenance of cellular diversity. Development 106:619–633

Hemler ME 1990 VLA proteins in the integrin family: structures, functions and their role on leukocytes. Annu Rev Immunol 8:365–400

Hinds JW, Hinds PL, McNelly NA 1984 An autoradiographic study of the mouse olfactory epithelium: evidence for long-lived receptors. Anat Rec 210:375–383

Hynes RO 1987 Integrins: a family of cell surface receptors. Cell 48:549–554

Jørgensen JM, Mathiesen C 1988 Continuous production of hair cells in vestibular sensory organs, but not in the auditory papilla. Naturwissenschaften 75:319–320

Korenbrot JI, Fernald RD 1989 Circadian rhythm and light regulate opsin mRNA in rod photoreceptors. Nature (Lond) 337:454–457

Land M, Fernald RD 1991 Evolution of eyes. Annu Rev Neurosci, in press

Mack A, Fernald RD 1991 Thin slices of teleost retina continue to grow in culture. J Neurosci Methods 36:195–202

Potten CS, Loeffler M 1990 Stem cells: attributes, cycles, spirals, uncertainties and pitfalls; lessons for and from the crypt. Development 110:1001–1019

Reh TA, Jones M, Pittack C 1991 Common mechanisms of retinal regeneration in the larval frog and embryonic chick. In: Regeneration of vertebrate sensory receptor cells. Wiley, Chichester (Ciba Found Symp 160) p 192–208

Ruoslahti E, Pierschbacher MD 1987 New perspectives in cell adhesion: RGD and integrins. Science (Wash DC) 238:491–497

Ryals BM, Westbrook EW 1990 Hair cell regeneration in senescent quail. Hear Res 50:87–96

Sichlau M, Paternostro M, Meisami E 1990 Two-dimensional models and morphometry of individual olfactory conchae in growing rats. Chem Sens 15:639

Thoms SD, Stocum DL 1984 Retinoic acid induced pattern duplication in regenerating urodele limbs. Dev Biol 103:319–328

Van Gelder RN, von Zastrow ME, Yool A, Dement WC, Barchas JD, Eberwine JH 1990 Amplified RNA synthesized from limited quantities of heterogeneous cDNA. Proc Natl Acad Sci USA 87:1663–1667

Watt FM 1983 Involucrin and other markers of keratinocyte terminal differentiation. J Invest Dermatol 81:100S–103S

Young RW 1967 The renewal of photoreceptor cell outer segments. J Cell Biol 33:61–72

Index of contributors

Non-participating co-authors are indicated by asterisks. Entries in bold type indicate papers; other entries refer to discussion contributions

Indexes compiled by John Rivers

Subject index